AF517733

New Vistas of Organic Farming

2nd Revised & Enlarged Edition

Mukund Joshi
Associate Professor,
Dept. of Agronomy,
University of Agricultural Sciences
GKVK, Bangalore - 65

Published by:

SCIENTIFIC PUBLISHERS (INDIA)

Jodhpur
–
5 A, New Pali Road
P.O. Box 91
Jodhpur 342 001
INDIA

Delhi
–
4806/24, Ansari Road
Daryaganj
New Delhi - 110 002
INDIA

Germany
–
Landstr, 240
Frankfurt 60388 -
GERMANY

E-mail: info@scientificpub.com
Website: www.scientificpub.com

2nd Revised & Enlarged Edition : 2016

ISBN: 978-81-7233-955-5

Typeset by Rajesh Ojha

Printed in India

FOREWORD

The awareness about organic farming is slowly pervading all the spheres of our society, including the health conscious consumers, who value the organic products as healthiest. The farmers, administrators, scientists, traders, exporters and even policy makers deal or attempt to deal with one or other aspects of organic sector, in the interest the health, trade, export or profitability. Most sections of the society accept the organic production as the most desirable pollution free food. Indeed, the interest in healthy pollution free food is increasing at an alarming rate.

Equally widespread perception, which is often debated, is the diagonally opposite issue that organic farming is not suitable to India, either due to heavy population of our country or tropical climate of India. Most sections of our society, excluding agricultural scientists and farmers are not even aware of the practical organic production, although they readily accept the organic products. Even among scientific community, many are not even aware of holistic meaning of organic farming. The debates by scientists and economic thinkers are have considered organic farming as use of manure, as they lack the holistic meaning of organic farming. In general organic growers have been inclined to practice organic farming either due to subsidy or reduced cost of cultivation. Most farmers are also not aware of the certification procedures and potential benefits of organic farming to the entire farm. Almost all sectors of society, including the farmers and many scientists hold large number of apprehensions doubting the usefulness or practicality of practicing the organic farming.

In such a scenario of ideological dichotomy, organic farming is slowly gaining momentum as an important future strategy. Even FAO and WHO sponsor the world wide promotion of organic farming. It is now practiced in as many as 139 countries, although organic area is small fraction of cultivated area. In this juncture, this book is an excellent attempt to show the holistic nature of organic farming, vastness of its potentialities and more natural option leading to sustainability in agriculture. The book has attempted to elaborately document the hill effects of continuing chemical agriculture besides showing the great potentials of practicing the organic farming. Interestingly, it has dealt with the history of organic farming as relevant to India, finding its roots in Indian civilization. But, a clear cut differentiation is brought out between modern organic farming and traditional agriculture, popularly held as organic by default. A detailed account is also presented about various organic options for pest control.

Very few books have presented the global scenario of organic farming. The author of this book has attempted to briefly describe the nature and spread of organic farming throughout the world, besides giving brief account of each country. This section is of special importance, as it can enrich the readers with the knowledge of organic situation in different continents and selected countries.

This book is an exemplary attempt to reveal the new vistas of organic farming, serving as a document many research workers and students can frequently refer. It is the result of vast practical experience and many years of research of the author along with deep insight to find lasting solutions to the problems of deteriorating soil fertility.

I wish that this book will be encouraged as important reference book in Indian agriculture for all scientists, administrators, students and farmers.

Dr. H.V. NANJAPPA
Dean (Postgraduate Studies)

PREFACE

Our nation has pursued the policies of intensive use of agro-chemicals in the last 30-40 years to meet the challenges of increasing the agricultural production. Level of consumption of agro-chemicals is even held as yardstick of agricultural development. Use of agro-chemicals along with other technologies like improved hybrids/varieties and irrigation has indeed elevated our country to self-sufficiency in food production.

But, in the journey of ever challenging agricultural development, we have reached a stage, where the basis of production itself is in perilous situation. Because, the use of agro-chemicals has damaged our eco-system and delicate balance between various components of eco-system. The biological basis of fertility imparts self-supporting feature in soil. Reducing organic carbon status on one hand and treating the soil as mere physical medium to supply the nutrients on the other - have ignored the biological basis of soil fertility. Similarly, the pest control by pesticides alone is akin to chemical invasion on eco-system. A pest is part of biological equilibrium in an eco-system and killing the pest by pesticides not only damages the eco-system but also kills predators and natural enemies of pests. A pest can express itself through biological survival/buoyancy mechanisms. These truths are conveniently undermined by advocation of increased pesticide use.

Use of fertilizers and pesticides had their designated aims of increased productivity and reduced damage due to pests respectively. But, the productivity of many crops has not shown proportionate improvement in the last 10-15 years - despite the increased use of fertilizers. Similarly, extensive use of pesticides has not reduced the losses due to pests. With these two facts in background - a stage has now reached to review whether promoting the use of agro-chemicals is appropriate strategy or not.

This book is written to enlighten the students and research workers that the use of agro-chemicals is perilous to our soils, eco-systems, water, foods, human health, animal health as well as productivity/pest control issues. The use of agro-chemicals has left us with unsustainable production systems. Factors leading to unsustainable agriculture are so pervasive and profound that urgent imperative need exists to achieve sustainable production - lest the food security will be in jeopardy in populous countries like India and China.

Therefore, the present day challenges in agricultural research include important and formidable aspects of attaining sustainability in various agro-eco-systems. The essential features of sustainability are more appropriately satisfied by principles and concepts of organic farming than any other. Organic farming deals with such methods, materials and practices - which are attuned to nature and natural processes. Hence, self-supporting concept of sustainability are more satisfied by organic farming.

It is appropriately surmised - with profuse justifications and technical evaluation of facts - in this book that sustainability could be achieved by organic farming in phased manner, without experiencing the negative impacts of disuse of agro-chemicals. An attempt is also made to present the concept of sustainability in holistic approach rather than fragmented approach. Although the empirical assessment of sustainability is in its nascent form, an account is also given to assess the sustainability in quantitative terms.

It is expected by the author that this book will be of technical value to researchers, teachers, students and administrators - who are interested to know the various issues connected to sustainability and organic farming.

MUKUND JOSHI

LIST OF TABLES

LIST OF FIGURES

CONTENTS

Chapter - 1

INTRODUCTION

Many claims have been made about spectacular achievements of Indian agriculture in the past three decades. The crop yields were increased by 50 to 150 per cent in food crops, 70 to 170 per cent in commercial crops, and 40 to 120 per cent in horticultural and plantation crops during this period. India has achieved self-sufficiency in food production more than twenty years back and the present level of food production is to the tune of 259 million tons. India has been exporting some agricultural goods like cotton and cotton garments, sugar, fresh fruits, besides the traditional export of spices and dry fruits. It is one of the major exporters of basmati rice as well as medicinal herbs. It is indeed a matter of pride to have achieved such an outstanding success in this brief period of six decades of independence.

The success of boosting up agricultural production has come through the release of high yielding, photo-insensitive varieties/ hybrids, which were highly responsive to chemical fertilizers and irrigation. The high yield potentiality of these varieties/hybrids necessitated an artificial supply of large quantity of nutrients from outside sources. As a sequel, the practice of fertilizer application became a part and parcel of our agriculture. This, in turn, promoted the large-scale production/import of inorganic fertilizers, leading to establishment of many fertilizer industries. Like any other new technology, the farming community resisted the use of fertilizers during early 1960's. But, the persistent national policies on promoting the fertilizer use, extensive field experiments to standardize the method and quantity of fertilizer application for different crops under varied situations and a network of field demonstration on the utility of fertilizers finally induced the farmers to use the fertilizers for almost all crops. From a meagre level of 0.66 lakh tons in 1950, the fertilizer consumption in India increased to 27.8 million tons in 2011. The global consumption of fertilizers has been to the tune of 184 million tons (FAO, 2011).

Increased crop yields by faster plant growth also promoted the large scale use of many other agro-chemicals such as insecticides, fungicides, bactericides, acaricides and rodenticides in order to save the crops from the damages by insect pests, diseases and other animals. The use of herbicides also became popular to control the weeds in irrigated areas owing to expensive labour and effectiveness of herbicides. Many growth promoters and growth retardants were also introduced commercially and their large scale use was aimed at improving the crop yields. As much as 2.4 million metric tonnes of pesticides are being used world over annually to save the crops from pests (Fishel, 2007). Out of this, 41822 tonnes are used in our country (Anonymous, 2010). Cotton and paddy crops account for 67.3 per cent of total pesticides consumption in India.

In the last decade (1991-2001), the global consumption of pesticides was reduced by 2.2 million tons. However, Asian continent recorded an increased consumption by 17 million tons, while European consumption was reduced by 4.26 million tons. American continent recorded constant level of consumption. In most developed countries, it has remained stagnant or reduced. In most developing and under-developed countries it has increased except African countries.

In future years, absolute dependence on the fertilizers for increasing the crop yields is not only dangerous due to adverse effects to our environment, health hazards to human and animal population but also untenable because the response to added fertilizers has been dwindling in important crops. It is also not practical to continue the use of fertilizers in next century, as naptha sources and other non renewable sources of fertilizer production are on the verge of exhaustion. **There is an urgent global need to find an effective alternative to fertilizer use, which can ensure increase in crop yields and at the same time cause minimum damage to the eco-system and human health**. Sustainable development and resource management are inevitable options to lead the globe further towards the prosperity. In the similar manner, natural alternatives to the use of pesticides are also to be identified and strengthened, as innumerous cases of pollution due to pesticides are reported. In this direction, organic farming takes a key role in the resource management of a given eco-system towards the sustained development of agriculture.

Our country has blindly followed whatever developed countries have found and introduced in their societies. The use of fertilizers, pesticides and other agro-chemicals is no exception to this. The blind perception that whatever is found in developed countries is good to us in unequivocally a blunder. As and when they realize that the new found material/practice is not good to their society and thereafter when they promote the ideas against their inventions/discoveries, the possible maladies are realized in India and efforts to discourage their use are made. This is testified truly in organic farming. Discoveries on plant nutrition, technology on fertilizer manufacture as well as large scale promotion of synthetic pesticides during 70s and 80s in Europe were transferred to Indian agriculture as invaluable opportunities. This led to establishment of mammoth pesticide/ fertilizer industry in India and many other countries. However, in the last decade, many western countries are promoting the non-chemical organic farming, even in India, large scale efforts are being made to promote the same.

As the developed countries have now found that the use of agro-chemicals cannot be sustainable from the long term view of ecological balance, many Indian farmers and scientists are conceiving the ideas against, the use of agro-chemicals.

However, the price paid to achieve the green revolution in India is beyond imagination. In an eagerness to quickly fight against the hunger and overcome the food shortage, an overwhelming use of agro chemicals was promoted, apparently ignoring the ill effects of these agro-chemicals. No doubt, the spectacular success in increasing the crop yields was achieved. The potential dangers of the large scale use of these chemicals were never even perceived when they were introduced. In the last 10-15 years profound thinking on the dangers of agro-chemicals was made which led to zero chemical farming in the form of organic farming/natural farming.

Two schools of thought are found pervaded in respect of use/dis use of agro-chemicals. A small group of scientists, innovative farmers and some extension workers acknowledge the dangers of agro-chemicals-i and promote the cause of organic/natural materials in partial/ total replacement of agro-chemicals. A large and strong group of planners, administrators, policy makers, scientists, successful farmers and industrialists believe that the use of recommended levels of agro-chemicals is not dangerous and their continued use is indispensable for increased and sustained agricultural production. They also believe that any attempt to discourage their usage would seriously jeopardize the present level of food production and push the globe towards the worst calamities like famines and hunger apart from totally disrupting the global food production. The second group may be correct, in the short run thinking, because such a level of food production [2425 million tons food grains+ 608 million tons of other agril crops]may not be an achievable prospect in the absence of agro-chemicals. In the long term perspective of sustainable agriculture and to save the mankind from the future hazards of the agro-chemicals, however, the use of agro-chemicals needs to be discouraged. The necessity of forcefully advocating the reduced use of chemicals can be realized if the dangers posed by these agro-chemicals to our soil, environment, food animals, micro-organisms, and human health as well as to our eco-systems are clearly understood. Many direct damages to human and animal health by persistent pesticides in soil, water, air, food and even processed food materials have been recorded. The human and animal milk, the drinking water, fruits, vegetables, processed and unprocessed foods have been polluted by the use of agro-chemicals. Many more indirect harmful effects of deterioration of soil fertility, increase in the soil erosion, change in the micro climate, stagnation of crop yields are also recorded. Thus, a state of dilemma is reached about whether to continue the use of agro-chemicals even at the cost of so many hazards mainly to retain the present level the production or to promote the disuse of agro-chemicals and save the globe from future chaos. The future hazardous situations may include the severely polluted water and food products, seriously imbalanced ecosystems, endemic occurrence of new and present diseases to human beings and animals, reduced soil fertility and disruption of the basic agricultural system, apart from innumerable occupational problems during the use and manufacture of agro-chemicals. Agro-chemical industries have been particularly highly polluting the environment and water around them. The accidental leakage of poisonous gases during the manufacture of the agro-chemicals can be disastrous to the society. Bhopal gas tragedy is one among such major tragedies, worth quoting.

Organic farming might replace the chemical farming in coming years, more as an inevitable measure than as any new opportunity to boost the agricultural production. According to many scientists and farmers, organic farming is not in any way different from traditional Indian agriculture. But the new found meaning of organic agriculture involves many new principles and practices , not found in traditional agriculture [for example, biofertilizers, bio pesticides etc]. Though organic farming is indeed conceptually not new, a fresh look on the principles of organic farming might be necessary to respond to some basic questions such as :

1. Can organic farming sustain the present level of agricultural production?
2. Can organic farming be followed in every part of the world having different agro-climatic situations?
3. In what all ways the organic farming can be practiced?
4. Is it possible to replenish the nutrients removed by the crops through organic farming, especially in the perspective of large scale uptake of nutrients of present day hybrids and varieties.

5. Is it possible to increase the production level to meet the demands of ever increasing population?
6. What do we mean by certification in organic farming and why it is needed?

Many new ideas/products/systems are included in the modern concepts of organic farming, besides, the use of organic manure. For example, the use of neem in various ways to control the pests; the use of many plant extracts for control of pests/diseases; the use of bio-fertilizers etc. The use of organic materials of different origins in various manners compatible with given eco-system to replace the agro-chemicals partially or fully without affecting the sustainability of production should be the goal set for future years.

The organic farming is not a practice like fertilizer usage or irrigation. In its real perspective, **organic farming is a philosophy by itself.** Organic farming is bound to have far reaching consequ-ences. It is expected to influence every aspect of rural and urban life, besides agriculture. Organic farming is not only aimed at maintaining the sustainable agricultural production but is expected to drastically change the economics of growing the crops, to substantially reduce the pollution as well as encourage the mixed farming in steep diversion from the present day agriculture. The organic farming is aimed at increasing the agricultural production, without losing a close co-ordination existing between crop husbandry with eco-system, animal husbandry, soil health and other aspects of rural life.

The continuous and indiscriminate use of agro-chemicals has not only polluted our eco-systems and posed great hazards to human and animal health, but also spoiled the whole agricultural production system. Fertilizers and pesticides have indeed done their intended job of increasing the yields and protecting the crops against pests by costing us the basic medium of agricultural production. Before agro-chemicals were introduced, the soil had remained as a dynamic living medium capable to supply and retain the plant nutrients. As it had self-sustained capacity to produce the nutrients, its fertility status was maintained for centuries. The present problems of sustainability have arisen only after the advent of large scale use of agro-chemicals. Similarly, the crops had their natural system of resistance to pests. The pests had natural enemies such as predators/parasites. The whole system was running on a delicate balance of predators - parasites - pests and an in built resistance mechanism in every crop. The large-scale use of agro-chemicals has upset such a natural equilibrium in more than one way. The adverse relationship between excessive supply of nutrients and incidence of pests and diseases was hardly considered while recommending the use of fertilizers. The nutrient cycles involved in the natural production and supply of nutrients were also adversely affected by artificial supply of ready to use nutrients. This can be justified by following examples :

a. Suppression of microbial decomposition, when large scale fertilizers are supplied, due to changed C:N ratio

b. Antagonistic effect of one nutrient on availability of other nutrient (ex: High Ca status induces K deficiency; reduced availability of boron in presence of excess Magnesium; Deficiency of zinc in presence of higher phosphorus etc)

c. Reduced availability of nutrients (like zinc) in reduced pH

d. Effect of filler materials like talc, dolomite used in preparation of fertilizers on pH and other chemical reactions in soil.

Many more examples of disruption of natural process of nutrient cycles can be quoted. Reduced nodulation in pulses, reduced organic matter status of soils, increased fixation of phosphorus and potassium, induced deficiencies of micronutrients etc., are all due to disruption of natural process of nutrient production and their availability.

Similarly, the control of pests had its own natural equilibrium. Most species of the crops had their natural resistance built in their genetic makeup. As the crops and their varieties were introduced from one country to other for increasing the agricultural production, they developed susceptibility to the pests and diseases in the new situations. For example, *G. hirsutum*, introduced from USA to India was severely susceptible to pests and diseases in new situation. The present difficulty in controlling the pests of cotton is due to the susceptibility of *G. hirsutum* to cotton pests. It has also spread the pests in its present virulent form even to local species. Over generations, the natural resistance of many species and varieties is also reducing, as the pests are artificially controlled by the use of pesticides. This would naturally upset the capacity of such species and varieties to resist on their own. Several examples of loss of resistance in many crop plants such as cotton, paddy, maize, sorghum, pulses can be quoted. As the crop species and their varieties lose their resistance, newer varieties/hybrids with better resistance are sought instead of understanding the reasons for loss of resistance. It is also a myth to perceive that synthetic pesticides can fully control pests, as the pests are also living entities and have potentiality to build up resistance of their own to a given pesticide over a period of time. The use of insecticdes is also reported to have resulted in a large scale resurgence of pests, as the insecticides kill the non-target predators/parasites of the pest. Loss of predator/parasite population would naturally allow the pests to grow much faster and the pest would resurge. Similarly, the insecticides have been reported to cause serious mortality of beneficial insects such as pollinators, aquatic and wildlife besides creating many residue problems causing hazards to human and animal health.

It is discernable from the foregoing information that application of agro-chemicals - fertilizers and pesticides - was conceived, practiced, encouraged regardless of its effects on natural balance systems and the nature provided equilibria in soil plant-animal interactive medium. The use of agro-chemicals cannot be construed to have improved the agricultural scenario, as generally perceived, because it has disrupted the basic framework of natural eco-systems.

It is in this background, the organic farming gets increased importance. Organic farming is aimed at improving the agricultural production by retaining the natural equilibria in soil - plant - animal interactive medium without disrupting the nature provided resistance mechanism in a given species of the crop. It is also aimed to improve the *in situ* production of nutrients without disturbing the nutrient cycles provided in an eco-system.

The main argument against the adoption of organic farming is its relative inefficiency in resulting into sustainable and adequate crop yields immediately, although organic farming is the only way to attain sustainability in agriculture in the long run. The sustainable crop yield is indeed a prioritized agenda of the agricultural development in most developing countries. But, in long run perspective, sustainability of the whole production system (comprising ecological balance, soil fertility, natural resistance to pests and diseases) should be the aim of agricultural development rather than a short term goal of increasing the crop yields alone. In this context, the rediscovered theme of organic farming can reach this goal over a long period of 8-10 years gradually replacing the chemical farming. In the transition period, simultaneous use of both the systems may have to be inevitably advocated. The ways and means of improving the crop yields within the meaning of organic farming may have to be studied with a thrust on long term sustainability of eco-system. Only such measure will be able to achieve the desired growth in agriculture.

In the forthcoming chapters of this book, an attempt is made to provide detailed sketch of the concept of organic farming, ,an elaborate account of potential dangers of the use of agro-chemicals along with damages already caused, potentials practicing organic farming and its linkage to sustainability, as well as certification in organic farming. An attempt is also made to trace the history of organic farming from Indian perspective. At the end, global and Indian Scenario of organic farming is presented. The vastness of this subject and relatively less number of works reported on many aspects of organic farming has limited the scope of testifying the veracity of many perceptions about organic farming. But, the totality of this vast subject pervading all the aspects of agriculture is basic framework, in which this effort is made.

REFERENCES

Dhaliwal, G.S. and Pathak, M.D. 1993. Pesticides in the developing world : A boon in bane. In: Pesticides; their ecological impact in developing countries, pp. 8.

Dudani, A.T. and Sengupta 1991. Status of banned and bannable pesticides in India. Pub. Voluntary Health Association of India, New Delhi, pp. 68.

Food and Agriculture Organization 2001. Year book on fertilizers, 51 : 17-38.

Chapter - 2

EFFECTS OF CHEMICALS USED IN MODERN AGRICULTURE

2.1 GENERAL ISSUES

Modern agriculture was conceptualized in the past 5-6 decades with the sole objective of increasing the crop yields. Increase in the crop yield was achieved by the introduction of crops/their varieties, evolution of new photo insensitvie varieties/hybrids, use of fertilizers, pesticides, irrigation and many other cultivation practices. As the yield potentiality of the crops was improved by various breeding methods, the crops demanded the use of more agro-chemicals (Fig. 2.1).

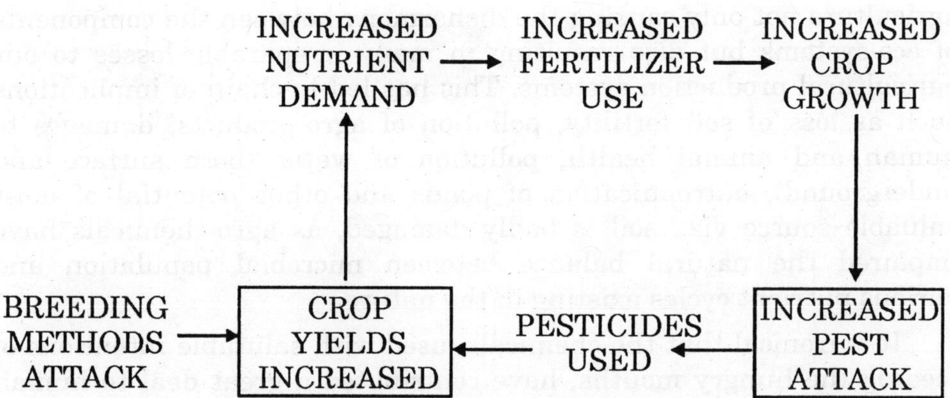

Fig. 2.1. Diagrammatic representation showing the necessities of using agro-chemicals.

Increased yield potentiality and the efforts to harness such potentiality naturally needed the supply of large quantities of nutrients artificially by using various inorganic chemicals. As the high yielding crops grew more vigorously, they were more susceptible to pests and diseases. This, in turn, necessitated the use of more pesticides to protect the crops. As the crops and their varieties were tried on new agro-climatic regions, as a part of introduction, their natural resistance was disrupted and the necessity to control the pests and diseases artificially became increasingly essential.

Thus, the chemicals were needed in modern agriculture both to increase the crop yields and to protect the crop from damages by pests and diseases. The agro-chemicals such as fertilizers, herbicides, insecticides, fungicides, bactericides, rodenticides, acaricides and growth regulators are now commonly used in the modern agriculture.

The dependence on agro-chemicals in the modern agriculture has been so much that it is hardly possible to imagine the agricultural development sans chemicals. In the pest 4-5 decades, as many as 150-200 types of chemicals were used to supply one or other plant nutrient and more than 600 types of chemicals were used to reduce the crop loss due to insects/diseases/weeds. Most national planning bodies, throughout the world, have regarded the level of use of agro-chemicals as a yard stick of agricultural development.

Notwithstanding the absolute necessities of increasing the crop yields in the perspective of global food shortages, these agro-chemicals infringed into naturally existing harmony between biotic and abiotic factors of crop production, besides acting as a source of pollution of water, food and air. Many toxic chemicals have entered the modern agriculture not only causing the disharmony between the components of eco-systems but also resulting in many irreparable losses to our agricultural production systems. This has led to chain of implications such as loss of soil fertility, pollution of agro-products, damages to human and animal health, pollution of water (both surface and underground), eutrophication of ponds and other potential of most valuable source viz., soil is badly damaged, as agro-chemicals have impaired the natural balance between microbial population and various nutrient cycles existing in the nature.

It is ironical that the chemicals, used with salutable intentions of feeding the hungry mouths, have contributed a great deal to human and animal miseries. A stage is now set to introspect whether the global agricultural development is in the right path or not. The use of agro-chemicals demolishing the basis of well balanced natural systems has led to the questions of sustainability in agriculture. An introspective rethinking is necessary to verify whether there is any need of so many toxic chemicals, causing series of health hazards, to improve the crop yields. While the population pressures are mounting on limited agricultural lands demanding the increase in the producti-vity, the increased awareness of hazards of agro-chemicals is discouraging their usage, which in turn may reduce the present level of productivity in short run. Indeed it is a formidable technical question to the agricultural scientists to solve this vexed problem of increasing the crop yields without depending on agro-chemicals. It is in this context, the organic farming is relevant as a part of sustainable agricultural system.

Since 1984, global condemnation of the use of toxic agro-chemicals in agriculture is on anvil as an effort to discourage the usage of such chemicals in agricultural production. Constant efforts are continuing to find alternatives for better way of farming without chemicals. The broader concepts of organic farming were perceived and presented by President of Norway during Rio-Earth Summit in 1992. The concepts such as zero chemical farming, natural farming, organic farming, bio-dynamic farming, nature farming and many other nomenclatures have been used to describe the new innovative or improved version of the old concept of organic farming. In essence, all these are similar so far as the use of naturally available materials to supply the nutrients and to protect the crop from damages. In fact, the organic farming is a better term to describe the phenomenon as it can explain the concept in its truthfulness with a stress on the use of organic materials than inorganic chemicals.

The large scale use of chemicals in agriculture has motivated the established of thousands of chemical industries throughout the world. These industries derive their existence on the sustained use of agro-chemicals. Besides causing direct pollution to air and water around their localities, these industries are solely responsible for innumerable ill effects in the agricultural eco-systems as well as chain of health hazards to animals and human beings. Notwithstanding the necessities of their existence, a question is now relevant to reconsider whether such an industrial development is a laudable social development.

Most of these chemical industries were established in the western countries. In the last 3-4 decades, the developing world imported these chemicals for their agricultural development. Of late, some developing countries such as India, China have established their own industries on the similar lines mostly in collaboration with their western counterparts. The dangers of agro-chemicals on the human society and eco-systems were recognized in western countries earlier to developing countries. Despite the vociferous efforts against the use of toxic chemicals in agriculture in their own countries, the developed countries adopted a veiled protectionist attitude in promoting the use of such chemicals in the developing and underdeveloped countries. It is distressing that developing countries have become a dumping yard for agro-chemical industries in the world. These industries derive their existence by their sales in developing countries. By adopting expensive advertisement strategies, by entering into many collaborative ventures, as well as by directly exporting the chemicals to such destinations, the developed countries have been continuously exploiting the situation for their benefit and to the detriment of the developing countries. This is glaringly evident in case of many chemicals banned in the developed countries but simultaneously promoted in the developing countries. Many examples of this type can be cited. The use of DDT was banned in USA in 1978. But its use for agricultural purposes was not banned till 1985 and its use for the public health continues unabated till today in India. Similarly, the use of EDB was banned in USA in 1983, but it is very widely recommended and promoted as the most effective insecticide against the store grain pests in India till today.

It is relatively easier for the developed countries to promote the sales of their agro-chemicals in the developing countries owing to less stringent laws, difficulties in executing the laws, poor literacy, unscientific and unsafe vending system as well as practice of using the agro-chemicals indiscriminately. In fact, the media advertisements, governmental concessions in terms of subsidies, ineffectiveness of legislation, delayed legal reprisals and any more lacunae of Indian society have been directly and indirectly beneficial for promoting such chemicals. Indiscriminate and unscientific use of agro-chemicals by the farmers has further helped in expanding the demand for these chemicals. The safety in using these agro-chemicals has been almost reduced to nullity in India, as an illiterate farmer is able to purchase quintals of poisonous agro-chemicals and is able to use it in the manner he likes, without regard to any safety measures. Our agricultural extension system has completely ignored the safety factor. The farmer not only endangers his health and the health of eco-system but also unknowingly becomes the cause of series of health hazards to the entire society. Neither the marketing system has any control over method of use, nor are the farmers aware of potent dangers. Any manufacturer would naturally make use of this opportunity to sell as much as he can in such a scenario.

In the forthcoming pages of this chapter, an attempt is made to give a detailed account of the effect of these agro-chemicals on soil, water, crops, food and food products, animal health, human health as well as the entire eco-system. Though the level of pesticide consumption in our country is as low as 0.30 kg ha^{-1} as compared to Netherlands (9.8 kg ha^{-1}), Portugal (5.3 kg ha^{-1}) and USA (2.8 kg ha^{-1}) it is necessary to curb the use of these chemicals at this stage itself to save our society from future chaos.

2.2 GLOBAL TRENDS IN USE OF AGRO–CHEMICALS

The chemicals used in modern agriculture are classified into following groups :

1. Fertilizers : To supply the nutrients for increasing the crop yields.

(a) Nitrogenous fertilizers

(b) Phosphatic fertilizers

(c) Potassic fertilizers

(d) Chemicals supplying calcium, sulphur, magnesium and iron

(e) Chemicals supplying manganese, copper, zinc, chlorine, boron, sodium, cobalt, vanadium and silicon

(f) Complex fertilizers supplying two or more nutrients but converted into pellets

(g) Mixed fertilizers

2. Pesticides : To reduce the losses due to pests such as weeds, insects, pathogens, animals.

(a) Herbicides

(b) Insecticides

(c) Fungicides

(d) Acaricides

(e) Bactericides

(f) Rodenticides

(g) Nematicides

3. Growth regulators : To promote or regulate the crop growth

(a) Defoliants

(b) Root promoters

(c) Retardants/inhibitors

(d) General growth promoters

4. Soil amendments

They may be gypsum or lime.

2.2.1 Fertilizers

The commercial use of fertilizers in agriculture started in early 1940s, though experimentally it was proved that the plant nutrients can be supplied to roots to improve the growth much before 1930s. In this brief span of 70 years, the use of fertilizers increased so rapidly that perhaps no other technology in agriculture has spread so fast. Though the use of fertilizers was started with some reluctance in the initial years, the dramatic results of fertilizers attracted the farmers throughout the world. The fertilizer usage became an easiest tool in farming to increase the yields. India could not remain exception to this worldwide phenomenon and the large scale commercial use of fertilizers commenced in India in late 1960s.

Globally, the consumption of fertilizers witnessed sharp increases in the decade 1970-1980. The technological improvement in the physical properties of fertilizers (like hygroscopicity, pelleting, and granular texture) promoted their easy transport and handling. The rate of consumption of fertilizers did not remain same in all the countries. The developed countries commenced their use earlier, reached their peak early. The consumption of fertilizers in those countries has slowly reduced from 1990s. But the developing countries commenced their large scale use of fertilizers in 1960s and their consumption is expanding till now. Among the developing countries, Asian countries consume maximum fertilizers, followed by Latin American countries. The major difference between the Asian countries and the developed western countries, as regard to the ill effects of the fertilizers would be the type of cultivation. In the developed countries, the agriculture being mostly extensive (as the pressure of population on the land is not so high) it is more convenient to plough back the crop residues in the soil along with the fertilizers. The rate of removal of nutrients is less due to low cropping intensity, but rate of addition of nutrients is high(both in terms of fertilizers and crop residues). In the developing Asian countries, the agriculture is mostly intensive and the cropping intensity is generally higher. It is hardly possible to plough back the crop residues to the soil. The rate of removal of nutrients is high due to high cropping intensity, but the rate of addition of nutrients is lesser. Hence, the level of organic carbon in former situations would be much higher than in the soils experiencing the intensive cultivation.

The world fertilizer consumption touched its peak of 184 million tons during 2011 (FAO 2011) (Table 2.1). Compared to consumption during 1961, it registered a growth of 153 million tons working out to annual growth by 3 million tons annually. In the last decade, the growth was to the extent of 4.6 million tons per annum. However, the negative growth was noticed in Europe, whose consumption started declining since early 1990s. In most continents, growth was witnessed in all fifty years, especially in the last decade.

Table 2.1. Trends in World fertilizer consumption (million tons)

Countries	1961	1971	1981	1991	2001	2011
Global	31.2	73.3	115.1	134.6	137.8	183.7
Africa	0.72	1.85	3.53	3.53	3.94	5.90
Americas	9.0	19.5	27.7	28.7	34.9	44.3
Asia	3.8	12.7	31.7	58.8	73.5	109.5
Europe	16.7	37.7	50.6	41.8	22.1	21.2
Oceania	1.0	1.6	1.6	1.8	3.3	2.0
China	0.73	4.6	15.2	29.7	33.4	56.9
India	0.34	2.6	6.1	12.7	17.3	27.8
USA	7.6	15.6	19.4	18.8	19.6	30.0

Source : FAOSTAT, 2011

The global trends in fertilizer consumption from 1961 to 2011 indicate that (Table 2.1).

- Almost 59 per cent of global fertilizer consumption is concentrated in Asia, particularly in India and China-which jointly account for 77 per cent of Asian and 46 per cent of global consumption.
- Europe has shown declining trends from 1991 till now and its fertilizer consumption has now pegged at 21-22 million tons. But, American continent has maintained consistently increasing consumption during this period.
- USA accounts for 68 per cent of total American consumption, indicating clearly that many Latin American and Caribbean countries consume very little fertilizers.
- Fertilizer consumption in Oceania has also declined in the last decade like Europe. But its consumption itself is as small as 12-18 per cent of Indian consumption, despite having large area under agriculture.
- Most developing countries in African continent have not registered growth in fertilizer consumption except S. Africa and Nigeria.

The Asian trends in fertilizer consumption indicate following :

- Except Japan, Korea and Saudi Arabia, most Asian countries have registered increasing trends of consumption.
- Steeper growth in fertilizer consumption was registered in China, India and Pakistan in the last 10 years. They incre-

ased their fertilizer consumption by 29, 44 and 57 per cent respectively.

- China itself consumes 52 per cents of combined Asian consumption of fertilizers, while India accounts for 25 per cent.
- Japan, Korean Republic registered an overall reduction of 27 and 43 per cent during the last decade.

It is evident from the above that Asia is the most popular fertilizer market for all the fertilizer manufacturers in world. India and China are particularly taking lead in their fertilizer consumption and may continue to lead further. Although increased fertilizer consumption increased the agricultural production, cumulative load of fertilizers on the cultivated lands over the past 20-30 years has culminated in many problems worth consideration. They are :

- Increased loss of nitrates has polluted the ground water and surface water. In general, 40-50 per cent of the applied nitrates are used by crops, taking into the consideration the soil type, method of application etc. The balance 50 to 60 per cent is either lost through water or volatilized to air. In submerged paddy, the loss of nitrates can be high as 66 percent of applied nitrogenous fertilizers.
- Increased acidity of the soils by fertilizer application. This is especially true when the fertilizers contain sulphur or phosphoric acid (e.g., super phosphate). They can retard the activity of microbes.
- Impairment of natural nutrient production systems in the soil, which are more dependent upon the decomposition of organic wastes.
- Imbalance set in by the use of N fertilizers in the nitrogen cycle- which essentially involve bacteria controlled processes such as nitrification, denitrification and ammonification.
- Large scale accumulation and fixation of some nutrients such as potassium and phosphorus in the soil. Phosphatic and potassic fertilizers are generally known to undergo fixation and only a part of them may be available to the immediate crop as a labile component.
- As all the fertilizers are applied to the top 30 cm soil layer, the continuous application can result into very high net cumulative enrichment of some plant nutrients like potassium and phosphorus as well as vast filler material.

- Application of fertilizers with excessive application of irrigation water has resulted in drain of applied nitrogen leading to nitrate pollution of surface and ground water.
- Application of some nutrients can suppress the availability of other nutrients, resulting into unavailability of such nutrients causing imbalance.

In India, the consumption of fertilizer has progressively increased over years from 0.67 lakh tonnes during 1951-52 to 27.8 million tonnes during 2011. Among different states, Punjab, Tamil Nadu, Andhra Pradesh consume more fertilizers to the tune of 221, 210 and 222 kg ha^{-1} (Anonymous, 2010) based on area under cultivation in these states. Without any regard to the detrimental effects of these fertilizers on the eco-system, the use of fertilizers is blindly promoted to achieve higher food production.

The fertilizer consumption in India has many peculiarities, which have enlarged the gravity of general problems. Most of such peculiarities are due to imbalanced and unscientific use of fertilizers. Some of them are :

- The farmers use the fertilizers an easiest and short cut method to improve the crop yields without having regard to their effect on soil fertility status as assisted by the soil tests.
- Most farmers presume that the fertilizers take care of all the crop requirements and valuable practice of adding the farm yard manure/ compost has been largely replaced by the fertilizer use.
- Wherever the soil tests are conducted, the test report is either ignored or not used by the farmers in its true spirit.
- Most of the farmers being illiterate, the fertilizers are hardly applied looking to the crop needs or recommendations and they are mostly unable to calculate the fertilizers actually required for the crops, as the recommendations are in terms of kg nutrients ha^{-1} and the available fertilizers are in specific formulations.
- The farmers use complex fertilizers without any consideration regarding the residues left by them for succeeding crops.
- As at least 30-40 per cent fertilizers are sold on credit, the farmers will not have choice of selecting the fertilizers during credit sales. The creditor dictates the type of fertilizers purchased by the farmers. The farmer is constrained to compromise with whatever he gets rather than what he wants.

- Very often, the fertilizers are likely to supply either more or less of nutrients to the crop than actually required. The known formulations such as 19 : 19 : 19; 10 : 26 : 26; 15 : 15 : 15; 17 : 17 : 17 or Dia ammonium phosphate (DAP) supplying 18 : 46 : 0 etc. are sold as tailor made for all crops in most situations. Such formulations do not conform to the requirement of most crops, forcing the farmers to use more or less nutrients. In most rural situations, farmer is forced to purchase what is available in his village than what he wants. The availability of fertilizers gets priority than the nutrient requirement and supply.
- Complex fertilizers are also commonly used for top dressing, which can only lead to increased leaching losses, wastage of nutrients, fixation etc., as the crop is hardly able to absorb all the nutrients from slowly releasing complex fertilizers, in a short time before being lost.
- Without having regard to various application losses, the fertilizers are broadcasted by most of farmers even under rainfed conditions leading to very poor fertilizer use efficiency.
- The farmers have almost presumed that any fertilizer can satisfy all the crop needs, though none of the fertilizers can supply all the crop nutrients.
- The fertilizers are used in place of application of farm yard manure or other organic manures. Most farmers have abandoned the age old and time consuming practice of preparation of farm yard manure due to presumption that the farm yard manure is unnecessary. Those who use the manure, use them in the unscientific manner leading to loss of nutrients from manure.

The continuous and indiscriminate use of fertilizers had resulted into irreversible manifold problems. The damage due to fertilizers on the soil, water and environment are more predominant in developing countries than the developed.

Thus, the fertilizers have done more harm to our soils than building up their productive capacity but temporarily increased the crop yields. Besides damaging various biotic and abiotic factors of the soil, the use of fertilizers has created problems of water pollution resulting into consequences such as eutrophication. According to UNEP (1991) 30-40 per cent of world's lakes and reservoirs are eutrophic and most often, the waters reaching these lakes and reservoirs are enriched with phosphorus and nitrogen. Eutrophication has not only changed the composition of aquatic flora and fauna, but has made their waters unworthy of use by human beings and animals.

2.2.2 Use of fertilizers and food production

Despite the manifold maladies created by the use of fertilizers, no one can ever deny a dominant role played by the fertilizers in the world food production. Fertilizers have almost yielded magical results in increasing the food production in the past five decades. The yields of many food have doubled/trebled. In absence of fertilizers, many populous countries like India and China would have faced much severe hungry deaths.

However, per capita food production has either changed as a function of fertilizer use or/and rate of change of population. This is evidenced by data presented in Table 2.2

Table 2.2. Per capita food production index in selected countries

Countries	1995	2000	2002	2005	2008	2011
Asia						
India	81.2	91.7	87.4	100.1	116.6	129.3
China	66.1	83.8	89.1	100.5	112.7	124.0
Bangladesh	66.8	89.5	90.2	102.9	120.4	132.1
Indonesia	77.8	78.3	86.1	98.1	113.0	124.6
Iran	69.3	77.8	89.8	103.0	98.5	114.1
Iraq	101.0	100.1	127.9	106.3	95.5	114.0
Israel	76.9	87.3	94.4	100.0	104.1	106.2
Japan	109.7	103.5	102.6	101.0	101.2	92.7
Philippines	74.6	84.1	90.9	100.2	113.5	115.1
Turkey	81.9	94.2	92.5	101.7	105.6	120.1
Europe & Oceania						
Australia	86.2	105.3	94.0	108.0	102.6	108.3
Czechoslovakia	102.8	97.9	96.7	98.3	101.5	97.6
Denmark	96.9	98.4	99.3	101.5	103.8	102.0
Finland	94.9	98.7	100.8	101.9	100.0	98.9
France	99.1	103.8	104.8	96.7	96.4	99.2
Germany	91.3	99.4	97.7	99.6	104.3	105.3
Greece	103.2	106.3	101.1	102.8	94.0	86.9
Hungary	85.2	89.0	90.3	95.0	103.8	88.2

Countries	1995	2000	2002	2005	2008	2011
Ireland	98.2	104.4	99.1	99.1	97.0	103.4
Italy	95.2	100.2	94.7	100.4	98.2	95.2
Poland	90.5	91.8	94.7	98.8	102.2	103.6
U.K.	109.1	104.3	101.3	100.0	103.0	104.1
Africa						
Nigeria	66.7	80.9	84.4	99.7	104.2	104.0
South Africa	71.0	95.8	97.0	102.4	118.9	116.1
Sudan	52.4	67.0	78.8	99.8	110.4	109.8
Tanzania	67.2	71.3	95.3	96.1	104.9	128.2
Uganda	76.0	88.4	97.3	100.0	104.6	114.7
Zambia	69.5	84.9	85.0	97.3	103.6	141.6
Namibia	91.0	84.9	84.6	103.6	91.1	91.3
Mozambique	77.3	90.8	94.2	94.1	99.4	119.5
Morocco	55.6	71.8	84.2	93.3	103.2	126.3
Ethiopia	55.9	70.0	85.3	101.7	113.0	135.6
Egypt	67.0	84.5	86.7	97.7	117.3	113.9
Americas						
U.S.A.	83.9	96.0	94.1	99.4	107.1	105.0
Canada	78.3	91.5	82.4	101.5	106.8	109.3
Cuba	88.2	123.5	118.5	93.8	96.4	96.5
Mexico	78.5	88.4	92.5	98.5	105.1	105.0
Argentina	64.5	90.6	100.3	104.4	102.5	120.9
Brazil	66.2	76.8	86.5	99.4	116.7	127.0

Source: World Bank, 2011; 20005 figures are = 100

The countries, having negative growth in per capita food production are Greece, Japan, Italy, Hungary, Poland and Cuba. Most such countries have very low growth in population and negative growth in per capita food production is due to reduced fertilizer consumption. Positive growth is recorded in countries such as China, Tanzania, Brazil, Zambia, Egypt, Ethiopia and India due to increased fertilizer consumption, despite increased population. Many European countries with negative growth in per capita food production can still manage to maintain their food supply because of low population pressure. But many African countries (e.g., Namibia) with negative growth in per capita food production will suffer in coming years due to aggravation of already dilapidated food situation.

The world food production is thus, fully dependent upon the fertilizer consumption. The excessive dependence of food and agricultural production on fertilizers on one side, serious dangers posed by their use on the environment, soil, human and animal health on the other side have led to vexed challenge about the continuing the use of fertilizer use.

This should not mean that the fertilizer use has to be pushed up from the present level. Global policy, in the past, had been to push up the fertilizer usage as a straight and easy remedy for food shortages. Because of such policies, many miseries are now surfaced and detected. Fertilizer usage can no more be an easy and straight solution to the future food shortages, as the response to added nutrients has diminished in most crops. The food production has to be increased by promoting the ecologically sound production systems and halting the processes of resource depletion.

Instead of adopting the easy bur ecologically dangerous solution of supplying readily available nutrients in the root zone by fertilizers, the sustainable systems will have to be developed to make the soil to produce, retain and supply the plant nutrients by eco-friendly practices. The soil should be able to supply the nutrients by *in situ* production methods and store the plant nutrients by making use of natural processes in given eco-system. This is possible only by practice of organic farming leading to sustainable agricultural development.

2.2.3 Pesticides

An estimated 2.4 million metric tonnes (MMT) of pesticides are used in world agriculture to save the crops from pests (Fischel, 2007). Nearly 17 per cent of this is constituted by the insecticides and 40 per cent of the total pesticides consumed are made up of herbicides (Table 2.3). Around 10 per cent of the total pesticides consumed comprise of fungicides while sundry pesticides account for 33 per cent.

Table 2.3. Consumption of pesticides in world (%)

Group region	World	Asia-Pacific	India
Insecticides	17	76	72
Herbicides	40	13	11
Fungicides	10	8	12
Other's	33	3	5

(David, 1992a; ADB, 1987 and Anon., 2001a, Fishel, 2007)

Though the chemicals such as sulphur were reportedly used as insecticide as early as 77 A.D., large scale use of chemicals for the control of pests started about seven decades back. Unlike fertilizers, more pesticides are used in developed nations than developing ones. Nearly 75 per cent of total pesticide consumption is used by developed countries (Anonymous, 1998) covering 25 per cent of global geographical area. Out of this, Europe and North America account for 27 and 30 per cent of global pesticide market, while Japan accounts for 12 per cent. But, three continents Asia, Africa and Latin America put together account for 30 percent market.

This heavy imbalance is pesticide use is evidenced by the pesticide load in different countries. It is as high as 9.4 kg ha^{-1} yr^{-1} in Netherlands (Table 2.4), 5.3 kg ha^{-1} yr^{-1} in Portugal, 2.5 kg ha^{-1} yr^{-1} in Germany and 2.8 Kg ha^{-1} yr^{-1} in USA as compared to 0.3 kg ha^{-1} yr^{-1} in India and 0.13 kg ha^{-1} yr^{-1} in Africa (Edwards, 1986). However, in recent years some countries of Latin America like Costa Rica and Columbia have recorded exceptionally high pesticide load of respectively 51 Kg ha^{-1} yr^{-1} and 16.7 Kg ha^{-1} yr^{-1}. In most of the nations, use of pesticides is concentrated on few crops (Dhaliwal and Pathak, 1993). For example, in USA 93 per cent of area under corn sorghum, soybean and tobacco receives very little pesticides while, relatively a smaller area under other crops receive the higher load of pesticides. Similarly, in China, chemical pest control is most commonly adopted in paddy than other crops.

Table 2.4. Levels of pesticide use in selected countries

Country/area	Pesticide use (kg ha^{-1} yr^{-1})
Netherlands	9.4
Portugal	5.3
Germany	2.5
USA	2.8
Austria	2.4
Thailand	1.40
Indonesia	0.60
India	0.30
Latin America	0.22
Oceania	0.20
Africa	0.127

(Anonymous, 2000)

While the herbicides and fungicides account for larger share of pesticides in developed countries, the insecticides constitute lion's share in developing countries. Large scale use of herbicides is an indication of acute shortage of agricultural labour in those countries. But in developing countries of Asia-Pacific region, the insecticides account for 3/4th share of total pesticides consumption (Table 2.3).

Among insecticides, the organophosphates and carbamates account for major share in developed nations, while, use of organo-chlorines is still predominant in developing countries. As the consumption of pesticides in developed countries is more, the world consumption of pesticides is dominated by organophosphates (36%) followed by pyrethroid (25%), carbomates (21%) and organo-chlorines (8%) (Szczepanski, 1990).

In Asia-Pacific region, China accounts for 51.3 per cent of pesticide market, while India consumes 13.9 per cent of the pesticides sold in this region (Table 2.5). Except Korea, Malaysia, Thailand, Sri Lanka, Fiji and Papua New Guinea all other countries in this region consume considerably large quantity of insecticides than herbicides.

Table 2.5 Pesticide market in Asia-Pacific region

Country	Insect-icides	Herb-icides	Fungi-cedes	Others	Total	% of market in Asia Pacific region
China	90.0	5.0	4.0	1.0	100	51.3
India	80.0	7.0	10.9	2.0	100	13.9
Korea	39.0	22.0	32.0	7.0	100	9.0
Indonesia	74.0	15.3	9.4	1.4	100	7.9
Malaysia	13.0	75.0	3.0	9.0	100	5.5
Pakistan	91.2	4.5	2.2	2.0	100	4.7
Thailand	51.0	39.0	9.6	–	100	3.6
Sri Lanka	51.3	38.1	6.3	4.4	100	0.6
Bangladesh	94.5	2.5	1.7	1.7	100	0.5
Burma	87.5	–	6.3	6.3	100	0.1
Papua New Guinea	15.0	60.0	20.0	5.0	100	0.1
Fiji	42.3	53.8	7.7	–	100	0.1
Western Samoa	33.0	67.0	–	–	100	0.1
% of market	75.8	13.4	8.4	2.4	–	100.0

(ADB, 1987)

In India, 80 per cent of the pesticides are insecticides, while, fungicides and herbicides contribute only 10 and 7 per cents, respectively. The consumption of insecticides began in India principally for malaria control and later expanded gradually to agriculture. The share of agricultural consumption has risen to reach 61.5 per cent by 1983-84 (CERC, 1989). Nevertheless, the use of pesticides for non-agricultural purposes, mostly for public health, is substantial. Among various crops having major pesticide consumption cotton consumes 44.5 per cent of insecticides used in India followed by rice (22.8%) (Dudani and Sengupta, 1991). Among other crops, pigeon pea consumes 12.8 per cent of total insecticides, while all other crops including sugarcane receive meager quantity of pesticides. Among different states.

Table 2.6. Consumption of pesticides in states of India

State	Consumption (MT) 2005	2009	Cultivable area (million ha)	% of total consumption	Per ha consumption (kg/ha)
Andhra Pradesh	1997	1015	14.2	2.4	0.07
Bihar	875	828	10.9	1.9	0.08
Gujarat	2700	2750	9.9	6.5	0.27
Haryana	4560	4070	3.7	9.7	1.1
Karnataka	1638	1647	11.9	3.9	0.14
Kerala	571	631	1.8	1.5	0.04
Madhya Pradesh	787	645	22.2	1.5	0.03
Maharashtra	3198	4639	21.4	11.0	0.22
Orissa	963	1588	5.3	3.5	0.30
Punjab	5610	5810	4.1	13.9	1.42
Rajasthan	1008	3527	20.6	8.4	0.17
Tamil Nadu	2211	2335	7.8	5.5	0.30
Uttar Pradesh	6671	9563	17.6	22.9	1.26
West Bengal	4250	4100	5.6	9.8	0.73
Other States/ Union Territories	1050	913	7.6	3.0	0.29
Total	39773	41822	164.6	100	0.254

(Anonymous, 2010)

The pesticide consumption in India has increased in recent years (Table 2.6) from 39773 tons in 2005 to 41822 in 2009.This increase is ascribed to higher consumption in the states of Uttar Pradesh,

Maharashtra, Punjab, Rajasthan and Orissa even though consumption was reduced in the states of Andhra Pradesh, Haryana and West Bengal.

The consumption of pesticides in India accounts for only 3 per cent of global pesticide market. Out of 600 distinct technical grades of pesticides, only 134 have been legally permitted for use in India (David, 1992 b). The pesticide load on Indian farm land is as less as 0.3 Kg ha^{-1} yr^{-1}. But, 45 per cent of total pesticides consumed in India reach cotton area of 8 million ha which accounts for less than 5 per cent of cultivable area in our country. Thus, the consumption of pesticides in India on cotton is well comparable to western countries (Bami, 1989), while other crops except paddy receive very low quantity of pesticides. Only 48 per cent of total cultivated land in our country receives pesticides (Gupta, 1989). Endosulphan mancozeb and malathion account for 50 per cent of insecticide consumption in our country, even though they were banned in other countries long back.

Although average pesticide load in India is as less 0.3 kg/ha/yr, Uttar Pradesh, Punjab, Haryana have pesticides load in the range of 1.1 to 1.42 kg/ha. But, Bihar, Kerala and Madhya Pradesh have a pesticide load of as less as 0.03-0.08 Kg ha^{-1} yr^{-1}.

As such a large quantity of pesticides and fertilizers are consumed every year to protect the crops or to increase the crop yield, it is increasingly necessary to know in detail about their effects on soil, water, air, ecological balance, pollution, animal and human health. Such a study becomes more relevant in the context of some pesticides being persistent in the environment with the potential bio-accumulation and bio-magnification. Some of the pesticides are known to be carcinogenic, tumorogenic and teratogenic with large scale ramifications on human and animal health. The bio accumulated pesticides in the human and animal bodies may act as source of many medical problemms, besides causing abnormalities to the innocent young ones. The use of agro-chemicals on the crops has caused many irreversible effects on the soil health, underground water contamination as well as hazardous pollution recorded in various unprocessed and processed foods.

In many crops, the present level of use of fertilizers has resulted in stagnation of yields. Any incremental yield can be expected only by additional dosage, evidencing weakened inherent capacity of soil to supply and store the nutrients. Similarly, the use of pesticides has not offered the solutions to pests fully. Despite the use of large quantity of pesticides, crop losses due to pests are not successfully controlled. Gupta (1989) reported a crop loss of Rs. 6000-00 crores in our country.

This raises the question as to whether the use of pesticides is the right solution to protect our crop or not. In surmise, the original purposes of using the fertilizers and pesticides (to increase the crop yields and to protect the crops from pests) are not fully achieved despite their large scale use.

2.3 FATE OF AGRO-CHEMICALS

Before studying the impact of various agro-chemicals on soil, air, water, human and animal health, a clear understanding of their fate on application to crops is needed.

2.3.1 Fertilizers

Different fertilizers used in agriculture meet different fate, when applied to soil or foliage. In general, their fate can be summarized as follows :

2.3.1.1 Nitrogenous fertilizers

Nitrogenous fertilizers would supply either NH_4+ ions or NO_3- ions or both depending upon the source. They are able to supply ammonium (e.g., urea, complex fertilizers) or nitrate ions (e.g., calcium ammonium nitrate, calcium nitrate, potassium nitrate) on dissolution. Ammonium ions may be adsorbed to the plant roots or to soil particles and later be available to plant roots (Fig. 2.2). Alternatively, they may be fixed inside the expanding clays or leached to subsoil if cat ion exchange capacity of the soil is less. Ammonium ions are nitrified later and added to the nitrate pool. Ammonium ions are also utilized by heterotrophs in the soil, which may later release nitrates to the nitrate pool. The added ammonium ions are subject to constant transformation to nitrate form by nitrifying bacteria if not utilized by the plants directly or not leached to subsoil. The nitrate form of fertilizers directly releases the nitrate ions, which may be utilized directly by the plants through the soil solution or to very limited extent through the anion exchange. Whenever, the soil is saturated with nitrates, the unutilized nitrates are usually leached to subsoil, or lost with surface or ground water or may be immobilized into organic forms by ammonifying bacteria or subjected to denitrification process leading to loss of nitrogen in the form of nitrites/nitrous oxide. When the nitrogenous fertilizers are added, the soil is saturated with the nitrate/ammonium ions and constantly subject to the processes of nitrification, ammonification or denitrification through the action of microbes. The nitrogen fractions not utilized by the plant roots are bound to be lost in some form or other unless fixed by clay minerals or immobilized by microbes.

Two conclusions can be drawn in N-dynamics.

(a) Supply of NH_4/NO_3 ions must be in accordance with the capacity of plant roots to take them up. Otherwise, major part of added nitrogen is likely to be lost in some way.

(b) Available NH_4/NO_3 ions may be immobilized, if soil has adequate microbial population.

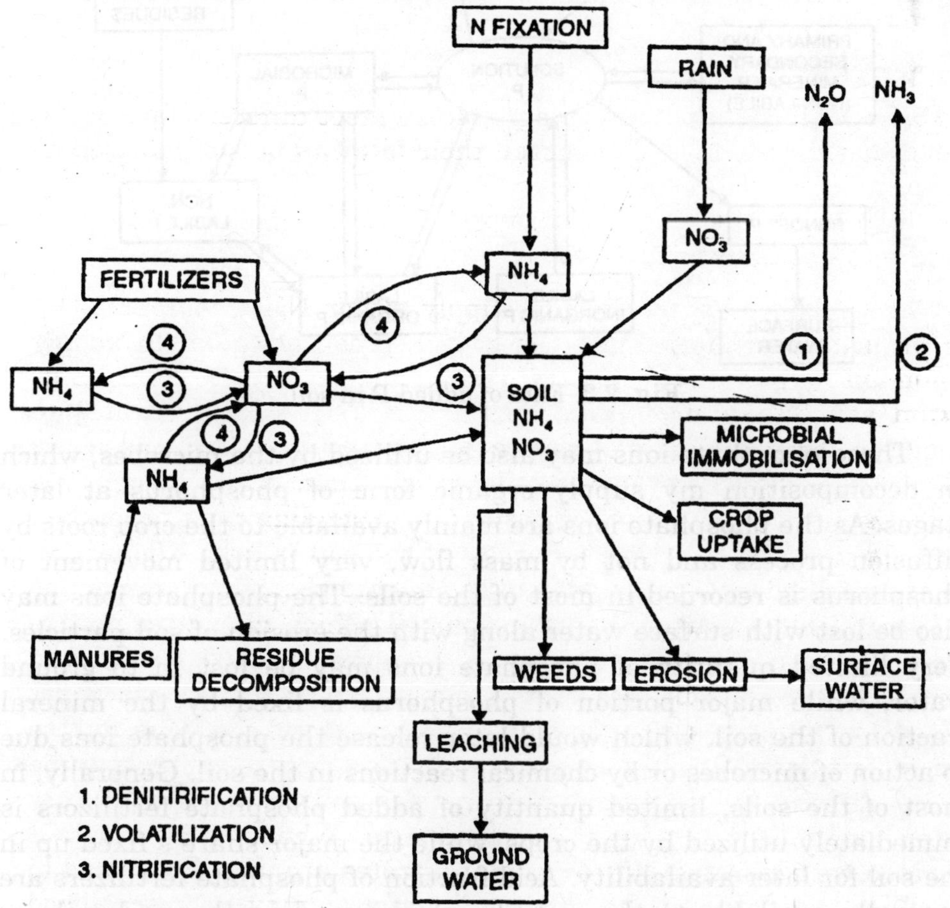

Fig. 2.2. Dynamics of nitrogen in soil-plant production system.

2.3.1.2 Phosphate fertilizers

Phosphorus is principally supplied to crops as super phosphate or rock phosphate or other complex fertilizers. These fertilizers contain phosphates in water soluble form or acid soluble form. On dissolution with water, the water soluble fraction supplies HPO_4+/H_2PO_4–ions to the soil solution. These ions are later absorbed by the crop roots directly or precipitated as Ca, Al or Fe phosphates or adsorbed by the soil particles depending upon the anion exchange capacity of the soil (Fig. 2.3).

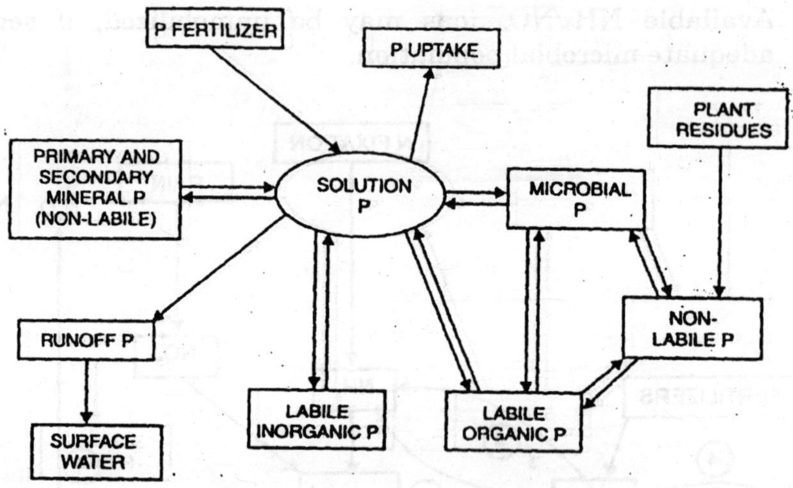

Fig. 2.3. Fate of added P in soil

These phosphate ions may also be utilized by the microbes, which on decomposition my supply organic form of phosphorus at later stages. As the phosphate ions are mainly available to the crop roots by diffusion process and not by mass flow, very limited movement of phosphorus is recorded in most of the soils. The phosphate ions may also be lost with surface water along with the erosion of soil particles. Very limited quantity of phosphate ions may be lost in to ground water, while major portion of phosphorus is fixed by the mineral fraction of the soil, which would later release the phosphate ions due to action of microbes or by chemical reactions in the soil. Generally, in most of the soils, limited quantity of added phosphate fertilizers is immediately utilized by the crops, while the major share s fixed up in the soil for later availability. Acid fraction of phosphate fertilizers are generally available to the soil phosphorus pool in the acid soils or alternatively fixed up in the soils as soil mineral fractions.

2.3.1.3 Potassic fertilizers

Potassium is supplied to the crops by fertilizers such as muriate of potash or sulphate of potash or complex fertilizers. On dissolution, these fertilizers release K+ ions to the soil solution. These ions may be directly utilized by the crops through the processes like mass flow or diffusion or adsorbed to plant roots/soil particles for later utilization by the plant roots. (Fig. 2.4) Potassium ions may also be lost down to deeper layers of the soil or to ground water or surface runoff. A part of applied potassium may be fixed in expanding clays, which may also be available to the roots at later stages, as K concentration in the soil solution is gradually reduced.

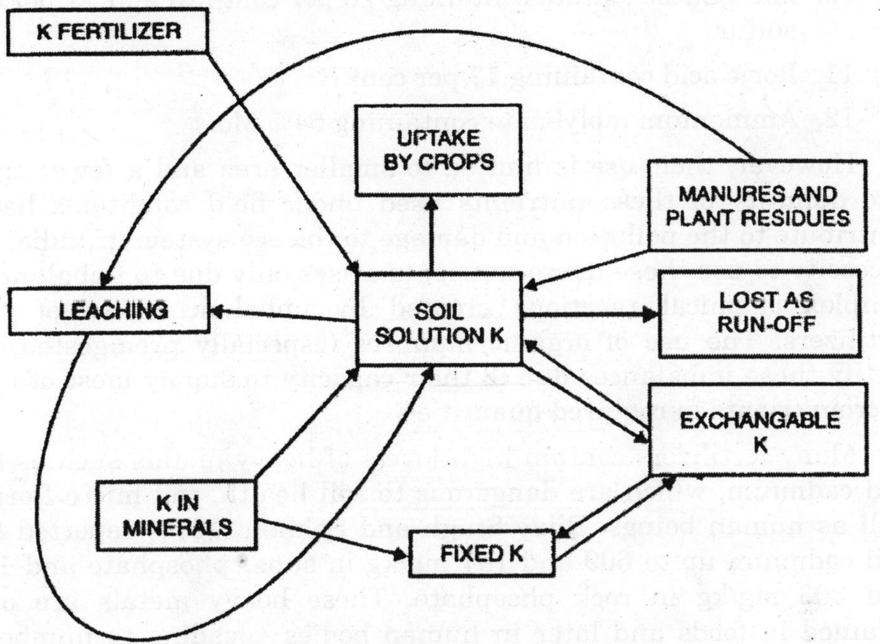

Fig. 2.4. K equilibria and cycling in soils.

In general, only a part of added potassic fertilizer is directly or indirectly taken up by the crops, while the remaining part is bound to enter the surface or ground water unless utilized by the microbes.

In addition to these nutrients, the crops also require calcium, magnesium, sulfur, iron, manganese, zinc, copper, boron and molybdenumnum for their growth. They are supplied through chemicals such as :

1. Super phosphate containing 20 per cent calcium and 11 per cent sulphur
2. Lime containing 25 percent calcium
3. Gypsum containing 28–32 per cent calcium and 18 per cent S
4 Ferrous sulfate containing 19 per cent Fe and 18 per cent sulfur

5. $MgSO_4$ (Epsom salt) containing 10 per cent Mg and 13 per cent S
6. Potassium sulfate containing 18 per cent S
7. Urea + sulfur containing 10-20 per cent S
8. Zinc sulfate containing 18 per cent S and 36 per cent Zn
9. Granular sulfur containing 68-90 per cent S
10. Manganese sulfate containing 26 per cent Mn and 11 per cent sulfur
11. Boric acid containing 17 per cent B
12. Ammonium molybdate containing 54% Mo.

However, their use is limited to smaller area and a fewer crops. The quantity of these nutrients used under field conditions hardly contribute to the pollution and damage to the eco-system in India. The necessity to use these micro-nutrients arises only due to imbalance in complex chemical reactions created by unbalanced use of NPK fertilizers. The use of organic manures (especially predigested) can rectify these imbalances due to their capacity to supply most of these micronutrients in required quantities.

Many fertilizers contain high levels of heavy metals such as lead and cadmium, which are dangerous to soil health, soil micro flora as well as human beings. Bijay Singh and Sekhon (1977) reported lead and cadmium up to 609 and 187 mg/kg in super phosphate and 1135 and 303 mg/kg in rock phosphate. These heavy metals are often retained in foods and later in human bodies - leading to number of health complications.

The foregoing information on the fate of added fertilizers clearly reveals that only a part of added fertilizers is actually used by the crops and the major share of applied fertilizers is subject to various processes of chemical or microbial reactions, which would release these nutrients later and expose them to losses through erosion, runoff water or ground water. The complex chemical reactions involving mineral and organic forms of these nutrients are governed by several alterable parameters such as moisture availability, pH, organic matter status of the soil, microbial population, temperature and uptake by the plants. Adverse balance in any of these parameters may drastically increase the extent of loss of nutrients. It is not safe to assume that all the added fertilizers are used up by the crops and calculations of nutrient requirement based on such assumption may lead to speculate either low or unbalanced nutrient requirement. This may result into low of balance between biotic and abiotic components of the soil, besides, leading to substantial loss of nutrients.

2.3.2 Pesticides

The pesticides are principally applied as emulsified concentrates, granular forms, dusts and fumigant. All these forms are subject to different pathways of degradation (Fig. 2.5), depending upon the method of application, concentration, atmospheric humidity, temperature etc.

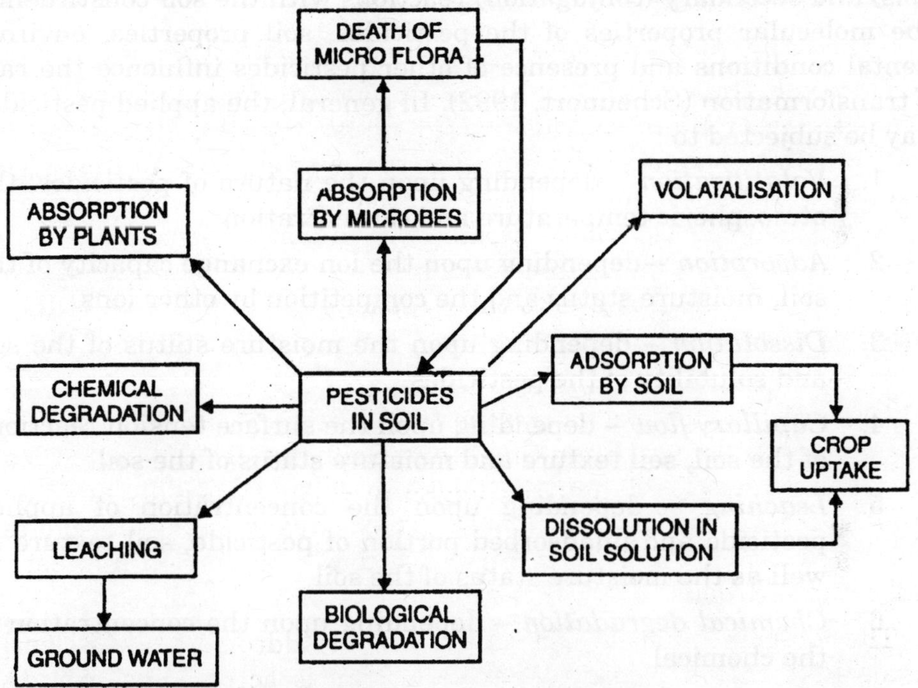

Fig. 2.5. Fate of pesticides in soil

The emulsified concentrates are applied as sprays directed to reach the foliage of the crop plants. But, major part of so applied emulsified concentrates reach the soil and very limited quantity of applied spray solution can be absorbed by the foliage in a limited time before being slipped away from the leaves. Apart from this, a major part of the spray solution directly reaches the soil and non-target area due to drift. The dust formulations also meet similar fate to reach more non-target area than targeted area. The granular form and fumigants are applied directly to the soils, where a part of them is useful to kill soil borne pests and useful insects and major part is lost in soil solution.

In any case, it is clear that only a small part of the applied pesticides reaches the target pests and major part of them reaches the undesirable targets such as soil, air and soil fauna. According to one estimate, only 0.1 per cent of the applied pesticides reach the target pests, while, the remaining amount of pesticides act as source of pollution of air, water, soil or take part in the metabolic activities of microbial populations and other soil fauna.

The applied pesticides undergo various abiotic and biotic transformation/reactions comprising primary (oxidation, reduction, hydrolysis) and secondary (conjugation) reactions with the soil constituents. The molecular properties of the pesticides, soil properties, environmental conditions and presence of other pesticides influence the rate of transformation (Scheunert, 1992). In general, the applied pesticides may be subjected to :

1. *Volatilization* – depending upon the nature of pesticides, the atmospheric temperature and concentration.
2. *Adsorption* – depending upon the ion exchange capacity of the soil, moisture status and the competition by other ions.
3. *Dissolution* – depending upon the moisture status of the soil and solubility of the pesticides.
4. *Capillary flow* – depending upon the surface tension reactions of the soil, soil texture and moisture status of the soil.
5. *Leaching* – depending upon the concentration of applied pesticide and unadsorbed portion of pesticide, soil texture as well as the moisture status of the soil.
6. *Chemical degradation* – depending upon the concentration of the chemical.
7. *Absorption by plants* – depending upon population of weeds and crop plants.
8. *Microbial degradation and accumulation* – depending upon microbial population, pH, temperature and aeration.

It is evident that the pesticides and their metabolites are subject to various dissipation pathways such as absorption and retention by the crops, retention by soil particles, movement through surface runoff often along with the flowing water/soil particles, diffusion through vapour phases and hydrodynamic transport through water column ultimately leading to loss along with the leaching water. It is unsafe to assume simply that the applied pesticides are degraded by various degradation pathways and hence pesticides do not harm the environment. Because, even after degradation, the pesticides serve as source of pollution to various components of the environment. Many pesticides falling in the groups such as aromatic hydrocarbons, hydro chlorine compounds (both aliphatic and aromatic) and substituted hydrocarbons show tendency of bio-accumulation in microbes, other animals and plants (Mackay, 1982).

The persistence of the applied pesticides in the environment refers not only to their existence in their original form but also in transformed forms after possible degradation. Thus, 20-70 per cent of pesticides applied in agriculture remain in the soil as persistent residue bound to soil colloids (Calderbank, 1989). Surface adsorption and covalent bonding are two main mechanisms of pesticide persistence in the soil. Ironically, the organic matter (which is a desirable component of the soil) is largely responsible in the initial process of binding the pesticides. With the passage of time, the bound residues tend to become more firmly bound and offer more resistance to the degradation process.

2.4 EFFECTS OF AGRO-CHEMICALS ON SOIL

The soil is a living, dynamic medium having a harmonious integration among its three components viz., physical, chemical and biological spheres. The top weathered layer of the soil harbors innumerable organisms (bacteria, fungi, actinomycetes, protozoa, algae, annelids, arthropods and mammalians) living in harmony with physical and chemical spheres. The complex chemical reactions undergoing in the soil are continuously influenced by the activity of these organisms as well as various physical parameters such as pH, temperature, moisture, concentration of CO_2 and O_2 and soil texture. The nature has provided an excellent mechanism of adaptation to these organisms to changing environmental cycles such as carbon cycle, nitrogen cycle and hydrological cycle. In any soil eco-system, there is an integrated food chain being supported by life cycles of different organisms. They live and die to support and nourish other organisms of the food chains leading to food webs. In the process of their natural existence, they release complex organic chemicals, which are able to support other organisms making the whole system stable. The entire system will also influence and being influenced by the plants and animals living above the soil. The system supports the plant growth naturally and in turn expects the plant residues to be ploughed back for continuing the carbon and nitrogen cycles in their natural forms. Left to the nature, the entire system works on a stable but delicate balance. In a given climate, all the components of the soil ecosystem viz., plants and animals on the soil, biosphere inside the soil and chemical cycles support each other. The enrichment of soil by organic components through continuous integration of living and non-living components of the soil ultimately helps to build up the organic matter. The physical characters of the soil not only influence and support the living organisms but also are influenced by the organic matter content in the soil.

Such a delicately balanced but stable medium of soil is basic essentiality of mankind, as it is this medium which supports all the agricultural crops of the world. Prior to the use of fertilizers and pesticides, agriculture was practiced with a limited human interference in support of natural processes. The crop residues were either returned to the soil or allowed to be decomposed with the animal excreta and it was applied to strengthen and continue the natural cycles thus, maintaining the organic carbon status of the soil. In those days, the use of pesticides was not in vogue because varieties of the crops grown in a particular locality were subjected to natural selection and surviving ones had natural resistance to the pests and diseases.

With the increasing global population and enhanced industrialization, the agenda of increasing crop yields and protecting the crops became imperative. Artificial supply of the nutrients to the soil through fertilizers as well as introduction of crops/their varieties from one region to other thereby, witnessing the wide spread susceptibility to pests and diseases, necessitated the use of agro-chemicals in the modern agriculture.

The effects of agro-chemicals on the soil are manifold. They have many direct and indirect effects on the soil health. These effects can be studied separately for the three components of the soil viz., physical, chemical and biological. Often, some of these effects are found to be interdependent resulting into cumulative harm. The final deleterious effect of a particular chemical may be masked by some advantage. It should not be construed that such a chemical has no harmful effect on soil health, as the benefits of such advantages may be short lived and in due course they may expose the environment to harmful effect.

2.4.1 Physical components

Soil physical component principally comprises the soil texture, soil structure, soil air and its composition, soil pH and soil organic matter. Except soil texture, all the other physical parameters are reported to experience one or the other deleterious effect by the use of agro-chemicals.

2.4.1.1 Soil structure

The evidence available on the effect of agro-chemicals on the soil structure is scanty and often conflicting. Essentially, the soil structure is a process of formation of soil aggregates by the flocculated soil particles. The extent of clay content and the nature of adsorbed cat ion influence such aggregate formation. Ca and H clays bring about better aggregation than Na, Mg or K clays. The clays saturated with Na, Mg and K minerals keep them dispersed and are not able to form soil aggregates. In general, the long term use of fertilizers such as sodium nitrate or ammonium sulfate can lead to reduced stability of the soil aggregates (Daji, 1988). Many potassium fertilizers are also reported to have various proportions of NaCl, which contains 40 per cent sodium (Tisdale *et al.,* 1985). In spite of potential damage to the soil structure by the use of such fertilizers, their usage is increasing in developing countries year after year in contrast with their declining use in developed countries. For example, in India 536 thousand tonnes of ammonium sulphate and 1188 thousand tonnes of muriate of potash was consumed during 1994-95 (Anno., 1995). Similar trend is recorded in many other developing countries.

2.4.1.2 Soil pH

Of all the physical factors, pH appears to be very important factor affected by the use of fertilizers.

When nitrogenous fertilizers such as ammonium sulphate, ammonium chloride, diammonium phosphate and urea are added to the soil, ammonium ions are released which would react with oxygen to form free hydrogen ions in presence of nitrifying bacteria.

$$2NH_4^+ + 4O_2 \rightarrow 2\,NO_3 + 4\,H^+ + 2H_2O$$

(in presence of nitrifying bacterial)

The continuous use of such fertilizers in the soils with poor lime status leads to development of acid soils (Daji, 1988). The relative acid forming characters of various nitrogenous fertilizers are recorded by Walcot *et al.* (1965) (Table 2.7).

Notwithstanding the benefits of nitrogenous fertilizers in increasing the crop yields, the increasing trend in the consumption of nitrogenous fertilizers in our country is a matter of serious concern from the point of their acid forming characters.

The consumption of acid forming fertilizers such as urea, ammonium sulphate, ammonium chloride, calcium ammonium nitrate, diammonium phosphate was recorded to be of the order of 15810, 591,

139, 629 and 3480 thousand tonnes during 1993-94, respectively in India. The cumulative effect of the usage of these fertilizers over a long period has reduced the soil pH in general. The application of these fertilizers, in fact, demands the application of lime for amelioration of acidity (Donahue *et al.,* 1990; Dahama, 1997). But, the lime application along with acid forming fertilizers was never recommended and practiced in our country.

Table 2.7. Relative acid forming characters of nitrogenous fertilizers

Nitrogenous fertilizers	**Final pH after 5 years**	**Lime needed (kg $CaCO_3$/kg of N)**
Soil only (no fertilizers)	5.7	0.0
Ammonium sulphate	4.0	12.7
Ammonium chloride	4.4	7.9
Urea	4.9	6.0
Ammonium nitrate	4.8	6.1
Urea form (slow release)	5.0	3.8
Anhydrous ammonia	5.1	11.3
	(initial soil pH 5.8)	

(Wolcott *et al.,* 1965)

The phosphatic fertilizers are also reported to increase the acidity of the soil. Triple super phosphate and single super phosphate are known to release the phosphoric acid and acidity the site of application (Tisdale *et al.,* 1985).

It is often pointed out that the soil acts as buffering agent and it will resist the sharp changes in pH by the presence of aluminum ions. But, the total amount of clay and organic matter and the nature of clay will determine the extent to which the soils are buffered. For example, ultisols and oxisols containing 1 : 1 type clays are less efficient in buffering than alfisols and mollisols containing 2 : 1 clays. Similarly, sandy soils with lesser clay content have poor buffering ability than clayey soils. All the soils poor in organic matter would certainly not be able to buffer the acidity as compared to soils rich in organic matter. The enrichment of soil organic matter by use of digested organic manure may mitigate the effects of acid forming fertilizers. This means that applied fertilizers are effective only in presence of organic matter.

In practice, the fertilizers are recommended and used in all types of soils without any regard to their buffering ability. It is unsafe to presume that all the soils are able to buffer the resultant acidity on application of fertilizers. It is such unsafe assumption that has led us to continuously recommended, use and promote the fertilizer consumption.

The acidity increased due to continuous use of fertilizers may have many direct and indirect effects on the soil health. Some of them are :

(a) The acidity in the soil may affect the effectiveness of herbicides such as paraquat, diquat, 2,4-D, nitrilies, phenols as their adsorption is reportedly more in acidic soils (Haris and Warren, 1964).

(b) The nitrifying bacteria, primarily responsible for conversion of ammonia nitrogen into nitrates are known to be active in the pH range of 6.5 to 7.5 and their activity is adversely affected in pH below 5.5 (Daji, 1988; Russel, 1975). When nitrification is slowed down due to acidity in soils, the leachates from such soils contain more ammonia molecules and may contaiminate the ground water.

(c) Nitrogen fixing bacteria are reported to be very sensitive to pH. A pH less than 6.0 is known to drastically reduce the rhizobium population (Rice, 1977; Russel, 1975) and activity of Azotobacter (Daji, 1988).

(d) Ironically, the acidity induced by the fertilizers may reduce the availability of many nutrients such as phosphorus, boron, copper and zinc, as they are known to react with other elements/compounds below a pH of 5.5 to 6.0 and become unavailable.

(e) If the pH of the soils is less than 5.5, the solubility of elements such as manganese and iron increases to such a level that their excess concentrations are harmful to the plant growth.

In general, it is surmised that in most of organic and mineral soils, the maximum nutrient availability is in the pH range of 6.5 to 7.5 (Lewis and Davis, 1970). Any reduction in pH less than 6.5 due to acid forming fertilizers will reduce the availability of many nutrients.

2.4.1.3 Soil organic matter

Soil organic matter acts as a store house of plant nutrients, besides increasing the exchange capacity and water holding capacity of the soils. It provides the energy to microbes, stabilizes the structure of the soil and acts as buffer against bringing out rapid changes in the soil pH.

The application of fertilizers may not reduce the organic matter status directly but the status of organic matter may decide whether the added nutrient is available to crops or not. It has major role in the cat ion exchange capacity of soils, which is major mechanism of nutrient availability. The presence of organic matter also increases the solubility and availability of phosphorus by :

(a) Formation of phosphohumic complexes which are more easily assimilated by the plants.

(b) Anion replacement of phosphate by humate ion

(c) Coating of sesquoxide particles by the humus to form protective cover and reduce the phosphate fixing capacity of the soils.

Logically, the addition of fertilizers must increase the organic matter status of the soils as more biomass is produced with the use of fertilizers and it should add up organic matter on decomposition. This logic is possible when the biomass is returned to the soil. Even when the biomass is returned to the soil, the soil organic matter may or may not increase depending on temperature, microbial population, aeration and pH. In general, the temperature climate favors the retention of organic matter formed, while in tropical climate, it is more easily lost due to faster decomposition and in severe cold condition the organic matter may not be formed at all.

In a tropical climate of India, the continuous return of biomass to the soil is essential for maintenance of organic matter status to ensure the proper availability of added nutrients. Continuous use of fertilizers not combined with return of crop residues can be dangerous to the soil health, though in short range it may improve the crop production.

2.4.2 Chemical component

Considering the present status of the fertilizer usage and the method in which its usage is promoted, it appears that the chemistry involved in the soil reactions is grossly over simplified and is marred with several convenient assumptions. The fertilizers, in general, might have increased the crop production, as the sites of fertilizer application would certainly enrich the availability of added nutrient. For this reason, the fertilizer usage has been regarded as an easy and quick method to increase the fertility status of the soil.

But, the complexity of the chemical reactions when a fertilizer is added to the soil deserve an overall consideration in the totality of all the reactions involved rather than single convenient reaction. As most fertilizers are applied through the soil, they would unfailingly disturb the whole chemical environment of the soil, besides, supplying particular nutrients. The applied fertilizer may affect the chemical nature of the soil or induce imbalances in other chemical reactions or mask the other chemical reactions or mask the availability of other nutrients in one or more following ways:

1. A fertilizer, supplying one nutrient, may inhibit the supply or availability of other nutrients. Conversely, a nutrient present in a soil may inhibit the availability of applied fertilizers. Many such examples can be cited (all the below mentioned information are obtained in Tisdale *et al.,* 1985).
   (a) As Ca and Mg compete with K for entry into the plant body and for site on the clay mineral, all the added potassium is not expected to be available to the crops (P : 268-269).
   (b) The presence of K will often restrict the adsorption of NH_4+ on clays as they can mutually compete with one another (P. 135). As a result, more NH_4+ ions are available for nitrification in the soil solution or may be leached if not nitrified.
   (c) Nitrogen fertilizers can influence manganese availability in several ways and many chloride salts such as KCl, NaCl and $CaCl_2$ have been reported to promote the manganese availability (P : 375-376). Excessive manganese leads to toxicity to the plants.
   (d) Excessive phosphorus or molybdenum in soil or deficiency of potassium or zinc may induce iron deficiencies (P : 370).
   (e) More NO_3 in soil may induce iron chlorosis (P : 370).
   (f) Application of nitrogen fertilizer may aggravate copper deficiencies (P : 363).
   (g) When calcium nutrition is excessive, there is greater requirement of boron (P : 355).
2. Many added nutrients may be fixed, immobilized or transformed in the soil and affect their availability to the crop plants. Some examples of this nature are:
   (a) Potassium is known to be fixed in the expanding type of clays such as montmorillonite and later available to crops (P : 263).
   (b) Phosphorus is retained and fixed in kaolinitic clay soils due to higher amounts of hydrated oxides of iron and aluminum (P : 206).

(c) Soils with vermiculite and illite clays have a capacity to fix NH_4 in their expanded lattices (P : 135).

(d) Several microbes are reported to utilize NH_4 nitrogen and fix them in organic forms, which may be later released after their decomposition (P : 114).

(e) Most of added sulphur in the surface horizon of well drained agricultural soils is transformed into organic forms, which may be later available to the crops (P : 298).

(f) Applied phosphorus and sulphur may be transformed into soluble sulfates and phosphates in calcareous soils (P: 299).

(g) NH_4 + ions, released out of ammonical fertilizers may be lost to air by volatilization as NH_3 gas if not adsorbed to clays or nitrified to NO_2.

(h) Due to activity of denitrifying bacteria, sizeable nitrate nitrogen is converted to gaseous nitrogen or nitrous oxide and lost to atmosphere.

3. The availability and release pattern of nutrients from a fertilizer source is highly correlated to the moisture status in the soil. Excessive moisture in the soil may drain away the nutrient ions or their converted compounds to deeper layers to contaminate the ground water. Soil moisture deficit may restrict the movement of nutrients and create a conducive situation for leaching or transforming fixation. For example :

(a) Excessive moisture in soil leaches excessive nitrates and ammonium ions (P : 703).

(b) Inhibited movement of K ions due to moisture stress may lead to fixation or transformation into unavailable forms (P : 263).

(c) Submerged condition of the soil leads to formation and excessive release of iron compounds causing to toxicity to the crops.

(d) Leaching of soil boron to deeper layers of sandy soils under heavy rainfall situations (P : 704).

(e) Excessive availability of manganese under moist conditions due to its higher solubility in reduced condition (P : 704).

4. In the guise of adding nutrients, a large quantity of filler material is added to the soil every year. Considering the global consum-ption of fertilizer to the tune of 137 million tonnes (1998-99), the addition of filler material cannot be regarded as chemic-ally irrelevant in the context of complex chemical reactions in the soil.

This is especially relevant in the light of repeated addition of fertilizers over several years. For example, the most popular filler material - talc - does contain calcium and although it is regarded as inert, such a large quantity of calcium added every year will certainly alter the soil chemical reactions adversely. Though no systematic study is reported on the effect of filler material on the soil reactions, excessive calcium can certainly be expected to come in the way of K, Zn and Fe availability.

Thus, the fertilizers added to the soil with a laudable motive of increasing the crop production may result into complexities in availability, reaction, transformtions and fixation of nutrients as well as mutually dependent interactions. To add to these complexities, the changing nature of the soil, varying climatic factors, natural micro flora of a given ecosystem and impact of changing hydrological cycles - all affect the fate of added fertilizers. It is grossly oversimplified influence if one concludes that a given quantity of fertilizer would make a fixed quantity of nutrient available to the crop in all the situation. The commercial practice of recommending and using the fertilizers based on simple mathematics of (soil available + added) = (crop removal + other losses) has rendered the soil unhealthy and harmed the natural capacity of the soil to produce, supply and retain the plant nutrients.

2.4.3 Biological component

The most relevant living component of the soil ecosystem is the biological component comprising soil flora and soil fauna. The present system of flooding the agriculture with numerous agro-chemicals including fertilizers is followed with least regard to the biological component of the soil. The dynamics of the life process in and above the soil have never received the attention they deserve while planning the use of agro-chemicals. The nature dependent cycles such as carbon cycle in conjunction with hydrological cycle indeed provide the basis for natural production of most of plant nutrients, if properly managed and maintained. But, the modern agriculture has used large quantities of agro-chemicals only to disrupt these natural processes severely jeopardizing the biological component of the soil ecosystem. Commercial use of fertilizers and other chemicals have set in large scale imbalance in the normal life process of soil fauna and flora. Some chemicals may increase the population of a particular group of bacteria (ex. Application of urea encourages nitrifying bacteria). The imbalance in most microbial processes as well as their population dynamics due to application of agro-chemicals are well documented.

Besides using the fertilizers, attempts are made on commercial scale to use chemicals such as N-serve (Nitropyrin), Terrazole, ATC (4-amino-1, 2, 4-triazole hydrochloride) DCS (N-2, 5-dichlorophyenyl) succinamic acid), Guanyltiourea (1-amido-2 thiourea or ASU) DCD (Dicyandiamide) to act as inhibitor of natural nitrification process. Such nitrification inhibitor are designed to restrict the activity of bacteria - Nitrosomonas. Most of such commercial attempts are made in USA and Japan. Similarly, bacteria called *Bacillus megatherium varphosphaticum* termed as phosphobacterins are used in USSR to increase the availability of native and applied soil phosphorous (Tisdale *et al.,* 1985). They are now used all over the world including India. Addition of Rhizobia, Azotobacter to promote N fixation has been an established practice in agriculture world over. These instances are examples of direct interference in the biological component of soil by man for further increasing the crop production. The use of chemical nitrification inhibitors, or adding bacterial medium may also change the composition of microbial population and set in imbalances in the natural processes.

2.4.3.1 Details of micro flora in soil

The soil contains microbial population including algae, fungi, actinomycetes, bacteria, protozoa as well as fauna such as annelids, arthropods and sometimes even mammalian members. A brief description of these organisms with a specific reference to the soil and the soil chemical processes is useful before discussing the actual impact of the agro-chemicals on them.

Algae

Algae are grouped into four classes:

1. Cyanophyceae (Blue green algae)
2. Chlorophyceae (Green algae)
3. Xanthophyceae (Yellow green algae)
4. Bacillariaceae (Golden brown algae)

Of these, the members of Cyanophyceae are most frequently found in the soil, while the members of Bacillariaceae are scantily available. Though all cyanophycan members are autotrophs, at least eight species of Cyanophyceae are found to fix nitrogen from the atmosphere for their nourishment and thereby are found to have agricultural importance (Table 2.8).

Table 2.8. List of species of Cyanophycaea

Species found in India	Species fixing N
Anabaena	*Anabaena*
Aphanocapsa	*Aulosira*
Aphanotheca	*Calothrix*
Campylonema	*Chroococcus*
Chroococcus	*Cyclindrospermum*
Cylindrospermum	*Nostoc*
Crinalium	*Plectonema*
Fischerella	*Tylopothrix*
Gloeocapsa	
Lyngbya	
Microcoleus	
Nodularia	
Nostoc	
Oscillatoria	
Plectonema	
Scytonema	
Stynonema	
Synecochoccus	

Fungi

They are thallophytes with saprophytic or parasitic food habits mainly living on the dead plant debris or live plants. Of the numerous genera of fungi found in the soil, the most commonly found fungi belong to genera such as *Alternaria, Aspergillus, Cladosoporium, Fusarium, Penicillium, Rhizopus, Trichoderma* and *Zygorhynchus.* Some less abundant species found in soil belong to genera such as *Absidia, Acroostalagnus, Bortrytis, Cephalosporium, Monilia, Oospora, Saccharomyces, Spicaria, Verticillium.*

The chief activity of the soil fungi is decomposition of organic compounds - both nitrogenous and non-nitrogenous - that are added to the soil in the form of manures and plant residues. They decompose the carbohydrates, organic acids, proteins, fats, lignins, resins, tannins and also take part in humus formation. Some fungi such as *Phoma* have N fixing characters. Soil fungi can also assimilate nitrates and ammonia into insoluble mycelial proteins (also called fungal immobilization). Besides all these beneficial functions, many soil fungi are parasitic and are responsible for number of plant diseases.

Of the saprophytic beneficial fungi, some are of great importance in the decomposition of organic compounds in the soil (Table 2.9).

Table 2.9. List of decomposing fungi

Carbohydrate decomposing fungi

Starch decomposing	*Aspergillus, Fusarium, Mucor, Rhizopus polyporus*
Gums and mucilage decomposing	Basidiomycetes
Hemicellulose decomposing	*Alternaria, Aspergillus, Chaetonium, Fusarium, Heliminthosporium, Penicillium, Polyporus, Rhizopus, Trichoderma*
Cellulose decomposing	*Aspergillus, Melancomium, Sardaria, Alternaria, Monilia, Sporotrichium, Botrytis, Monnosporium, Stachybotrys, Chaetomium, Mycogone, Stemphylium, Cladosporium, Penicillium, Trichoderma, Fusarium, Pyronema* and *Verticillium*
Protein decomposing	*Fusarium, Aspergillus, Mucor, Citromyces, Penicillium, Monilia, Trichoderma*
Lignin decomposing	*Stereum, Polyporus, Fomes, Polystictus, Xylaria, Torula, Poria, Agaricus*
Humus forming	Humus formation process is promoted by ligno-protein complexes, which are formed by many fungi including species such as *Alternaria, Aspergillus, Cladosporium, Dematium* and *Hymicola.*

Actinomycetes

They are unicellular organisms like bacteria but produce non-septate mycelium like fungi. Of the four families of *Actinomycetes* two families viz., *Actinomycetaceae* (genera *Actinomyces* and *Nocardia*) and *Streptomycetaceae* (genera *Stroptomyces, Micromonospora* and *Thermoactinomyces*) are of agricultural importance. While *Micromonospore* and *Thermoactinomyces* are found on the compost heaps and mud of fresh lakes/ponds, *Nocardia* and *Streptomyces* are found in arable soils.

Despite the fact that they are more numerous than fungal population, they contribute very little towads the soil organic matter. Nevertheless, their role in the decomposition of organic substantances cannot be completely ignored.

Bacteria

Bacteria are the most abundant group of microorganisms in the soil. They are more numerous than all other microorganisms put together in the soil (Table 2.10). Each gram of soil may contain 0.1 to 100 million bacteria. The bacterial population varies considerably from season to season, day to day. Usually, top 2 cm of weathered soil does not have bacteria, while they occur in maximum frequency at a depth of 8 to 10 cm.

Table 2.10. Distribution of microorganisms in India soils (x 10^6)

Soils	Number per gram of soil in 15 cm depth (in millions)		
	Bacteria	**Actinomycetes**	**Fungi**
Deep balck, broach	1.29	0.10	0.03
Medium black, Poona	14.94	0.29	0.009
Medium black, Karjat	1.58	0.40	0.02
Alluvial, Anand	20.11	0.26	0.03
Lateritic, Calicut	1.06	0.47	0.12
Lateritic, Mulki	0.78	0.11	0.10
Red, Bangalore	1.34	0.62	0.008

(Daji, 1988)

The bacterial population and its activity are influenced by composition of soil, air, supply and withdrawal of moisture, nature and quantity of substrate, soil reaction, soil temperature, soil organic matter and type of the soil. In general, higher the organic matter higher is the bacterial population. As more and more bacteria are responsible for production of more organic matter, increase in the organic matter would further increase their population. Large number of groups of bacteria is of agricultural importance. Unlike fungi, only a few bacteria cause plant diseases and the beneficial groups of bacteria are more than harmful groups. Classified according to their functions, the agriculturally important soil bacteria are grouped in Table 2.11.

Protozoans

They are unicellular flagellate or non-flagellate or ciliate organisms with saprophytic food habits. Important genera of the soil protozoans fall in the classes such as *Rhizopoda* (genera like *Bodo, Cercomonas, Heteromito, Tetramitus*) and *Liliophora* (genera like *Colpodo, Vorticella, Gastrostyla)*.

Table 2.11. Bacterial species of agricultural significance

Agricultural significance	Bacterial species
(a) Nitrogen fixing	
Non-symbiotic	
Anaerobic	*Clostridium*
Aerobic	*Azotobacter, Aspergillus*
Symbiotic	
Aerobic	Rhizobium, Agrobacterium
(b) Ammonifying	*Bacillus, Clostridium, Micrococcus, Pseudomonas*
(c) Nitrifying (aerobic)	
Nitrate producing	*Nitrosomonas, Nitrococcus*
Nitrite producing	*Nitrobacter*
(d) Denitrifying (anaerobic)	*Bacillus, Micrococcus, Pseudomonas, Achromobacter*
(e) Cellulose decomposing	
Aerobic	*Angiococcus, Cytophago, Polyangium, Spororocytophaga*
Anaerobic	*Methanococcus, Methanobacterium, Methanosarcina*

Though their agricultural importance is not fully understood, their saprophytic food habit is associated with accelerating the N fixation by *Azotobacter* and NH_3 release by the ammonifying bacteria.

In addition to these, the soil harbours number of earthworms (Annelids), arthropods such as beetles, ants, flies, worms etc. In fact, a food web can be traced in every soil eco-system which comprises bacteria, fungi, actinomycetes, protozoans as well as higher forms like annelids and arthropods. It is a dynamic living network of all these organisms which co-ordinates with the other physical and chemical components of the soil to form a stable system.

The use of agro-chemicals to increase the crop production or to protect the crop from damages of pests has innumerable direct and indirect effects on the population of all these organisms. Though all of them have mechanisms to adjust to the changing natural cycles, they can hardly adjust to the chemical onslaught of agro-chemicals. A brief review is presented here below on the deleterious effects of agro-chemicals on the soil organisms. Both the list of agro-chemicals and their multi-dimensional effects on soil organisms are indeed innumerous. Following are few examples to indicate that soil bio-sphere can be an undesirable target of agro-chemicals.

When a pesticides is added to the soil, the microbes in the soil are subject to following effects:

1. They may be suppressed - both in terms of population and activity.
2. They may act on the pesticides and break them down.
3. They may alter the composition of the microbial population.

All these effects are dependent upon various abiotic factors such as concentration and type of pesticides, soil type and clay content, soil temperature, soil pH, soil moisture as well as biotic factor such as composition of microbial population.

The agro-chemicals are reported to have inhibitory, stimulatory or neutral effects on micro-organisms, both in terms of number and their activity (Agnihotri *et al.,* 1981). The effect is dependent on typs of micro-organisms, type and doses of pesticides, environmental conditions and also the time factor. Many pesticides are also known to be broken down by microbial activities thus, on some occassions, microbes decide the effectiveness of agro-chemicals than being affected by the agro-chemicals.

Innumerable number of reports of inhibitory effect of agro-chemicals on the microbial population and their activity are documented.

2.4.3.2 Effects on N-fixing bacteria

Balasubramanian (1981) indicated, in general, that the recommended levels of pesticides adversely affected the growth, respiration, carbon and phosphorus metabolism of important N-fixing bacteria like *Azotobacter chroococcum, Rhizobium* spp and *Pseudomonas solanacearum*. Many other research workers have also recorded inhibitory effect of pesticides on growth and activity of N fixing bacteria (Balasubramanian *et al.,* 1974; Oblisami *et al.,* 1973; Swaminathan and Prasad, 1978). Martensson (1992) has recorded adverse effect of herbicides such as bentazone, chlorsulfuron, monochlorophenoxyacetic acid and glyphosate on rhizobial population. He also recorded the inhibition of nodule development by pesticides such as mancozeb. Growth of *Nostoc muscorium* was significantly reduced by application of 2-4 mg/l of benthiocarb with total death of algae at 8 mg/l (Bhunia *et al.,* 1992). They also recorded 36 per cent and 65 per cent reduction in the activity of *Nostoc muscorium* by the application of 10 and 20 mg/l of methyl parathion, respectively. Dutta *et al.* (1977) had earlier recorded similar reduction in the activity of symbiotic bacteria due to the use lindane. Aldicarb at 5 ppm and higher concentrations is reported to have inhibited the biosynthesis pathways of *Azotobacter chroococcum* resulting in its reduced growth (Balasubramanian and Narayanan, 1980).

As against this general suppressive effects of pesticides, a report on stimulatory effect of simazine used at the rate of 50 to 300 micrograms/g of culture or 1 g of soil on the nitrogen fixation by *Azotobacter chroococcum* is also made (Martinez-Toledo *et al.,* 1991). However, at higher concnetrations, it had negative effect. Kale *et al.* (1989) have reported that carbofuron had no significant inhibitory effect on *Rhizobium* and *Azotobacter.*

The nature of effects of applied pesticide (inhibitory, stimulatory or neutral) on N-fixing bacteria is decided by number of factors such as type of pesticide and its concentration, composition of microbial population, pH, aeration and temperature. Since most pesticides are reported to have caused one or other suppressive effects on N-fixing bacteria at recommended rates, it can be concluded that nitrogen fxing process is highly sensitive to pesticides. Extensive care should be exercised to avoid the ill effects of the pesticides on N-fixing process. A general, recommendation of seed treatment with pesticides before or after the inoculation with N-fixing bacterial culture is highly questionable practice in the light of foregoing discussion.

As all the micro-organisms would co-exist in the soil with their mutually dependent life-cycles, the impact of applied chemical need to consider the increase/decrease in populations of different microbes rather than the effect on one particular type of microbe. In this context, the addition of N fertilizer saturating the soil with NH_4+ ions would certainly alter the composition of microbial population. Dolgova *et al.* (1988) have reported a reduction in bacteria utilizing organic N while nitrifying/denitrifying bacteria population were reportedly increased. Similarly, the composition of total bacterial population is expected to change when a pesticide curbs the activity of one type of microbe.

2.4.3.3 Effects on other micro-organisms

Number of reports are made on the suppressive effects of pesticides on various micro-organisms, other than those of N-fixing characters. They are summarized in Table 2.12.

Number of useful annelids are reported killed due to use of pesticides. Clements *et al.* (1991) reported that continuous use of phorate (study of 20 years) killed earthworms and other soil invertebrates from first year itself. Pesticide treatment with phorate greatly reduced organic matter content and infiltration rate and pH of soil. In the above studies, the effects of pesticides on non-pathogenic micro-organisms have been investigated, besides, the toxic effects on phytopathogens. Non-target toxic effects of many pesticides, which are widely recommended and used in agriculture, need to be considered seriously in further promoting their usage. The ignorant usage of pesticides without any regard to their non-target toxic effects would ultimately culminate in drastic reduction in the microbial activity in the soils.

Table 2.12. Effect of pesticides on micro-organisms

Pesticide used	Type microorganisms	Reference
Carbaryl	Fungal population	Gaur and Bardia (1970)
Sumithion	*Sclerotium rolfsii*	Reddy and Anilkumar (1975)
BHC	Soil bacteria	Gaur and Bardia (1970)
Heptachlor	Soil bacteria	Chandra (1964)
Endrin, Malathion	Soil bacteria	Swaminathan and Sullia (1969)
Diazinon	Soil bacteria	Sathpathy (1974)
Malathion	*Bacillus thuringensis*	Pishoka and Timchenko (1979)
Paraquat and Alachlor	Soil bacteria and fungal population	Sahid *et al.* (1992)
Chlorsulfuronn	*Macrococcus roseus* *Azotobacter agalis*	Burnet and Hodgson (1991)
Sulphometuron	*Pseudomonas aeuruginosa* *Basillus cereus* *Saccharomyces cerevisiae*	Burnet and Hodgson (1991)
Prephenphos	Denitrifying bacteria nitrifying bacteria	Martinez-toledo *et al.* (1992)
Phorate	Nitrifying bacteria	Bhaskar Rao (1975)
Phorate and Disulfoton	Soil bacteria	Srinivasalu and Rangaswamy (1973)
Diuron, Nitrofen	Soil bacteria	Prabhakarsetty and Hosmani (1980)

Micro-organisms have now been regarded as major agents of break down of pesticides. The extent and the type of microbial population in the soil may ultimately decide the fate of applied pesticide as well as the extent of pollution due to their metabolites in surface and ground water. Number of research workers have documented the microbial degradation of pesticides. They are summarized in Table 2.13.

The presence of one insecticide may have positive/negative impact on the usefulness of other insecticide. For example, Ramanand and Sethunathan (1990) have recorded complete inhibition of degradation of carbofuran by *Arthrobacter* when benthiocarb and mancozeb were used. The microbial degradation of pesticides may lead to toxic metabolites, which in turn may be harmful in many ways. They may have inhibitory effect on other micro-organisms such as useful fungi (Raghu *et al.,* 1975 and Sidramappa *et al.,* 1973).

Table 2.13. List of pesticides broken down by micro-organism

Pesticide broken down	Type of micro organisms involved	References
Chlordane	*Mocardiopsis* spp. (actinomycete)	Beeman and Matsumara (1981)
Thiram	Soil fungi	Raghu *et al.* (1975)
Fensulphothion	*Pseudomonas acaligenes*	Sheela and Vasant-harajan (1977)
2, 4-D	Soil bacteria	Sandman *et al.* (1988)
Alachlor	Aerobic/anaerobic bacteri	Anonymous (1988)

Several other workers have reported the stimulative effects of pesticides on many phytopathogens. For example, *Pseudomonas solanacearum* - a causal organism of tomato wilt has been reported to have prolifertated at low concentrations of 5 ppm of aldicarb (Balasubramanian and Gita Nilkanthan, 1978) although they obtain inhibitive effect at 10 ppm and higher concentrations.

Chemical degradation of pesticides is also facilitated by presence of clay minerals, organic matter, soil pH and soil CEC. The rate of dissipation of pesticides was more in the soil applied with pig manure and cattle manure due to increase in populations of bacteria, actinomycetes and fungi (Muthukrishnan *et al.,* 1981).

Thus, the interactive effects of quantity, dosage, timing of application of pesticides with number and type of micro-organisms in presence/absence of organic matter would ultimately decide whether a chemical has suppressive/stimulatory effect or change the composition of microbial population or itself is broken down to metabolites. In any case, the applied pesticide can not be expected to bring about only desired result (control of pests) but it can also bring about may other undesirable effects and alter the basic composition and activity of microbial population.

In summary, the physical, chemical and biological components of the soil are certainly exposed to so many harmful effects of agro-chemicals that the natural features of the soil ecosystem are irreversibly disturbed. Inter-relations between these components and the complex chemical interactions undergoing in the soil cannot be easily ignored, when agro-chemicals are so widely recommended and used. The repeated use of agro-chemicals ignoring these harmful effects has already damaged many natural processes in the soil ecosystem significantly and induced the disharmony. Such a scenario of the use of agro-chemicals has also resulted into the undisputable damage to the natural inbuilt fertility of the soil, besides polluting our environment profusely in terms of air, water and food contamination. There is an urgent need to curb the use of dangerous chemicals in agriculture, without lessening the food production. In the recent years, many naturally available materials have been identified for the pest control. They are not only eco-friendly, cheap and naturally available but are able to effectively control the pests without much damage to the ecosystem. They also do not have potentiality to cause the environmental and food pollution, as they are bio-degradable into simple harmless compounds with less/no persistence values. Besides these naturally available chemicals, number of other non-chemical methods has been identified for control of pests. They include measures such as use of natural enemies of pests, use of physical traps, effective planning of the cropping system to avoid crop pests etc. These attempts offer long term solutions to our agricultural problems without damaging the soil in any significant way. It is pertinent to mention here that such solutions might not be available now for all the problems of pests in agriculture. Possibly, some such measures might have met with partial or less success as compared to conventional chemical method. But, in the long term interest of sustaining the agricultural production system, it is necessary to pursue such methods in future years and persistent efforts can lead to more effective and eco-friendly methods of pest control. It is also necessary to perceive that the damage already caused to the soil eco-system in the course of 15 to 20 years cannot be easily and quickly reversed, even if the new natural methods are pursued. It may take long time to reverse the whole situation and build up a healthy and fertile soil.

2.5 EFFECT OF AGRO-CHEMICALS ON ECOLOGY OF AQUIFERS

The chemicals used in agriculture are not only contaminating the soil, water, air and microflora in the vicinity of their use, but have outreached the vast geographical area around them through water. Non-point contamination by agro-chemicals has such wide spread ramifications that many of the flowing and standing water bodies thousands of kilometers away from the site of application have exhibited the contaminants in their waters. Besides causing pollution to the streams, rivers and the tributaries, seas, ponds, wells, estuaries, lagoons, reservoirs around agriculturally active zone, the pollution is now traceable in arctic oceans where the pesticides were never used. Persistent pesticides are carried over a long distance and accumulate in the water bodies often with increased concentration. Though the contamination of water is due to many other human activities such as mining, deforestation, industrialization and urbanization, the agriculture also contributes substantially to the contamination of water in a major way.

No where in the world, an agricultural land receives water only to the level of field capacity either by rain or by irrigation, except when the crops are irrigated by drip irrigation. The excess water has to find its way out of agricultural land as run off or enter the underground water to reach the drainage. Wherever the erosion is minimized by leveling, bunding etc., the surface run off losses may be less, but corresponding increase in the seepage and resultant increase in the ground water are inevitable. It is this path of water, through which agro-chemicals applied to the soil enter the hydrological cycle to create many ecological imbalances.

Two prominent effects of agro-chemicals on the ecology of water bodies can be identified.

1. Eutrophication of water bodies and their impact on aquatic eco-systems and related food chain.
2. Contamination of water bodies and the change in their flora and fauna.

2.5.1 Eutrophication

The ultimate natural structures receiving the discharge of all the contaminants carried by the water are low lying structures such as pond, estuaries, reservoirs or lakes. Eutrophication is a concept, wherein excessive minerals and other chemicals including pesticides are flown from an agriculturally active zone to a water body leading to change in the quality and composition of aquatic eco-system. The massive use of agro-chemicals has to culminate into eutro-phication of water bodies in the area around the site of their application.

The deposition of agro-chemicals principally consists of nitrates, phosphates, sulphates, carbonates, bicarbonates, chlorides, residues of herbicides, insecticides, fungicides and nematicides as well as various metabolites of these pesticides. The level of deposition depends directly upon the extent of these chemicals being used and the method of application, soil type and type of crop in the catchment agricultural area of water body in question. The unscientific method of irrigation, as practiced in developing countries, has indeed aggravated the situation of pollution through water because :

1. Availability of irrigation has led to increased use of agro-chemicals.
2. Indiscriminately applied irrigation water has great potentiality of carrying the agro-chemicals to the water bodies.

The manner in which irrigation water is applied to the soils in many developing countries including India is far from scientific requirement. In most command areas as well as area irrigated by tube wells, jack wells etc., the irrigation water is applied to fields almost without any regard to water holding capacity of the soil, drainage, crop requirement and various other considerations. Wrongful method of irrigation has indeed led to two way losses. Wastage of costly nutrients through drainage water as well as wastage of water itself. Both of these result in eutrophication of water bodies down the stream, leading to change in the composition of aquatic eco-system.

The eutrophic water bodies experience many ecological imbalances including the concentration of chemicals. Year after year, the concentrations of the accumulated chemicals increase both due to evaporation and due to recharge by new contaminated flow of water every season. The changed concentration of chemicals may render the water bodies unfit for many aquatic organisms or may change the composition of aquatic flora/ fauna. The populations of zooplanktons and phytoplanktons may change the whole aquatic environment suitably altering the animals they support. The entire food chain and the composition of food web may ultimately change due to all these changes. Similarly, nitrate accumulation in aquifers leads to excessive growth of moss, leading to decrease in oxygen which affects the growth and development of other species of flora as well as population of fauna in the related food chain of such aquifer.

Many reports have been made around the world about the eutrophication of water bodies by the nearby agricultural lands. Some of them are :

1. Delivery of excess phosphorus from the non-point source of cultivated muckland to Lake Ontario in Canada (Longbucco and Rafferty, 1989).
2. Aquifers in arid zone receiving nitrate rich water from upstream cultivated land in Australia (Barnes *et al.*, 1992).
3. High HCH detected in Shiga prefectures on Japan (Ishamuru *et al.*, 1987).
4. Unconfined aquifers recording high phosphorus up to 20 mg/1 (Handa, 1991).

5. Heady discharges of nitrogen made in North Sea by Norway and other riparian countries (Sodal *et al.*, 1989).
6. Decline in aquatic macrophytes and the change in composition of aquatic fauna in Netherlands due to eutrophication (Smolders *et al.*, 1995).
7. Eutrophication of Norwegian Skagerrak coastal region leading to reduced oxygen concentrations around the coast (study period 1927-1993) due to sedimentation of phytoplanktons at bottom (Johaannessen *et al.*,1996).
8. Environmental impact of using excess P leading to P erosion ultimately converting a fresh lake into eutropihic (Karim, 1992) in Thailand.
9. Nitrate pollution leading to eutrophication of Indian water bodies (Kansal *et al.*, 1992).
10. Coastal lagoons in Mexico accumulating high levels of Heptachlor, DDE, Endosulphan leading to eutrophication (Albert, 1996).
11. 30 to 40 per cent of world's lakes and reservoirs are eutrophic (UNEP, 1991).

2.5.2 Contamination of water bodies

The surface and drainage water from an agricultural land carries with them large number of chemicals including various metabolites of pesticides. The composition of such contamination depends directly upon the extent of use of such chemicals in the catchment. Such cotaminated water forms the part of flow of stream and then river etc. Alternatively, such contaminated water may be a part of ground water. Thus, the flowing as well as ground water are afflicted by the increased concentration of agro-chemicals. This would force upon the eutrophication of final water body receiving surface waters. Besides this, the changing chemical composition of flowing water also brings about substantial changes in the plants and animals supported by such streams.

Number of research workers have reported about carriage of agro-chemicals by the flowing water. Few of them are :

1. Matukhno (1988) recorded 42.91 per cent more N and 8 to 89 per cent more P in river water found due to cultivation.
2. Leaching of surplus P from nearby crop land accounted for high dissolved P in USA (Longbucco and Rafferty, 1989).

3. Alachlor found in river water, tap water, run-off water up to 270 micrograms/1 (Chesters *et al.,* 1989).
4. In the regions of intensive cultivation of Moscow, the water courses have recorded high nitrates, phosphates, K, Ca far exceeding WHO limited (Bashkin, 1989).
5. In Temuco and Rapel river basins of Chile, water and sediment samples from various places contained geotoxicants comprising of metalochlor, atrazine, triazines, benomyl (Dutka *et al.,* 1996).
6. Atrazine found in water flows of Des Plains wetlands of USA exceeding the US federal drinking water standard (3 micrograms/l) (Alvord and Kadlec, 1996).
7. Recovery of Atrazine and isoproturon from drainage water in Australia (Angus *et al.,* 1996).
8. Offside runoff losses of metolachlor and metribuzin applied to soybean crop (Webstern and Shaw, 1996).
9. Shallow ground water and tile drain water from San Joaquin valley in California recording herbicides like Atrazine, Bromacil, 2,4-D, Diazinon DCPB, EDB, Diacamba and Diuron (Domagalski and Dubrovsky, 1992).

In the western countries, water in the river courses and water bodies are also affected by acidification due to acid precipitation (acid rains) or due to acidic discharges through rock deposits. Acid rains are due to direct impact of large scale burning of fossil fuels and release of excessive nitric oxide, sulphur dioxide to atmosphere which are later converted to nitric acid and sulphuric acid during precipitation. Instead of water, the precipitation of diluted acid is received in many countries during the so called rainly months. The run off and drainage water from such lands receiving acid rains would logically be acidic and these fluids are carried through water courses and are led to water bodies. As a result the soil, water course and water body are likely to be subjected to acidity. Recently, Verma (1992) has recorded such rain acidification in India too due to formation of nitric acid by action of nitric oxide on rain in his study of data from 10 background air pollution monitoring (BAPMON) stations. In future years, the consequences of rain acidification may take worse shapes and would certainly affect the eco-system drastically.

Acidicity of water bodics and the water courses would mean a clear shift in the chemical reaction under going inside the water, change in number and composition of micro flora, change in composition and reduction of most of aquatic animals, besides, causing number of indirect effects.

With or without eutrophication, with or without underground water pollution, the pollution of our waters by increasing proportions of hazardous chemicals is a matter of serious global concern. The concentration of chemicals may be less in the water bodies at present. But, the concentration of these chemicals may reach alarming proportions, if the agro-chemicals are used on much larger area. In respect of some pesticides, their concentration in aquifers may not be an indicator of danger. Because, even low concentration of many organochlorine pesticides are dangerous as they have bio-accumulative and bio-magnifying characters.

Incessant charging of our waters with such dangerous chemicals may encourage the transfer of many bio-magnified chemicals to other aquatic or terrestrial organisms in the food chain. The classical case of Michigan sea lake, whose fishes were eaten by many pregnant women in Michigan city in USA and their off springs have not only carried the bio-magnified pesticides but are subjected to many psychological and physiological problems.

The extent of use of pesticides in the developing countries is far less than western countries. Nevertheless, the experience of our western counterparts should serve as fore-warning to the entire globe. The miserable experiences recorded in many western countries may recur in other parts of the world in which aggravated form. It would be unwise and unconcerned outlook to ignore such experiences and continue to use pesticides, as if the pesticides are the only solutions to agricultural pest problems.

2.6 EFFECT OF AGRO-CHEMICALS ON GROUND WATER POLLUTION

The human interference in the agricultural eco-system in terms of application of hazardous agro-chemicals to increase the crop yields or to protect the crop has undisputedly affected the ground water quality. This is substantiated by profuse data recorded across the world. Ground water quality is indeed affected by many other human activities such as mining, urbanization, industralisation burning fossil fuels, various types land uses, besides agriculture. Modern agriculture is inseparably dependent upon agro-chemicals and this has affected the ground water quality in much dominant manner than others (Keeney, 1989). However, the public awareness in respect of pollution of ground water due to agro-chemicals is at lower ebb as compared to pollution due to other agents. Many vociferous protests have been raised by many voluntary organizations concerned with the environmental pollution and about the water pollution due to non-agricultural agents. The pollution due to agro-chemicals is yet to be made aware of globally. This is especially true in developing countries, where it is generally acknowledged that the public awareness is at its lowest level due to various reasons such as poor literacy, poor legislation and many issues. In India, particularly the awareness of undergound water pollution due to the use of agro-chemicals is yet to be realized by many scientists, administrators and planners leves apart the general public.

The pollution due to use of agro-chemicals on ground water is mainly due to fertilizers and pesticides.

2.6.1 Pollution due to use of fertilizers

The pollution due to use of various fertilizers can be discussed mainly on nitrates and phosphates, as the literature on potassium are other nutrients as a source of pollution is meager.

2.6.1.1 Pollution due to nitrogenous fertilizers

Profuse evidences in the recent past suggest that nitrogenous fertilizers in any form ultimately contribute to a great extent to nitrate contamination of both surface run off and ground water. Both nitrate and ammoniacal fertilizers would ultimately lead to such contamination, as ammonical form of nitrogen is a part of dynamic nitrogen cycle in the soil and is always subjected to continuous transformation into nitrate due nitrifying bacteria. A source of nitrates contaminating the ground water is better explained by the concept of nitrate saturation, which is defined as the availability of ammonium and nitrate ions in the agriculturally active vadose zone in excess of total combined plant and microbial demand for nitrogen at any given point of time (Aber *et al.*, 1989). The nitrogen demand by any crop or microbial population at a given site is continuously spread over a period of time and the application of demanded nitrogen in a single dose or even split doses (2-3) can certainly result into undesirable nitrogen saturation. At this point, if the irrigation water or precipitation is received by the soil profile, no management practice including split application can ensure that the adverse nitrogen saturation is avoided. The extent of adsorption of ammonium ions by the soil particles is by far meager as compared to total ammonium ions released when N fertilizers are applied (even in case of split doses). The extent of crop uptake is always less than the saturation of nitrogen ions in the soil profile immediately after application. In the latter part of the crop duration, the crop may experience negative nitrogen balance and may experience nitrogen deficiency. But, by this time major part of applied nitrogen would have been subjected to leaching and would have been a part of underground water. Judicious management of nitrogen may minimize nitrate contamination (Russelle and Hargrove, 1989). Total elimination of nitrate accum-ulation in soil profile can hardly be possible in practical agriculture. Proper use of nitrogenous fertilizers involve accurate determination of nitrogen need of the crop and microbes at different stages of the crop growth, varying levels of losses due to volatalization and timing of nitrogen application to match with time related need (Peterson and Frye, 1989). In the commercial agriculture, however, none of these can be indeed controlled. Recommending the split doses of nitrogen on adhoc basis is indeed based on crop response than actual quantification of nitrogen losses.

The extent of nitrate contamination depends upon factors such as amount of overloading the soils with N fertilizers (Pustaji, 1989), extent of rainfall distribution (Hall *et al.,* 1989), type of soil texture (Jorgensen *et al.,* 1989) as well as extent and method of irrigation. Characterstically, the nitrate contamination is more in underground water, while surface runoff may be more contaminated by ammonical nitrogen (Matukhno, 1988), as nitrate ions being negatively charged are more easily lost to deeper layers of the soil than ammonium ions, which carry positive charge and have access to adsorption by clay minerals.

The extent and characterization of contamination of ground water by nitrogenous fertilizers are not only decided by the depth from which the water is tested but very strongly influenced by other factors of nitrogen cycle such as bacterial population oxidizing the ammonium ions into nitrate ions (Barnes *et al.,* 1992), extent of denitrification resulting into loss of nitrogen into gasses such as nitric oxide, nitrous oxide and elemental N. Increased nitrate concentration in the soil profile may change the composition of bacterial population (Dolgova *et al.,* 1988) as bacterial population utilizing organic N was found to be reduced, when soil profile had higher nitrate concentration, while nitrifying and denitrifying bacteria grew at the expense of other micro-organisms.

The loss of nitrogen in the form of ammonium ions may also contaminate the water, if the soil is over saturated with ammonium ions beyong its CEC. Most irrigated soils receiving heavy doses of nitrogen fertilizers experience such a situation. But, predominant type of N contamination continues to be due to nitrates.

The ground water contamination by nitrates has series of implications, most important being the contamination of drinking water.

Number of research workers have acknowledged about nitrate contamination in driking water wells (Hallburg, 1989 as well as Moody, 1990 in USA; Barnes *et al.,* 1992 in Australia; Mascher and Marth, 1991 in Australia; Pustaji, 1989 in Hungary). Nitrates can contaminate both on farm and off-farm water resources (Crutchfield, 1989), as the ground and surface water supplies are subject to non-point contamination leading to contamination of lakes, streams and estuaries etc. and change the chemistry of these water bodies. Ultimately, all the water bodies, lakes, rivers and even seas are over loaded with nitrates. Two tributaries of Desna river in Ukranian region experienced 42-91 per cent more N compared to their original nitrate concentration (Matukhno, 1988). Sodal *et al.* (1989) have cited similar experience in Norway, where international agreements were necessary to be made to reduce the nitrate discharge into north sea by 50 per cent. Lola *et al.* (1988) have reported the nitrate contam-ination to be the reason for eutrophication of Ivankova reservoir. Fleischer and Stibe (1989) have recorded severe loss of fish and marine population due to coastal eutrophication caused by exceesive use of nitrogenous fertilizers in Sweden.

More concerning ill effect of nitrate concentration in underground water is its effect on human health. World Health Organization has set the acceptable limit of nitrate N in the drinking water as 10 ppm (the limit of NO_3, is 45 ppm). Nitrate nitrogen concentrations in the vadose zone and below the agricultural fields typically are in the range of 5-100 ppm with frequent detections up to 20-40 ppm (Bouwer, 1990). Mascher and Marth (1991) have reported from Australia that 21 per cent of the underground samples they tested had more than 30 ppm of nitrates and 10 per cent of them had more than 50 ppm. Similarly, Barnes *et al.* (1992) from Australia and Pustaji (1989) from Hungary have also reported that 1/4 to 1/3 of drinking water wells in the rural areas have nitrate levels beyond the safe standards.

Existence of nitrates in natural water like rain, river water (at source), snow and naturally formed ice is a common phenomenon. It ranges from 0.2 to 3 mg NO_3 /l (equivalent to 0.8 to 12 mg NO_3 – N/L). Reports by Sarin *et al.* (1992) and Handa *et al.* (1991) have shown that natural waters in India (like Himalayan snow, Bhagirathi, Alkananda and Ganges at Rishikesh) contain up to 1.0 mg NO_3/1. But, the level of nitrate pollution in ground water is aggravated due to human activity like agriculture and urbanization.

First extensive report of nitrate pollution in India was made by Handa *et al.* (1975), who concentrated on the problem of nitrate pollution in his later works (Handa, 1986; Handa *et al.,* 1991). Since then, many workers have reported about nitrate pollution from several parts of India (Mehta *et al.*, 1990; Kansal *et al.*, 1992; Bulusu and Pande, 1990). In their detailed research paper, Agarwal *et al.*, (1999) have correlated nitrogenous fertilizer consumption in different states to average NO_3 level in ground water in respective states. For example, Haryana with 91 kg N-fertilizers/ha/year has 99.5 mg NO_3/l and Orissa with 8.5 kg N fertilizers/ha/year has 14.8 mg NO_3/l in their respective ground waters. They have even classified the country in to low risk states (Himachal Pradesh, Madhya Pradesh, Orissa and Maharashtra), moderate risk states (Uttar Pradesh, Bihar, West Bengal, Andhra Pradesh, Gujarat, Karnataka and Tamil Nadu) and high risk states (Punjab and Haryana).

The extent of nitrogenous fertilizers used annually to our crops can certainly indicate that the extent of nitrate pollution in the lakes, rivers and streams in the regions below our agriculturally potential zones. The nitrates in the drinking water, if an excees of the safe limits, can increase the possibility of cancer through the reduction of nitrates to nitrites in the oral cavity or stomach and the reaction with amines to form nitrosamines, which are most potent carcinogens in combination with aflotoxins B (Mirvish, 1977). He also reported that higher concentrations of nitrates can cause tumors in any the parts of human body. He has recorded geographical correlation between nitrate intake and gastric cancer.

Nitrogenous fertilizers, so commonly recommended and used in the modern agriculture, have caused many serious implications on the human health and the ecosystem, besides increasing the crop yields. In fact, the reduction in the consumption of fertilizers in entire Europe, Japan and other developed countries (Table 1.1 and 2.1) is the result of identification of such potent dangers due to nitrates. The concerns about nitrate contamination in the drinking water are already voiced in most developed countries such as UK, France, Germany, USA, Canada, Japan, Russia, Ukrain, Australia, Spain, Italy and Poland. But in a developing country like India, neither awareness about the nitrate pollution in the drinking water is created both in rural and urban areas nor any attempt is made to trace back the ill effects of nitrates on our society and eco-systems. On the contrary, the nitrogen consumption is on increasing trend every year with no realisation of potent dangers of nitrates.

Bouwer *et al.* (1990) have suggested that ground water contamination by agricultural chemicals can be minimized by using best management practices for crop production, including the low input sustainable agriculture. It is necessary to design and manage the irrigation systems for zero or minimum percolation beyond root zone. In this context, experience in Israel deserves seriousd consideration. Shevals and Waldman (1989) from Israel have reported that relative contribution of nitrogen fertilizers to the ground water pollution decreased over years from 1970 to 1984, when N consumption increased from 116 kg ha^{-1} to 156 kg ha^{-1}. They attributed this to better fertigation techniques. Many more strategies include *in situ* microbial production of nitrates which will be gradual and according to the crops need, growing deep rooted crops to remove nitrates from deeper layers, better irrigation methods.

2.6.1.2 Pollution due to phosphatic fertilizers

The phosphorus is known for its poor movement in the soil columns. Nevertheless, it is reported to be a cause of contamination mostly by runoff water. Longabucco and Rafferty (1989) have reported surplus phosphorus leached in sub-surface runoff water entering Lake Ontario. They recorded 72 per cent of dissolved reactive P and 39 per cent of total P entering the lake due to P losses in the upstream agricultural soils. Similarly, Ohmura and Kurokawa (1990) have reported that 60 per cent of applied phosphorus appearing in suspended state on river surfaces during snow melt season in Japan. Matukhno (1988) has also reported the increase in P concentration of tributaries of Desna river in Ukrain by 8-69 per cent. In India, Handa (1991) has reported about the phosphate contamination of ground water in unconfined aquifers up to 0.2 mg/L.

2.6.2 Pesticides contamination in ground water

Persistence of pesticides in soils is very closely correlated to the pesticide contamination of ground water. If the pesticide is able to persist in the soil by absorption or adsorption or transformation, it may not leach down to ground water. Otherwise, it or its metabolites will certainly contaminate the ground water or run off water.

Innumerable number of reports are recorded about the ground water and surface water contamination by various pesticides. Croll (1991) reported the occurrence of atrazine, simazine, mecoprop, isoproturon, chloroturon in the surface water in the concentrations up to 11.5 mg/L in Anglian water region of UK, while he could also report a more mobile herbicide atrazine to contraminate ground water at a low concentration up to 0.1 mg/L. Oki and Giambelluca (1989) have reported contamination of ground water by fumigants such as DBCP, EDB in Hawaii islands. Smith *et al.* (1990) have detected atrazine up to 350 mg/ L after 19 days of application at 60 cm depth of the soil, while the ground water concentration up to 90 mg/L was recorded by them after six months of application. Stover and Guitjens (1990) detected extremely toxic insecticide aldicarb- in ground water. Recent sampling in USA/ Canada indicate the occurrence of atrazine and its metabolites as principal toxic contaminant in ground water (Belluck *et al.,* 1991). Similarly, Alachlor and its metabolites are also demonstrated to contaminate the ground water (Anonymous, 1988). High levels of aldicarb (more than 10 mg/ L) and sulfones and sulfoxides of aldicarb were recorded in as many as eight states of US (Moye and Miles, 1988).

Angus *et al.* (1996) were able to recover three per cent of applied atrazine in drainage water, while Alvord and Kadlec (1996) recorded atrazine up to 8 to 10 mg/L of drinking water, which exceeds US Federal drinking waters stands (3 mg/L). Similarly, Kloos (1996) recorded 0.56 ppb of DCPB and 0.25 ppb of EDB in well water of metropolitan areas of California. Similar reports are available in case of metolachlor (Webster and Shaw, 1996), Propazine, Simazine, Diuron, Diazinon, DCPB and EDB (Domogalski and Dubrovsky, 1992) and Alachlor (Bubler *et al.,* 1993).

That the contamination of surface and ground water by the agrochemicals is both point and non-point needs serious consideration. The sites receiving chemicals contaminate the water in the immediate vicinity and later to various non-point sites in irrigation/drinking water wells, lakes, estuaries, ponds, reservoirs, streams, rivers, sea etc. During such course, the contaminants not only cause health hazards to human beings and aquatic/terrestrial animals but may change the composition of aquatic micro flora and other animals. Some pesticides such as organo-chlorine insecticides exhibit the bio accumulation and bio-magnification at different trophic levels typically illustrated by Fig. 2.6.

The developing nations have relatively lower pesticide load (Table 2.4) as compared to developed nations. But public awareness about the use of contaminated water is almost negligible in developing countries. In India, it has not been possible to provide even filtered and chlorinated water to the entire rural population. Most states are simply not geared to consider proposals of purifying the drinking water to cleanse the dissolved the contaminants like nitrates and pesticides/their metabolites. In this background, further loading the underground water by agricultural contaminants would certainly lead to a point of no return as the scenario of public water purifying systems would either remain same or go to worse situation in future years.

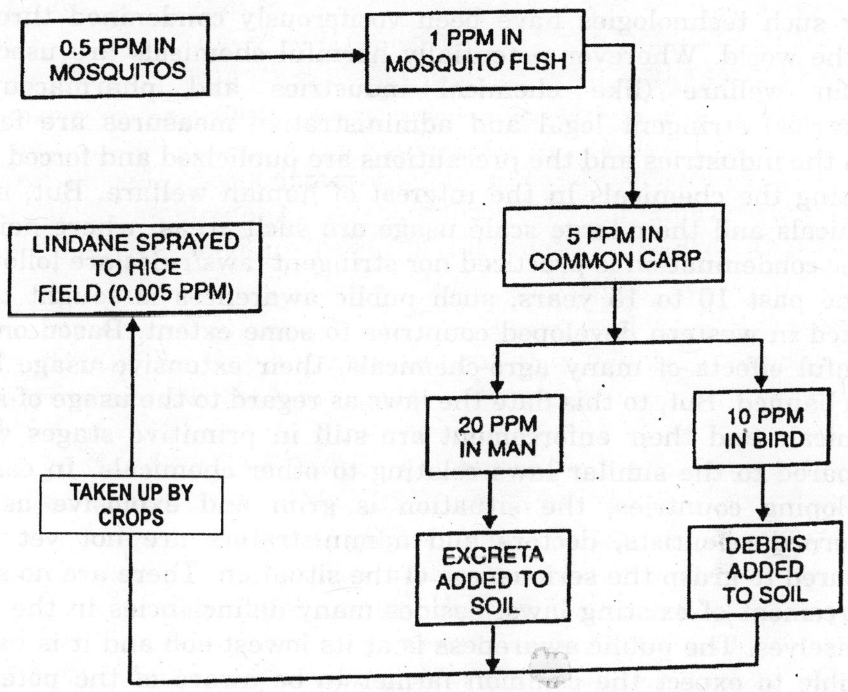

Fig. 2.6. Representation of bio-magnification in aquatic ecosystem

The pollution of ground water is also due to many other human activities such as mining, increased burning of fossil fuels, industrialization and urbanization. The extent of ground water pollution by these actviities is also substantial in the developing countries. Combined with such activities, agriculture is expected to further aggravate the ground water pollution even at present level of pesticides and fertilizer use. The situation is extremely grim and alarming. There is an urgent and compelling need to initially assess the extent of pollution of ground water by different human activities in drinking water wells, water supply reservoirs, lakes, rivers. It is also absolutely essential for the pollution control authorities to consider the issues arising out of pollution of drinking water due to agro-chemicals. Ways and means of minimizing the ground water pollution due to agro-chemicals must be recognized, discussed, improved and implemented in their true perspective.

2.7 EFFECT OF AGRO-CHEMICALS ON HUMAN HEALTH

Many technological improvements hazardous to human health have faced public condemnation in the history of mankind. Chemical weaponry, atomic warheads, ballistic missile technology and many other such technologies have been vociferously condemned through out the world. Wherever potentially harmful chemicals are used for human welfare (like chemical industries and pharmaceutical industries) stringent legal and administrative measures are forced upon the industries and the precautions are publicized and forced care in using the chemicals in the interest of human welfare. But, agro-chemicals and their large scale usage are such areas, where neither public condemnation is practiced nor stringent laws/rules are followed. In the past 10 to 15 years, such public awareness is sought to be created in western developed countries to some extent. Based on the harmful effects of many agro-chemicals, their extensive usage have been banned. But, to this date the laws as regard to the usage of agro-chemicals and their enforcement are still in primitive stages when compared to the similar laws relating to other chemicals. In case of developing countries, the situation is grim and explosive as the concerned scientists, doctors and administrators are not yet fully prepared to grasp the seriousness of the situation. There are no strict enforcement of existing laws, besides many definciencies in the laws themselves. The public awareness is at its lowest ebb and it is hardly possible to expect the common farmer to be aware of the potential dangers of the agro-chemicals on human health.

The agro-chemicals used in agriculture, in the guise of increasing the crop yields and protecting the crops, have indeed caused perpelexing situation in the mankind as far as health of human beings and animals is concerned. Unlike human and veterinary pharmaceuticals, the agro-chemicals cannot be straight away ingested into the plant bodies. An estimated 0.1 to 1.0 per cent of applied pesticides actually reaches the targeted pests leaving more than 99 per cent as pollutant to soil, water, air, crops, micro flora, eco-system to cause direct and indirect health problems.

2.7.1 Types of effects

The harmful effects of agro-chemicals on human helath are multi-faceted and multi-routed. They can be classified as :

Direct effect

(a) Routes of exposure	Ingestion, contact, inhalation through eyes, mixed
(b) Reasons of exposure	Accidental, intentional, professional, unintentional
(c) Results of exposure	Death, dermatitis, irritation, burning, inflammation, nausea, insomnia, loss of weight, loss of libido

Indirect effects (unintentional)

Indirect effects may be caused by entry through food, air or water.

(a) Through food

1. Direct ingestion of contaminated food
2. Through fod chain (may or may not be associated with bio-magnification)

   Terrestrial

   Aquatic

   Both

(b) Through air

(c) Through water

Results of Indirect exposure

Dermatitis, reproductive impairment, deformity of limbs, loss of weight, hereditary deformity, miscarriage, loss of remembrance, loss of balance, loss of reasoning, tumors, ulcers, cancer etc.

2.7.2 Implications of agrochemicals on human health

The proportion of ill-effects of agro-chemicals on human health can be better visualized if the extent of their usage and section of human population affected are briefly reviewed. Nearly, three million people across the world are subjected to pesticide poisoning each year, with 2.2 lakh deaths due to pesticides. The global consumption of pesticides is to the tune of 4.6 million metric tons. With a very high pesticide load on the society (up to 10.8 kg/ha/yr in Japan) in developed countries, the population is more exposed to indirect effects than direct effects as they follow stringent laws in the pesticide usage. In contrast, the developing countries experience a very low pesticide load (0.14 kg/ha/yr in India) on their society. But, they are more prone to direct effects owing to less stringent laws in force as well as indirect effects due to government and public negligence on mitigating such effects. For example, Sweden having 8.6 million population has reported 774 cases of poisonings and 5 fatal cases due to pesticide usage (WHO recorded world poisoning rates are 44 deaths/million and 600 poisonings/million) which are substantially low as compared to China having reported 3900 fatal cases and 43000 cases of poisoning cases (Ekstrom *et al.*, 1996). These statistics do not include unreported cases of direct effects as well as many indirect effects which may be substantially high.

The cumulative injury caused by the pesticides and other agro-chemicals both by direct and indirect effects can be so large that they exceed the much condemned ill-effects atomic explosions. The alarming feature about the pesticides is that the injuries of this magnitude are repeated every year, while the atomic explosions occur once in a while. Yet, the use of agro-chemicals have never received global condemnation as strongly as atomic plants. The beneficial effects of agro-chemicals to the mankind are almost questionable when their usage has large hazardous effect on public health, besides causing the increase in the crop yields. This perception is especially relevant in the context of prolonged ill-effects of agro-chemicals even after 2-3 decades of their usage (for example, organo-chlorines having entered our eco-systems 15-20 years back are reported to have remained till now and are also reported to have travelled to arctic oceans, whether they were never used).

In-depth study of various indirect ill effects of agro-chemicals indicates that the direct effects of death, irritation, inflammation, nausea, dermatitis are indeed of minor importance. the indirect effects are due to non-point contamination and have pervaded the entire globe. Irrespective of mode of entry (such as water, food, air or food chain) the pesticide contaminants have the potentiality to cause innumerable human physiological abnormalities often carried to their off springs, if mother is exposed to such contamination during the pregnancy. Such abnormalities include tumors, cancer, ulcer, reproductive dysfunction, mental illness, miscarriage, limb defomrities, loss of memory and loss of reasoning. Exposure to such indirect risks is unintentional and can be hardly curbed by laws. Many agro-chemicals reaching human beings through one of indirect media, have potentiality to bioaccumulate and biomagnify, as they enter food chains and food webs. The ultimate user is unaware of presence of pollutants. While the western society is struggling hard to contain the situation out of such indirect exposures, developing countries are yet to realize that there are potential dangers due to indirect effects of pesticdes. The entire globe is to be considered as one unit, while assessing the indirect effect of pesticides, because persistent agro-chemicals applied in any part of the globe may enter other parts through various ways such as export/import of food and other agricuoltural commondities or transportation of pesticides through water or through many food chains involving aquatic fomrs. In fact, the exposure to indirect effects is difficult to measure, but indeed 10 to 20 times larger than that of direct effects. Pollution due to agro-chemicals has deepened the already aggravated pollution scenario in the world. The cow's milk, fruits, vegetables, grains, meat, egg, processed foods and even human breast milk fed to infants are already polluted with agro-chemicals. The river water, pond water, borewell water and ocean water are found to have alarming concentrations of agricultural chemicals.

The reports of health hazards in the developing countries are not systematically documented as in case of developed countries. Sometimes, the frequency of direct effects itself may not be acurately accounted because many effects such as skin irritation, eye irritation, inflammation are never referred to doctor by the subjects. The public awareness about the health in rural areas in developing countries including India is so poor that minor symptoms are either ignored or self-treated with available local medicine. Hence, forthcoming discussions are based upon information available in western countries. Wherever the reports of contamination from the developing countries are available, they have been made use of suitably.

The fatal reports due to agro-chemicals should not be the only criterion to propagate the idea of their disuse. They have remained in our eco-system, in our bodies, in our food; they cause direct and indirect deleterious effects on our health. Many have carcinogenic, mutanegic tumrogenic and teratogenic potentiality. These pesticides have remained as a source of several human illness, besides causing implications such as neurosis, hepatic and neural complications, deformities of foetus, complications in pregnancy and delivery, still births, loss of libido etc. The kind of scenario, these pesticides along with other industrial pollutants have pushed our mankind into, can be better realized by the general increased incidence of tumors, cancers, foetal defomities in human beings in the last 30 to 40 years. Hence, the pesticides are the source of innumerable medical problems to the public than causing fatal consequences alone. As many pesticides have remained in our eco-system, they have potentiality to cause such toxicolgical consequences for a few more decades. The strong advocation against the use of insecticide finds relevance in this perspective.

The vast information on human health hazards due to the use of pesticides in various developed and developing countries is briefly reviewed with a stress on persistent organo-chlorines commonly recommended and widely used in developing countries.

2.7.3 Chlorinated hydrocarbons (organo-chlorine pesticides)

These pesticides were commonly used in most developed countries to control the sucking and leaf eating pests. Owing to their cheap availability and effectiveness in the control of wide range of pests, they became very popular and very soon spread to all other parts of world. Consequent to finding that these pesticides are highly persistent in the enviroment and due to their bio-accumulative nature they were banned in most developed countries. As a sequel, their consumption has been almost negligible in most developed countries, while the developing countries have continued their usage. The persistent nature of these chemical have resulted in their wide spread condemnation, though they are still the cheapest and effective pesticides. These chemicals have now been found to have both bio-accumulative and bio-magnifying natures. Most of these chemicals and their metabolites are now traceable not only in agricultural eco-systems but also in our food, drinking water, human and animal bodies as well as many manufactured food stuffs. They have already entered most of the food chains across the globe. What is most concerning is that they can be traced now in many developed countries too, even after decades of the ban on their use. This should serve as premonition to the entire world that even if all their users ban them now, their harmful effects can be noticed by our future generations.

The problems of use of organo-chlorine pesticides is a world wide issue with time stretched medical problems. Following issues deserve serious consideration to decide whether their usage has to be continued in the control of pest or not.

(a) Why at all their usage be continued in developing countries, when most developed countries have banned them, owing to their persistent nature and potentiality to cause health hazards.

(b) How and why their continued usage has been recommended for the control of mosquitoes as a part of public health management in a country like India, although their usage has been banned for agricultural use.

(c) What is the technical justification in banning some organo-chlorine compounds such as DDT and BHC but simultaneously encouraging the use of many other organochlorine pesticides such as endosulfan, Heptachlor, Lindane, Aldrin etc.

Each one of the above issues deserve a detailed consideration. Though it is beyond the scope of this book to draw discussions on these issues, it is pertinent to surmise that use of all the organo-chlorine pesticides need to be stopped in toto by all the countries, ignoring their effectiveness to control our agricultural pests. Now, there are many alternative and effective pesticides for better control of pests. The continued use of organo-chlorine pesticides can be an open invitation to ever lasting and perplexing health problems to human beings and animals of mankind.

The health problems related to the use of organochlorine pesticides are many fold and pervaded through out the globe. They mostly cause indirect ill-effects on human health. Profuse evidences are recorded in the research literature, which commonly points out that their continued use is not only dangerous to human health but also result into vexed environmental malady with absolutely no solution in the sight.

Though many organo-chlorine pesticides are used in agriculture, most important among them are DDT, BHC, Lindane (Gamma BHC), Heptachlor, Aldrin, Dieldrin, Endosulfan, Chlordane and Endrin. The health hazards due to these chemicals have been briefly reviewed below:

2.7.3.1 DDT (Dichloro diphenyl trichloroethane)

DDT on degradation in the eco-system or on metabolism in an organism is converted to followng metabolites, each one of which have persistence values and pose health problems:

DDE	-	Dichloro diphenyl dichloro ethylene
Dicofol	-	Dichloro diphenyl dichloro trichloro ethanol
DDCN	-	Dichloro diphenyl acetonitrite
DDD	-	Dichloro diphenyl dichloroethane
DDMU	-	Dichloro diphenyl chloroethylene
DDMS	-	Dichloro diphenyl chlorethane
DDNU	-	Dichloro diphenyl ehtylene
DDNS	-	Dichloro diphenyl ethane
DDOH	-	Dichloro diphenyl ethanol
DDA	-	Dichloro diphenyl acetic acid
DDM	-	Dichloro diphenyl methane
DBH	-	Dichloro benzohydrol
DPB	-	Dichloro benzophenone
PPCA	-	P-chloro phenyl acetic acid
PCBA	-	P-chloro benzoic acid

Though the direct health hazards due to the use of DDT are very few, the indirect effects on human health are innumerable. Most such recorded indirect health effects are due to the capability of DDT and its metabolites to accumulate in the anumal bodies and cause physiological abnormalities after a long spell of time. They also exhibit a dangerous phenomena of bio-magnification, wherein the concentration of accumulated chemical increases at each trophic level and can enter human body in its concentrated form via food chains. Some salient health hazards due to use of DDT are:

(a) The metabolites of organo-chlorines have a potentiality to act as endocrine disruptors. They are known to have caused cryptorchidism (undescended testicles) in UK (Chilvers *et al.*, 1984) and hypospadias (abnormal penis development) (Giwercman *et al.*, 1993).

(b) As DDE is oestrogen like compound because of its feminizing and emasculating effects, reduction in sperm counts in human beings from 113 x 10^6 / ml to 66 x 10^6 observed between 1938 to 1990 has been ascribed to continued exposure to and deposition of DDE in human adipose tissue (Carlsen *et al.*, 1992).

(c) Teratogenic effect of mother's exposure to PCBs (Poly Chloro Biphenyls is a common abbreviation given to the residues of organo-chlorines in the ecosystem) have adversely affected cognitive, motor and behavioral development of her offspring (Jacobson and Jacobson, 1993). At the age of four years, such children, who were exposed to in-utero PCB contamination, had difficulty with verbal and numerical memory. Mother's breast milk with PCB concentration of more than 3.5 ppm can further create loss of muscle tone, abnormal reflexes in infants (Rogan *et al.*, 1986).

(d) PCBs are known to be disruptors of immune and nervous system in human beings (Colborn and Clement, 1992).

(e) The concentration of testesterone was found to be inversely proportional to the concentration of DDE in mammalians (Subramanian *et al.*, 1987). The reduced concentration of testosterone is known to reduce the fundamental masculine activity of producing sufficient and healthy sperms.

(f) Increased frequency of complicated pregnancies, premature births, new born infants with low weight were associated with DDT and its metabolites concentrated in mother's body (Komarova, 1983; Medved, 1971).

(g) Tumorogenic effects are associated with accumulation of DDT and its metabolites in human adipose tissue (Spynu, 1989).

(h) DDT and its metabolites are associated with pathology of nervous system, carcinoma and liver cirrhosis (Agarwal, 1980).

The DDT/PCBs are found in the bodies of aquatic animals in most part of the world. They have been found in arctic whales and Antarctic fishes, though DDT was never used in these regions. This reveals that problem of DDT/PCBs is not restricted to the region where DDT is used but all pervaded through out globe. Continuous pollution of oceans and consequent entry of DDT/PCBs in to the bodies of lower aquatic animals and later bio-concentration in higher carnovorous animals has led to such a consequence. Terrestrial animals are also equally exposed to such contamination. Man is the end user in many food chains and he is exposed to very high levels of DDT/PCBs due to such bio-magnification phenomenon. The problems due to DDT/PCBs in the environment and food chain have attained global importance because people of every country are exposed to residues of DDT in one way or other (like food chain, imports, direct use) irrespective of whether they use DDT in their country or not. Number of reports are made through out the world about accumulation of DDT/PCBs in human adipose tissue (as they are lipohilic), human milk. Some of such reports are :

(a) The children, whose mother's milk fat PCB exceeded 1.25 ppm, accumulated PCBs in their bodies at the very young age (Jacobson and Jacobson, 1993). Such children had difficulty with verbal and numerical memory.

(b) During the lactation, the infants were exposed to PCB levels up to 27 mg/kg (b.w.d.) (Tilson *et al.,* 1990).

(c) The mothers who ate Michigan Lake fish for a long time before conception transferred their accumulated PCBs and DDE to their foetus (Rorcevic *et al.,* 1987).

(d) Organo-chlorine insecticide residues (DDE) in human adipose tissue up to 18.36 mg/g (extracted on lipid basis) are recorded in Mexico (Albert *et al.,* 1983).

(e) DDE/DDT levels in human milk up to 10.35 mg/g (e.l.b.) in Mexico (Albert *et al.,* 1981).

(f) The umbilical cord blood was found to have 0.03 mg/g (e.l.b.) of DDE, while mother's milk had its level up 6.3 mg/g in sample survey at Yaqui valley, Sonara, Mexico (Gracia and Meza-Montenegro, 1991).

(g) Higher levels of DDT/DDE were found in children's adipose tissue than adults due to predominance of milk products and fat in children's diets (Spynu, 1989).

(h) DDT was detected in most of the blood samples from Israel, USA and UK and its level was up to 50 mg/g (Davies, 1973).

(i) While Dale *et al.* (1965) reported 23 ppm DDE in blood samples of people around Delhi, Ramachandran *et al.* (1974) have reported its concentration up to 21 ppm even after nine years. In Punjab, it was found to be in the range of 10 ppm (Kalra, 1977). It was found that these levels are far more than those found in other countries as well as exceed USEPA and WHO limits.

(j) Occupational exposure to DDT has been reported to have resulted into accumulation of DDT level up to 1359 ppm (Poland *et al.,* 1970) in human body.

(k) DDT spray operators have been found to have accumulated 1.272 mg/l of their blood in Gujarat and occupational exposure to DDT was recorded to have contributed up to 35 mg/kg of their body (Agarwal, 1980).

Several recent reports are made from different parts of India indicate that human adipose tissue, blood and cerebro spinal fluid accumulated DDT. Chawla *et al.* (1978), Gupta and Kalra (1983) have reported DDT concentrations in human adipose tissue. Bhatnagar *et al.* (1992) have reported that blood samples of Indians contain 10 times higher DDT content than other countries. Similarly, DDT concentrations are found in human milk in a range of 0.04 to 2.25 ppm (Siddiqui *et al.,* 1981; and Narasimharao, 1990).

2.7.3.2 BHC (Benzene Hydrochloride), Lindane, HCH

Benzene hydrochloride, Lindane (Gamma BHC) and Hydrochloro Hexane (HCH) form an important group of organo-chlorine chemicals widely used next only to DDT. BHC was introduced by the industry to replace DDT claiming lesser environmental persistance. But, time has proved it as wrong. BHC, Lindane and HCH are now found to be highly persistent in the environment.

The residues of BHC, Lindane and HCH are found through out the world like DDT. As their metabolites also have entered food chains, they also pose an equally complicated and globally pervaded problem as that of DDT. Some of the health hazards with the use of BHC, Lindane and HCH are summarized as follows :

(a) BHC is classified as possible carcinogen by USEPA (United States Environment Protection Agency) (USEPA, 1986). Sufficient evidences of its carcinogenicity in other mammalians are available but not yet recorded in human beings.

(b) Exposure of human beings to seed wheat contaminated with BHC in Turkey from 1955 to 1959 caused epidemic of BHC induced porphyria turcica with symptoms ranging from disturbed porphyrin metabolism, cutaneous lesions, hyper pigmentation (Cam and Nigogosyan, 1963). In children under one year, pink sores were observed followed by 95 per cent mortality (USEPA, 1985).

(c) It has inhibitory effect on lymphocytes in human beings (Rouz *et al.,* 1979).

(d) Liver tumors are reported to have been developed at an intake rate of 52 to 100 mg/kg (Thorpe and Walker, 1973).

Like DDT and its metabolites, BHC is also known to bio accumulate and bio-magnify through food chain. It is known to accumulate in the body/milk. The potential risks due to BHC are more due to its capacity to bio-accumulate and bio-magnify in animal bodes and enter human body in its concentrated form through food chain. Besides, the residues of BHC reported in fruits, vegetables and grains, it finds its entry into ground water, later transferred to terrestrial and aquatic animals. Number of reports are made about the accumulation of BHC/HCH in human bodies. Some of them are summarized as follows (BHC accumulates in human tissues as HCB, alpha HCH beta HCH, gamma HCH).

(a) Beta HCH accumulation in human adipose tissues up to 0.33 mg/g (lipid basis in Mexico (Albert *et al.,* 1983).

(b) Beta HCH concentration up to 1.63 mg/g and HCB concentration up to 0.21 mg/g in human milk recorded in Mexico (Albert, 1981).

(c) Alpha HCH and gamma HCH concentration up to 0.03 mg/l in umbilical cord blood (Gracia-Bnuelo and Meza-Montenegro, 1991).

(d) Total BHC concentration up to 6.2 mg/kg in human renal fat samples from South Africa (Wiese, 1976).

(e) Beta BHC of 27.8 ppb in human blood serum (Kurtz, 1983).

(f) HCH concentration up to 1.19 mg/kg was found in human adipose tissue in France (Davies, 1973).

(g) Total HCH concentration up to 30 ppm in human fat samples in Ludhiana (Chawla *et al.,* 1978).

(h) HCB concentration of 9.2 ppm in human milk fat in Swiss and German women (Spindler, 1983).

(i) Lindane is found to accumulate in humn milk (Siddiqui *et al.*, 1981) and also reported to have been transferred to human foetus through placenta (Pardovski *et al.*, 1977).

There are number of reports available on Indian studies regarding BHC accumulations in various parts of human body. Agarwal (1980) and Rahupathy and Kuttalam (1992) have reported BHC accumulation in human blood. BHC was found in cerebrospinal fluid and in umbilical cord collected from new born after 24-48 hours of child birth (Kalra, 1983 and Siddiqui *et al.*, 1981). BHC concentration up to 4.06 ppm was found in adipose tissue samples from Ludhiana (Chawla *et al.*, 1978). Similarly, BHC concentration were found in the range of 0.014 to 0.83 ppm in human milk. Acute poisoning due to HCH have also been recorded in India (Kalra, 1977). Spontaneous abortions and abnormal deliveries are reported to have been induced by higher levels of HCH accumulation in blood (Siddiqui *et al.*, 1981).

2.7.3.3 Heptachlor and heptachlor epoxide

Heptachlor is an organo-chlorine insecticide which was banned through out Europe, USA about 20 years back. But, it is continued to be used in Asia for the control of soil borne insects such as termites on an exhaustive scale. With a half life of more than one year in soil, it is much more persistent organo-chlorine chemical than even DDT and BHC. Edwards (1966) suggested a range of 3 to 5 years for disappearance of 95 per cent of applied Heptachlor, while some have reported its persistence up to 13 years (Stewart and Fox, 1971). But, before it is being degraded, it is converted to heptachlor epoxide, which is also equally persistent and poisonous.

It has a very high bio-accumulative and bio-magnificable value in the eco-system and human beings are mostly exposed to it by food chain or occupational / operational exposure. As heptachlor and its metabolite migrate slowly and bind to soil very strongly, they are not found in the underground drinking water as a contaminant. Entry of heptachlor via food ingestion is the common mode of entry. As heptachlor has been detected at a very low levels in air (1 ppt), it is not posing any danger as air pollutant.

Although there are number of reports of mammalian toxicity, no evidence exists to date to indicate that low level exposure to heptachlor is dangerous to human health. However, acute exposure (via ingestion, dermal or inhalation) of heptachlor is reported to have affected central nervous system, leading to salivation, muscle tremors, convulsions (Dodey and Kammer, 1953). Inhalation and dermal exposure during the manufacturing and use of heptachlor are reported to be associated with blood dyscrasias (eganemia and leukemia) Infante *et al.*, 1978).

Many reports are made to show that human beings have tendency to bio-accumulate heptachlor epoxide in adipose tissue, brain adrenals, lung, heart, liver, kidney and spleen (Curley *et al.*, 1969). It has been detected in human adipose tissue in Great Britain (Abbott *et al.*, 1981), Brazil (Wassermann *et al.*, 1972), Japan (Curley *et al.*, 1973), Israel (Wassermann *et al.*, 1974), USA (Burns, 1974; Green *et al.*, 1980). Luquet *et al.* (1974) have concluded that human beings have tendency to accumulate more heptachlor than animals. Human beings are exposed to ingestion of heptachlor epoxide through food chains of both terrestrial and aquatic types. As the chemical is persistent and is shown to exhibit carcinogenicity and teratogenicity in many mammalian animals (Fendick *et al.*, 1990) serious concerns have been expressed by scientists across the world that heptachlor epoxide may cause health hazards to human beings to at a later date.

Like other organo-chlorine compounds, heptachlor is reported to accumulate as heptachlor epoxide in human milk in the concentration ranging from 0.02 mg/kg (Mussalo-rauhamma *et al.*, 1988) in Finland and up to 35 mg/g (Takaahashi *et al.*, 1981) in Hawaii. Heptachlor has also found in eggs, cheese and cow's milk (Albert, 1996) and has potentiality to accumulate in various food items even after processing.

2.7.3.4 Chlordane

It is a broad spectrum insecticide primarily used for termite control. It is highly persistent, poorly hydrolysed with a half life in soil up to four years. The residues may persist for 14 years or even longer. It also has property of bio-accumulation and bio-magnification.

Chlordane has been classified as a probable human carcinogen (USEPA, 1987). Chlordane at a dietary levels of 25 and 50 ppm for 18 months results in high incidence of hepatic carcinoma (Epstein, 1976). However, chemical is known to non-teratogenic and non mutagenic. Blood dyscrasias such as aplastic anemia (Klemmer *et al.*, 1977; Infante *et al.*, 1978) refractory megablastic anemia (Furie and Trubowitz, 1976), acute stem cell leukemia, acute lumphoblastc leaukemia and myleomonocytic leukemia (Infane *et al.*, 1978) are reported to be due to dermal or inhalatory exposure to chlordane.

Acute or chronic exposure to chlordane is known to affect central nervous system, whose effects such as irritability, salivation, laboured respiration, tremors, convulsions are reported (Dodey and Kammer, 1953). The major routes of exposure to chlordane have been found to be drinking water, dermal contact, inhalation and food. As it has potentiality to bio-accumulate, it shows large possibility of entering the human body through food chain of terrestrial/aquatic nature.

2.7.3.5 Endrin

It is an important organo-chlorine cyclodiene insecticide once widely recorded and used in western countries. In the developing countries also it was very widely used. Of late, the use of endrin has been banned in western countries as well in some developing countries such as India. Some developing countries continue its use even today to control wide range of insect pests.

Being highly persistent, it bio-degrades poorly and has potentiality to bio-accumulate and bio-magnify. It enters human body through contaminated water and food (direct contamination or indirectly through food chains). It is bound to soil particles firmly and allows small portions to migrate slowly with water flowing through the soil. In mammals, it is found to be eliminated by various metabolic pathways. It has a half life in human bodies varying from 1 to 2 days.

Apart from number of deaths by swallowing endrin deliberately (Tewari and Sharma, 1978) it can cause headache, dizziness, abdorminal discomfort, nausea, vomiting, insomnia and mental confusion (USDHHS, 1978). Food contamination by endrin resulting into unintentional ingestion have alsdo been reported. Bread made out of wheat contaminated by endrin during shipment contained up to 150 mg endrin/kg. Such ingestion through bread caused severe poisoning of 59 people in Wales (Davis and Lewis, 1956).

It has not exhibited carcinogenic potentiality and mutagenic potentiality. However, in a study by Albert (1996) it has shown teratogenic potential. He found 0.0219 mg/l of blood in umbilical cord and 0.0998 mg/l in blood of three months old body fed by months exposed to endrin in Mexico.

2.7.4 Effect due to Carbamates

As compared to organo-chlorine compounds, carbamates such as aldicarb, carbaryl, carbofuran, zineb, maneb, mancozeb and ziram are found to be moderately persistent. They are used as insecticide, acaricide, nematicide as well as fungicide. On application, they are rapidly metabolized to sulfoxides and sulfones. They are reported to be degraded by anaerobic/aerobic bacteria with a half life up to six weeks. They may be found in underground water and may contaminate the drinking water.

Apart from the risks of exposure through direct and deliberate ingestion, unintentional ingestion through food are reported to have caused complaints such as diarrhea, abdominal pains, vomiting, nausea, excessive perspiration, blurred vision, headache, convulsions and temporary loss of limb functions (Goes *et al.,* 1980). Industrial exposure to carbamates may lead to the complaints such as nausea, dizziness, weakness, convulsions, salivation and vomiting (Sexton, 1966 and Haines, 1971). Severe poisoning by carbofuran may lead to muscular twisting, chest tightness and coma (Tobin, 1970). Acetylcholine esterase depression in blood may lead to all such symptoms.

Most of the carbamates have shown no evidence of carcinogenicity, mutagenicity nor any teratogenicity. However, tumorogenicity has been exhibited by carbamates such as thiram, zineb, ziram and maneb (Spynu, 1989). A marked increase in aberrative cells were found in a culture of human peripheral blood leucocytes, when treated with zineb and ziram.

2.7.5 Effects due to organophosphates

Organophosphates were introduced in agriculture as a replacement to organo-chlorines, as the former were considered to be less persistent in environment than latter. Many organophosphates degrade faster biotically and abiotically and pose lesser problem of persistence. They normally do not bioaccumulate and biomagnify like organo-chlorines, thereby reducing the indirect risks. Nevertheless, the dangers of wide scale use of organophosphates are still inevitable because of the following reasons.

- The pesticide residues on economic part such as fruits are conveniently transferred to human body before degradation.
- The metabolites of organophosphates may pose problems of contamination with ground water and even drinking water.
- Health hazard due to accidental, professional or intentional exposure to chemicals.

For these reasons, many organophosphrous chemicals such as phorate and methyl parathion are placed under `extremely hazardous pesticides' category by WHO (WHO, 1988). World Health Organization have fixed permissible maximum daily intake for different organophosphates in case of contaminated foods (WHO, 1987).

2.7.6 Nitrates and nitrites

Nitrogenous fertilizers used on agricultural lands has undoubtedly led to nitrate contamination of drinking water. Excessive nitrate concentration in drinking water is reported to have caused number of health hazards to human beings.

Lethal dose of potassium nitrate for an adult ranges from 54 to 154 mg/kg (Burden, 1961). The toxicity of nitrates in human beings is due to reduction of nitrate to nitrite. By reacting with blood hemoglobin, nitrite forms methemoglobin, which does not transport oxygen to tissue leading to asphyxia (USEPA, 1985a). Methemoglobin levels in human blood above three per cent is defined as methemoglobinemia (Skrivan, 1971).

Children and pregnant women are especially prone to higher levels of methemoglobin, as 278 cases of cyanosis were detected in USA, which were associated with nitrate contaminated water (Walton, 1951). Skrivan (1971) has also demonstrated that the women at 30th week of pregnancy are more prone to methemoglobinema due to nitrites in their bodies. Excess nitrates in drinking water has been associated with high risk for cancer, as the nitrates would be reduced to nitrites in oral cavity/stomach, which would react with amines to form nitrosamines, which ae known to be potent carcinogens (Mirvish, 1977). Many studies with mammalian animals have shown that these reactants of nitrites were carcinogenic and tumorogenic. Tumors were developed in all organs including stomach, oesophagus and nasal cavity by nitrosoamines (NAS, 1981). More than 75 per cent of N-nitrosocompounds are shown to be carcinogenic (Shank and Magee, 1981).

Thus, there is reasonable evidence to show that human beings may be exposed to potential risks of carcinogen by repeated and prolonged intake of nitrates in their drinking water.

2.7.7 Effects due to EDB (Ethylene Dibromide)

It is the most popular fumigant used through out the world to protect store grains as well as to kill soil insects through fumigation. Being a volatile liquid, the fumes of EDB escape easily to the atmosphere but leave considerable residues on the substrates. Further, it poses many problems like derma and ophthalmic toxicities. Major human health hazards of EDB are outlined hereunder:

(a) EDB entry into human body by major routes such as inhalation, dermal and oral methods- can affect adrenal glands, heart, kidney, liver, pancreas, testes and central nervous system. Of them, primary target organs are liver and kidney. Estimated lethal doses of human beings is 160 mg/kg, while systemic toxicity is reported with dermal application of even 16 mg/kg (Alexeef *et al.*, 1990).

(b) Contact of EDB with skin can produce severe irritation, blistering, while vapours can produce severe eye irritation (Letz *et al.*, 1984).

(c) 26 cases of dermal toxicity and four cases of eye irritation are reported by EPA (1981).

(d) Occupational exposure to EDB in air has produced nasal and throat irritation above 50 ppm (Alexeef *et al.*, 1990).

(e) Concentrations up to 28 ppm EDB in air have resulted into occupational fatalities (Let *et al.*, 1984).

(f) Subchronic human exposure to EDB resulted in inflammation of connective tissue and swelling of eyes and irritation to respiratory tract (Kochman, 1928).

(g) EDB binds DNA following invivo administration or in vitro incubation resulting into several mutagenic reactions. EDB is also indicated to be initiator of carcinogenic process (Alexeef *et al.*, 1990).

(h) Hardin *et al.* (1991) have shown that EDB can exhibit a clear teratogenic potential and reproductive toxicity. A dosage of 0.9 ppm on male humans resulted in sperm abnormalities (Schrader *et al.*, 1987), while a dosage of < 5 ppm has been reported to have reduced the fertility rates in men (Wong *et al.*, 1979).

(i) Number of instances to the exposure of EDB less than five ppm at the time of its application leading to reduced sperm counts (Ter Haar, 1980).

(j) It is known to be a potential carcinogen and mutagen (Alexeef *et al.*, 1990).

(k) Besides the entry into human body by direct exposure, it can also enter the human body through water or EDB protected food grains/ food stuffs. When used as soil fumigants, it can find its entry into underground water. Post harvest storage of barley, corn, oat, rice, rye, sorghum, wheat by fumigation with EDB is shown to have contaminant even after processing (Alexeef *et al.*, 1990).

2.7.8 Effects due to atrazine

Large number of reports show that atrazine is easily leached out with water unless the soil has rich organic matter. The ground water is contaminated by atrazine leached from such soils. Health problems of many aquatic and terrestrial animals drinking such water have been reported. However, human health problem due to contamination of atrazine in water have not taken alarming shape. The risks due to atrazine are restricted to occupational/professional/intentional exposure during its manufacture or use. Cantor *et al.* (1985), Hoar *et al.* (1985) and Gass (1991) have respectively reported lymphocytic pymphoma, colon cancer and gastrointestinal complaints due to such exposures to atrazine. However, such studies are sporadic and statistically not significant to show that atrazine poses problem to public health. Similarly, risk of primary ovarian tumors in the residents using atrazine contaminated water was derived by Donna *et al.* (1989). Such studies need further confirmation before declaring the herbicide as a cause of public health hazard. Intentional or mistaken swallowing of atrazne has been reported to have caused hospitalization (Loosli, 1995).

Thus, at present atrazine can not be declared to be a cause for hazards to public health, though it is a potent chemical to cause tumors and cancer to human beings if atrazine charged water is used continuously over years. No study about atrazine contamination of food and water is reported from India.

2.7.9 Effects due to phenoxy alkanates

Though many phenoxy alkonates such as 2,4-D, 2,4, 5-T, MCPA, MNPB, 2,4,DB are used in agriculture, most commonly employed herbicides among them are 2,4-D (2,4-Dichlorophenoxy acetic acid) in its salt or amide forms and 2,4,5-T (2,4, 5-Trichloro propionc acid).

These chemicals are mainly used as systemic, pre-emergence herbicides on various crops such as paddy, wheat, corn, pasture crops, sorghum, barley etc. They are commonly employed to kill broad leaved weeds. These are degraded in the environment and are not considered to be persistent compounds. They are readily degraded by the bacteria and also undergo hydrolysis. They have no capability to accumulate in the soils. They do not bioaccumulate in the plants/animals. 2,4-D and 2, 4, 5-T have been detected at very low concentrations in drinking water samples in USA and thus, are not potent chemicals of serious concern to cause indirect effects on public health.

However, occupational, industrial exposure and intentional intake of these chemicals are reported by many research workers. Total death due to intentional ingestion of 2,4-D amine (Nielson *et al.*, 1965) and occupational exposure of 2,4-D and 2,4,5-T resulting into nerve imbalances and malignant lymphoma (Singer *et al.*, 1982 and Hardell *et al.*, 1981) are reported.

Phenoxy alkones have not been shown to be carcinogens but mild mutagenicity and foetal toxicity have been recorded in mice (Curtney, 1977).

2.7.10 Effects due to alachlor

Alachlor is an important pre-emergence herbicide used on wide variety of crops for control of weeds. It is degraded in the soil by several mechanisms including bacterial decomposition (both aerobic and anaerobic) and metabolised rapidly by the crops. It has moderate mobility in light soils but is adsorbed to clay minerals in clayey soils. It has been demonstrated to migrate to ground water but does not bio-accumulate in the animals and plants. The occurrance of alachlor in the drinking water has been documented by USEPA (1985). The drinking water has been the major route of intake of alachlor in man.

In this background, the risks of public health due to alachlor are restricted to direct intake of alachlor due to accidental, occupational or intentional exposure as well as indirect ingestion of alachlor along with drinking water. Alachlor has been considered as probable human carcinogen (Anonymous, 1988). Cancer risk estimates for human beings s calculated to be due to exposure to alachlor levels in drinking water from 0.002 mg/l to 0.53 mg/l (USEPA, 1986). Alachlor is also demonstrated to be mutagenic and highly teratogenic (Shirasu *et al.*, 1980 and IRDC, 1980). Besides this, it has been associated with hepato-toxicity (Almed *et al.*, 1981).

Acquavella *et al.* (1996) and Leet *et al.* (1996) have reported many fatalities and cancer incidences among workers of alachlor manufacturing units. Because of potential carcinogenicity of alachlor, widespread concern have been expressed in many recent literature about alachlor pollution in drinking water supplies.

Many more health hazards of direct and indirect nature have been documented in respect of agro-chemicals not discussed above. Effect of some chemicals is not yet studied in depth. It may be out of the scope of this book to narrate the entire range of ill-effecs of all the agro-chemicals on human health. But, it is sufficiently clear from the foregoing information that most commonly employed agrochemicals have one or more direct/indirect effects on human health. Many more agro-chemicals now entering the market may be potentially more dangerous than previous chemicals. This is especially true in case of developing countries, which has been the popular pesticide market for many pesticide industries located in developed countries. Due to poor public awareness and difficulties in enforcing the existing laws, these foreign companies have ample scope to find an analogue of an old and possibly banned chemical and push them into markets of developing countries very easily through highly responsive mass media campaigns. This strategy is also adopted by many indigenous manufacturers too. Most popular example of this nature are BHC and Lindane. It took almost 15 years to ban BHC in India for public use in spite of known fact that it is highly persistent, dangerous, bio-accumulative and bio-magnifiable chemical. Most scientists are aware of the fact that lindane is gamma BHC and behaves almost similar to BHC. However, lindane is yet to be banned and it is being widely used in our agriculture. Similarly, DDT is banned for its agricultural use. But, Dicofol (its analogue) is marketed in India widely.

Most pertinent question would be that, are we producing and supplying more and more food in our country at the cost of our public health? It is for the planners and administrators of our country to consider whether our public health should be a priority or producing more at the cost of public health should be our national priority. Instead of possible hungry dealths (had we not used agro-chemicals to increase our food production) we have been able to give the food to all but weakened the very basis of existence both in terms of impaired public health and soil health. Where are we heading to? It may not be too late now at least to correct ourselves and go in the right direction to conserve natural capacity of the soil by producing food through sustainable systems of agriculture at the same time protecting the public health.

2.8 PESTICIDES IN INDIAN FOOD COMMODITIES

The storage of organo-chlorine pesticides in human tissue is directly proportional to their dietary intake of insecticide residues. The dietary intake of HCH in India in an average vegetarian diet is 19.59 μg and that of a non-vegetarian diet is 23.26 μg per day. This is 2-20 times higher than that in other countries of the world (Gupta *et al.,* 1982).

Many reports are already available in India showing alarming level of pollution in food commodities due to pesticides, which include food grains, vegetables, oils, fruits, milk, milk products, meat, chicken, egg, butter, ghee etc. The continuous accumulation of pesticides, through dietary intake, in human bodies is not detectable unless external manifestations are complained by patients. Such manifestations can be observed only after sufficient deposition of these bio-accumulative pesticide residues in human bodies.

Food grains

In as many as 85-100 per cent rice grain samples collected from Haryana, Maharashtra, Tamil Nadu, West Bengal, Bihar, Orissa, Kerala and Gujarat contained HCH residues in the range of 0.001-0.563 ppm (Balwinder Singh and Dhaliwal, 1993). Sorghum grain samples showed 38% contamination with HCH and DDT residues (Jadhav, 1986). Kalra (1989) has reported the DDT contamination in wheat flour. About 66 per cent of cereals and cereal products were found to be contaminated by organo-chlorine residues (Laxminarayana, 1980). Even pulse grains were also found to show detectable level of HCH and DDH (Jadhav, 1986).

Other food materials

Besides food grains, number of Indian workers have reported the presence of DDT, BHC, HCH, DDE from fresh vegetables , fruits, vegetable oil, spices, milk, milk products, meat, poultry eggs and fish.

These are reported in summarized way as follows :

S.N.	Food item	Contaminant	References
1.	Bhendi, brinjal, cabbage, carrot, potato, tomato, radish, cauliflower	DDT, BHC	Krishnamurthy and Reddy (1978), Gupta and Kalra (1983), Jadhav (1989), Banerji (1989), Narasimharao (1990)
2.	Grapes	DDT, Lindane, Malathion	Singh and Chawla (1979)
3.	Grapes	DDT, Endrin, Lindane	Laxminarayana (1980)
4.	Mango, Grapes, Sapota, Guava	DDT, HCH	Jadhav (1986)
5.	Banana, Apple, Mango, Plum, Sweet Lemon Guava	DDT, HCH	Lal *et al.* (1989)
6.	Coconut oil, Groundnut oil and Mustard oil	HCH, DDT	Thakre *et al.* (1969), Laxminarayana and Krishna Menon (1975), Battu *et al.* (1980), Srivastava *et al.* (1983), Dikshit *et al.* (1989)
7.	Groundnut oil, Sesame oil	HCH, DDT	Krishnamoorthy and Raghupathy (1990)
8.	Cotton seed oil	HCH, DDT	Talati *et al.* (1991)
9.	Block pepper, Ginger, Turmeric	DDT, HCH. Endrin, Dieldrin	Sullivan (1980)

10.	Ghee	DDT, BHC	Dhaliwal and Kalra (1978), Kalra and Chawla (1983).
11	Milk, butter, baby milk powder	HCH, DDT	Kumar (1991), Shah *et al.* (1992), Kathpal *et al.* (1992), Dhaliwal (1990)
12	Meat samples of goat, sheep, chicken, cow, buffalo	DDT, BHC	Kalra and Chawla (1983), Battu *et al.* (1984), Laminarayana and Menon (1972), Reghupathy and Kuttalam (192), Sharma *et al.* (1979).
13	Fish	DDT, BHC, dieldrin, lindane, aldrin, heptachlor	Bhinge and Banerji (1981), Battu *et al.* (1984), Ramesh *et al.* (1990)

Pesticide contamination in food is not easily detectable by appearance or taste. Pesticides in the food is not restricted to surface of food articles-but embedded into the molecules of fibre, carbohydrate, proteins as well as fats. The contamination of food articles may or may not alter the taste of food. Due to all these features, the conventional procedure of examining the food for contaminants is insufficient as far as pesticides are concerned. Foul smell, unnatural colour and deposition of foreign material are not found in most cases of pesticide pollution. A common man can be easily misled by such pollutants - as nothing odd is seen and felt while eating such food. Some cautions people wash the pesticide afflicted vegetabls/fruits. Although it is helpful to remove due or physical depositions from the surface, this practice may not remove the pesticides inside the fruits and vegetables.

Pesticides are not fully removed even while boiling, cooking or any type of processing. This is evidenced by traces of pesticides found in processed foods like flour, bread, pulp, extracted oil etc. Boiled milk, boiled eggs, cooked rice, baked rotis, cooked vegetables can retain a part of their original pesticide contamination.

There is really no easy way to escape from pesticide pollution through food articles - except following pesticide free cultivation. Even this method may not be fool proof, because a pesticide accumulated in soil may enter the plant bodies even when no pesticide is applied or pesticide ridden water may pollute soil causing pollutant to enter the system.

Very few attempts are made in India by any agency/organization till now to:

1. Train the public to identify and avoid pesticide contamination through food.
2. Tackle the food pollution by legislative methods
3. Carryout comprehensive research on medical/environmental implication of pesticide pollution.

Before any more dangers can occurs to aggravate the scenario of pollution of our food, it is necessary to take series of regulatory steps to reduce the deleterious effects as early as possible.

2.9 EFFECTS OF AGRO-CHEMICALS ON ANIMAL HEALTH

Use of agro-chemicals have world wide detrimental effects on entire biosphere and it is meaningfully testified by series of health hazards recorded in the entire animal kingdom from protozoans to mammalians. Non-target and non-point source pollution by various pesticides have a wide range of both short term and long term effects on the animal life. However, most of the detriment on the animals due to pesticides is unintentional. These effects can be classified as :

(a) Effect due to direct exposure

- Through herbivorous food chain
- Direct absorption
  1. Cutaneous
  2. Ingestion

(b) Effects due to indirect expsure

- Through ecosystem
- Through carnivorous food chain

Most pesticides, except herbicides are intented to kill some class of animals designated as pests. In addition to killing their target pests, they damage vast range of other non-target animals. While assessing the lethal effects of agro-chemicals on animals the commonly employed expressions are :

1. LD_{50} value	:	50 per cent of lethal dose of a chemical for killing an organism
2. LC_{50} value	:	50 per cent of lethal concentration of a chemical at which a known organism gets killed

3. ppm : Parts per million of a chemical in a solution

4. ppb : Parts per billion of a chemical in a solution

It is relevant to mention that detrimental effects on animal health are globally measured by the complete death of an organism. In case of higher animals, many more dangerous effects on their physiological processes might occur, but such effects have gone unrecorded.

The subject of effect of agro-chemicals on animal health is so exhaustive and the works reported on this subject are so many that it is hardly possible to review the whole work within the scope of this chapter. However, relevant research works are reviewed critically to justify globally well known detrimental effects especially on higher animals and many beneficial animals. An attempt is also made to trace their effects on other members of the food chain by indirect exposure.

2.9.1 Effects due to direct exposure

The applied pesticide many enter animal bodies either by direct cutaneous absorption or oral ingestion (*ex.*, exposure of earthworms to soil insecticides or killing of predators/parasites of pests while spraying insecticides against such pests) or through herbivorous food chain (*ex.*, a cow eating the grasses sprayed with insecticide or poultry birds feeding on grains contaminated with detectable levels of pesticides).

Most animals harboring in agricultural ecosystem such as ants, insects, mites, rats, earthworms, snails, cows, buffaloes, poultry birds, fishes, frogs, oysters, goats, sheep, horses, crows, parrots, sparrows are exposed to pesticides by such direct route. Many pet animals such as cat, dog, horse, rabbit are also affected by such exposures.

The extent of damage caused by the pesticides on these animal population is judged by LD_{50}/LC_{50} values tolerable by a given animal species and the concentration/quantum of chemical to which they are exposed. Many persistent pesticides many remain in the bodies of animals either as applied compound or as their metabolites and may cause innumerable ill effects during the course of their life and even beyond their life in their off springs.

2.9.2 Effects due to indirect exposure

Contrary to the popular view by a layman, the pesticides contaminate the environment/eco-system far away from the site of their application and act as source of health hazards to animals in new environment/ecosystem. The pesticide applied in an agriculturally active zone may reach some downstream ponds through flowing water and may kill the aquatic populations of such pond. Birds and higher mammals move from one ecosystem to other and may carry the pesticide residues in their body and their excreta may act as a source of pollution in the new ecosystem.

Indirect exposure may be due to entry via air, water, soil or contact. But more important indirect route of exposure is through carnivorous food chain. Most persistent pesticides bio-magnify in the food chains and ultimately cause health hazards to animals which are never exposed to the chemical in question directly during the course of their life.

In most developing countries including India, the rearing of useful/domesticated animals is often carried out without any managemental control over what they eat and drink. Such a situation can pose more dangers of both direct and indirect exposure to these animals.

The discussion on effect of important pesticides on animal health is made hereunder based on their economic/environmental importance. As the exhaustive review of effect of all pesticides on animal health is beyond the scope of this book, monocrotophos, chlorpyiphos, methyl parathion and DDT have been selected for discussion.

2.9.3 Effect of monocrotophos

It is a broad spectrum systemic insecticide used very commonly to control wide range of insects on crops such as cotton, paddy, sorghum, wheat, pulses, maize as well as various vegetables. It can kill pests and predators with equal efficiency. The toxicity to applied monocrotophos varies widely with species, stage of life cycle, route of exposure and the concentration of the chemical to which those animals are exposed.

Monocroptophos is rapidly degraded in soil and aquatic ecosystem and thereby its residues are not accumulated in an ecosystem (Guth, 1994). However, it is reported to be lethal to wide range of terrestrial and aquatic animals before degradation. Some examples are:

(A) Beneficial insects/animals are killed by monocrotophos in wide ranging situations either by direct exposure while being sprayed to targeted pests or by indirect exposure through soil and water before degradation. Such effects are summarized as follows :

Insects/Animals	LD_{50}/LC_{50} value	References
Honeybee	LD 0.14/μg/bee	Bai and Reddy (1977)
Earthworm	LC 35.2 μg/kg	Guth (1994)
Carabid beetle	–	Heinrichs *et al.* (1979)
Predators on cotton pests	–	Laster and Brazzel (1968) Lingers *et al.* (1968) Roach and Hopkin (1988)

(B) Birds are known to be highly sensitive to monocrotophos. Unintentional exposure to monocrotophos sprayed/ contaminated foods would kill the birds instantaneously as their LC_{50} values are as low as 0.19 mg/kg to 3.4 mg/kg. Wide range of avian population including Japanese quail, Mallard duck, domestic flea, Canada gose, Golden eagle, Sareech owl, Rock dove, house sparrow, parrot, crow, Bengalese finch, Red winged black bird and even poultry hens are known to be very sensitive to monocrotophos (Guth, 1994).

(C) Monocrotophos entering water bodies can kill large number of aquatic organisms and can results into commercial loss as well as environmental hazard. Number of reports are available to indicate hazardous effect of monocrotophos on aquatic animals.

Animal	Exposure (hrs)	Effect	Effective dose (mg/l)	References
Fresh water shrimp	96	LC 50	0.3	Johnson and Finley (1980)
Prawn	96	LC 50	1.9	Nagabhushan *et al.* (1983)
	48	LC 50	4.5	
Red cray fish	36	LC 50	1.3	Chang and Lange (1967)
Brown shrimp	36	LC 50	0.069	Butler (1965)
	24	LC 50	0.32	Butler (1965)
Oyster	–	Shell growth affected	>1	Butler (1965)
Rainbow trout Killifish Channel catfish Bluegill sunfish	–	100% killing	5–10 mg/l	Guth (1994)

(D) Monocrotophos is moderately toxic to many other smaller and bigger animals of both aquatic and terrestrial habitats.

Organism	Exposure (hrs)	Effect	Effective dose (mg/l)	References
Waterflea	24	LC 50	0.00024	Rawash *et al.* (1975)
	96	LC 50	0.0034	Sanborn (1979)
Tubificid worm	72	100% kill	4	Nagvi (1973)
South African Frog	96	LC 50	>100	Snowder and Chambers (1989)

2.9.4 Effect of chlorpyriphos

It is an important organo-phosphate insecticide very widely recommended to kill wide range of pests on many agricultural crops. It degrades by biotic and abiotic transformation in both terrestrial and aquatic environments (Racke, 1993). It is reported to have half lives of 1-7 days in water and 22-51 days in sediment.

Chlorpyriphos is reported to be lethal/toxic to various animals from planaria to aves. Chlorpyriphos received at 10 mg/kg on soil is reported to have decreased fungal density, increased ammonification and decreased nitrification. It is also reported that birds, fishes, earthworm, predators and other soil inhibiting insects are highly sensitive to even low concentration chlorpyriphos but amphibians show better tolerance, chlorpyriphos exhibits varying levels of toxicity to many fresh water invertebrates. High level of sensitivity to chlorpyriphos by beneficial insects/ animals are summarized as below :

A. Toxicity to beneficial insects/animals

Insects/Animal	LD 50 value (mg/kg)	References
Honey bee	0.054 –1.44	Barren and Woodburm (1998)
Lady bird beetle	0.036 – 0.50	
Earthworm	104–1174 mg/kg	Ma and Bodt (1993)

The toxicity of chlorpyriphos on other animals is summarized as below :

B. Toxicity to other animals/organism

Animals	Phyla	LC/EC	Values	References
Planaria	Planaria	LC	2-4.3	
Oligocheta	Annelida	EC	>36	

Snail	Mollusca	EC	> 94-200	Barren and Woodburn (1998)
Leech		LC	> 128	
Oyster		EC	34-1090	
Crayfish	Crustasea	LC	6-41	
Crab		LC	60-200	
Fairy shrimp		EC	<0.62	
Mysid		LC	0.035-0.056	
Tadpole shrimp		LC	4	
Blue crab		EC	5.2	
Brown shrimp		EC	0.20	
Grass shrimp		EC	0.29-1.50	
Pink shrimp		EC	2.4	
Midge	Insecta	LC	0.44	Barren and Woodburn (1998)
Black fly		LC	27	
Caddis fly		LC	0.3-0.9	
Damselfly		LC	8.4-14.3	
Dragonfly		LC	0.10	
Mayfly		LC	0.25-3.00	
Midge		LC	0.007-90	
Mosquito		LC	0.0011-7	
Stonefly		LC	7-3	
Fishes	Pisces	LC	0.5-100 μg/l	Ma and Bodt (1993)
Bluegill Silverside	Pisces	LC	1.7-10 μg/l	
Frog and tadpoles	Amphibia	LC	1-3000 μg/l	
Water fowl Quaails Chicken Pigeon Doves Cranes	Aves	LD	8.3–83 μg/l	Ma and Bodt (1993)
Blackbirds Starlings Sparrows	Aves	LD	13-75 μg/l	Ma and Bodt (1993)

2.9.5 Methyl parathion

It is commonly recommended organophosphate for control of sucking inects and leaf eating insects on wide variety of crops. Although it is not considered as highly persistent pesticide and it is readily degraded in an agricultural ecosystem, several reports are made about its free availability in surface and underground water. Its half life is reported to be 9-17 days in ponds (Crossland and Bennett, 1984), 5-6 days in rearing ponds (Sabharwal and Belsare, 1986), 0.6 hr in river (Smith *et al.,* 1978), 27-28 hrs in eutrophic ponds (Smith *et al.* 1978) and up to 168 hr in clear lake in California (Epperson *et al.* 1976).

If the pH is less than eight, it is reported to have half life in lake sediments up to 2138 hrs (Smith *et al.,* 1978).

All free swimming species of aquatic insects are reported to be affected at 100 ppb (Crossland, 1984) while many aquatic animals are reported to have very low toxicity to methyl parathion.

Species	Test period (Hrs.)	Toxicity LC 50	References
Rainbow trout	-	21-2.5 pprs	Palawski *et al.* (1983)
Water flea	48	0.14 pprs	
Fathead minnow	96	9.96 pprs	
Carp	96	7-13 pprs	
Blue gill sunfish	96	1.0 pprs	
Side swimmers	96	4.5 ppr	Meyer and Ellesieck, 1986
Grass shrimp	48	0.4 pprs	
Stone fly	96	0.3 pprs	

2.9.6 DDT

Residues of DDE are recorded in teal and mallard samples, respectively in concentration of 0.033 μg/g and 0.058 mg/g while these animals have recorded a PCB concentration in the range of 0.025 μg/g to 0.272 μg/g in their bodies (Botero *et al.,* 1996). Similarly, Kennish and Ruppel (1996) have recorded DDT and its metabolites in the range of 25-300 μg/kg weight of bodies of blue fish, blue crabs collected between the period from 1988 to 1991.

DDT and its metabolites have been found in arctic and Antarctic fishes (Subramaniam *et al.,* 1983), cetaceans of pacific ocean, Indian ocean (Tanabe *et al.,* 1986), in grey whales along western coast of N. America (Varanasi *et al.,* 1994) and in black footed albatrosses (Jones *et al.,* 1995).

Some investigators have reported many adverse health effects of DDT on seals and whales. Their inferences indicate clearly that DDT can cause wide ranging long term effects in seals and whales, which include reproductive, sexual abnormalities as well as aberrations in immune system of these animals. Their investigations are summarized as below :

Health effect	Animal	References
Reproductive dysfunction	Common seals	Reijnders. 1986
Immune suppression	Harbour seals	Ross *et al.*, 1994
Tumors	Beluga whales	De Guise *et al.*, 1994
Abnormal sexual development	Beluga whales	De Guise *et al.*, 1994
Impaired reproduction	White croakers	Cross & Hose, 1988
Behavioral abnormalities	Herring gulls	Fox *et al.*, 1978
Reduced sperm counts	Polar aquatic mammalians	Bacon *et al.*, 1992
Reduced population	Seals	Reijnders, 1986

2.10 REFERENCES

Abbott, D.C., Collins, G.B., Goulding, R. and Hoodless, R.A. 1981. Organochlorine pesticide residues in human fat in united Kingdom. *Brit. Med. J.*, 283 (6304) : 1425-28.

Aber, J.D., Nadelhoffer, K.J., Steudler, P. and Melillo, J.M. 1989. Nitrogen saturation in north eastern eco-system. *Bio. Sci.,* 39 (6) : 378-386.

Acquavella, John, F., Riordan, S.G., Margaret, A., Lynch, C.F., Collins, J.M., Ireland, B.K. and Heydens, W.F. 1996. Evaluation of mortality and cancer incidence amoung alachlor manufacturing workers. *Environ. Health Persp.* 104 (7) : 728-733.

ADB, 1987. Handbook on use of pesticides in Asia-Pacific region., Mauila pp. 65-92.

Agarwal, H.C. 1980. Pesticide residues in man in India, In : Pesticides residues in environment in India. Proc. of Symposium held at UAS, Bangalore in Nov., 1978, pp. 54-55.

Agarwal, G.D., Lankad, S.K. and Malkhed, T. 1999. Diffuse agricultural nitrate pollution of ground water in India. *Water Sci. Tech.,* 39 (3) : 67-75.

Agnihotri, N.P., Sharma, A.P. and Krishna Singh 1981. Influence of insecticides on soil micro-organisms and their biological activity. *Pesticides,* 15 : 16-24.

Albert, L.A. 1996. Persistent pesticides in mexico. *Rev. Environ. Contam. And Taxicol*, 147 : 6-70.

Albert, L., Vega, P. and Portales, A. 1981. Organo-chlorine pesticides in human milk samples from comara lagunera, mexico. *Pestic. Monit. J.* 15 (3) : 135-138.

Albert, L., Mendez, F., Cebrian, M. and Portales, A. 1983. Organo-chlorine residues in human adipose tissue in Mexico-results of preliminary study in three Mexican cities. *Arch. Environ. Health,* 35 (3) : 262-69.

Alexeef, G.V., Kilgore, W.W. and Li, M.Y. 1990. EDB : toxicology and risk assessment. *Rev. Environ. Contam. and Toxicol.,* 112 : 50-95.

Almed, F.A., Tegeris, A.S. and Underwood, P.C. 1981. Alachlor : Six month study in dog : unpublished study submitted to USEPA by Monsanto Co. study No. CDL : 246292, 246293, 247376.

Alvord, H.H. and Kadlec, R.H. 1996. Atrazine fate and transport in Das Plaines wetlands, Ecological modeling, 90 (1) : 97-107.

Angus, B.J., Stolting, H., Harris, G.L., Howse, R. and Rohn, K.C. 1996. Leaching of recently applied and aged residues of isoproturon and atrazine through soil core. *Chemosphere,* 33 (7) : 1297-1305.

Anonymous, 1988. Fate of alachlor in environment. *Rev. Environ. Cvontam, and Toxicol*. 104 : 9-20.

Anonymous, 1995. Fertilizer association of India News, 1995, pp. 1-102.

Anonymous,1998, World resources: A guide to global environment. Pub: World Environmental Library pp: 285-296

Anonymous, 2000. Pesticide information, 26 (3) : 5.

Anonymous, 2001a. Pesticide information, Annual issue, pp. 72-73.

Anonymous, 2000, Pesticide use use statistics countries compared, NationMaster.com

Anonymous, 2010a. Indian fertilizer scenario. http://fert.nic.in/sites/default/files/Indian-Fertilizer-Scenario.pdf

Anonymous 2010b. Pesticide statistics, http://ppqs.gov.in/IpmPesticides Cont. htm

Bacon, C.F., Jarmon, W.M. and Costa, D.P. 1992. Organo-chlorine and polychlorinated biphenyl levels in pinniped milk from arctic, Antarctic, California, Australia. *Chemosphere,* 24 : 779-791.

Bai, A.R.K. and Reddy, C.C. 1977. Laboratory toxicity to some insecticides in Apis Cerana India. *J. Apic. Res.,* 16 : 161-162.

Balasubramanian, A. 1981. Effect of pesticides on non-target soil microorganisms. In : Progress in soil biology and ecology in India Ed.: G.K. Veeresh, UAS, Technical Series, 37 : 305.

Balasubramanian, A., Shantaram, M.V., Emmimath, V.S. and Sidramappa, R. 1974. Lindone increases nodulation and yield in soybean plants. *Pesticides,* 8 (1) : 35.

Balasubramanian, A. and Gita Nilakanthan, 1978. Effect of Aldicarb on a non-target soil dwelling phytopathogen. In : Soil biology and a ecology in India. Ed. : CA Edwards and GKVK, Bangalore 65 Veeresh, UAS Tech. Series, 22 : 304-309.

Balasubramanian, A. and Narayanan, R. 1980. Influence of pesticide on growth and metabolism of beneficial non-target soil bacteria. In :

Pesticide residues in environment in India, Ed. CA Edwards, GKVK, Bangalore - 65 Veeresh and H.R. Kruger, UAS Tech. Series, 32 : 486-492.

Balwinder Singh and Dhaliwal, G.S. 1993. Pesticide contamination of fatty foods in developing contries. In : Pesticides their ecological impact in developing countries. Ed : Dhaliwal, G.S. and Balwinder Singh, Pub. Common Wealth Pub., New Delhi, pp. 65-91.

Bami, H.L. 1989. Pesticide pollution - rational assessment in India context. Paper presented at National symposium on impact and management of pollutants on crop productivity, Hissar held on Feb. 16-18, 1989, pp. 10-79.

Banerjii, S.A. 1989. Pesticide residues in food items available in Bombay market. In : Pesticide residues in food consumer education and research centre, Ahemabad, pp. 304-309.

Barnes, C.J., Jacobson, G. and Smith, G.D. 1992. The origin of high nitrate ground water in Australian arid zone. *J. Hydrology,* 137 (1-4) : 181-197.

Barren, M.G. and Woodburn, K.B. 1998. Ecotoxicology of chlropyriphos. *Rev. Environ. Contam. Toxicol.,*144 : 1-76.

Bashkin, V. 1989. Accumulation of agro-chemicals and their metabolites in surface and ground water in various agro-climatic regions. In : Toxic substances in agricultural water supply and drainage. *Proc. 2nd Pan Am., ICIO Region,* p. 64-79.

Battu, R.S., Chawla, R.P. and Kalra, R.L. 1980. Insecticide residues in market samples of vegetable oils and oil seed cakes from selected areas of Punjab. *Indian J. Ecol.* 7 : 1-7.

Battus, R.S., Gupta, S.C., Chawla, R.P. and Kalra, R.C. 1984. Residues of DDT and BHC in market samples of meat of various animals. *Indian J. Ecol.* 11 (2) : 177-182.

Beeman, R.W. and Matsumara, F., 1981. Metabolism of Cis and Trans chlordane by soil organisms. *J. Agri. and Food Chem.* 20 : 84-89.

Belluck, D.A., Benjamiu, S.L. and Dawson, T. 1991. Groundwater contamination by atrazine and its metabolites : risk assessment, policy and legal implications. ACS Symposium series, Pub. Minnesota pollution control agency, 459 : 254-273.

Bhaskar Rao, B.R. 1975. Studies on the effect of phorate on nitrification inhibition in soils of Karnataka. M.Sc. (Agri.) Thesis submitted to UAS, Bangalore, pp. 25-148.

Bhige, A.R. and Banerjii, S.A. 1981. Estimation of organo-chlorine pesticide residues in different organs of the pomfres fish available in Bombay. Scaranger, 11 : 25-30.

Bhunia, A.K., Basu, M.K. Roy, D., Chakrabarti, A. and Banerjee, S.K. 1992. Growth, Chlorophyll-a, N-fixing ability and certain metabolic activities of Nostoc muscorium-effect of methyl parathion and benthiocarb. *Bull. Environ. Cont. and Toxicol.,* 47 (1) : 43-50.

Bijay Singh and Sekhon, G.S., 1977. Impact of fertilizer use on environmental pollution in Punjab present status and fertilizer projections. *Fert. News.* 23 (2) : 7-11.

Botero, J.E., Meyer, M.W., Hurley, S.S. and Rusch, D.H. 1996. Residues of organo-chlorines in mallards and blue winged teal collected in Columbia and Wiscounsin, 1984-1989. *Archives of Environ. Contam. And Toxicol.* 31 (2) : 225-231.

Bouwer, H. 1990. Agricultural chemicals and ground water quality. *J. Soil and Water conservation* 45 (2) : 184-189.

Bouwer, H., Dedrick, A.R. and Jayner, D.B. 1990. Irrigation management for groundwater protection. *Irrigation and drainage systems*. 4 (4) : 375-383.

Bubler, D.D., Randall, G.W.; Koskineu, W.C. and Wyse, D.L. 1993. Alachlor and atrazine losses from subsurface till drainage of clay loam soil. *J. Environ. Qual.,* 22 : 583-588.

Burden, E.H.W.J. 1961. The toxicology of nitrates and nitrates with particular reference to potability of water supplies. *Analyst,* 86 : 429-433.

Burnet, M. and Hodgson, B. 1991. Differential effects of sulphonyl urea herbicide - chlorsulfuran and sulphometuron methyl on micro-organisms. *Archives of microbiology*, 155 (6) : 521-525.

Burns, J.E. 1974. Organo-chlorine pesticide and poly chlorinated biphenyl residues in biopsed human adipose tissues - Texas 1969-1972, Pestic. Monitoring, 7 : 122.

Butler, P.A. 1965. Commercial fishery investigation. In Effects of pesticides on first and wild life, USDI Fish and Wildlife Circle, 226 : 65-77.

Bulusu, K.R. and Pande, S.P. 1990. Nitrates- A serious threat to ground water pollution, BHD-JAC News, Quarterly J. CGWB, 5 (2) : 39.

Calderbank, A. 1989. Reviews of Environmental Pollution and Toxicology, Vol. 108-95.

Cam, C. and Nigogosyan, G., 1963. Acquired toxic porphyria cutanea tarda due to hexa chlorobeuzene, JAMA, 183 (2) : 88-91.

Cantor, K.P., Everett, G.D., Blair, A., Gibson, R., Schuman, L.M. and Isocson, P., 1985. Farming and no-hodgkin's lymphoma *Am. J. Epidamiol.*, 123 : 535.

Carlsen, Giwercman, A., Keiling, M. and Skakkebeck, M.E., 1992. Evidence of decreasing quality of samen during past 50 years. Br. Med. J., 305 : 609-613.

CERC, Consumer education and Research Centre, 1989. Pesticide residues in food-Ahmedabad, pp. 177-222.

Chandra, P. 1964. Effect of heptachlor on microbial activities in a loam soil. *Indian J. Microbiol.,* 4 : 59.

Chang, V.C.S. and Lange, W.H. 1967. Laboratory and field evaluation of selected pesticides for control of red cray fish in California rice fields. *J. Econ. Entomol.,* 60 : 473-477.

Chawla, R.P., Sharda, V.m., Joja, B.S. and Kalia, R.L. 1978. A survey of residues of organo-chlorine insecticides in biopsy human fat samples. In : pesticide residues in environment in India. Proc. Symposium held in Nov. 1978 at UAS, Bangalore pp. 54-55.

Chesters, G., Simsiman, G.V., Levy, J., Alhajjan, B.J., Fathulla, R.M. and Harkin, J.M. 1989. Environmental fate of Alachlor and metolachlor. *Rev. Environ. Pollut and Toxicol.* 110 : 63.

Chilvers, C., Forman, D., Pike, M.C., Fogleman, K. and Wadsworth, M. 1984. Apparent doubling frequency of undescended testes in England and Wales 1962-1981, Lancet : 330-332.

CIBRC, 2000. Central Insective Board and Registration Committee on http://cibrc.nic.in/pestconsum.asp

Clements, R.O., Murray, P.J. and Sturdy, R.G. 1991. The impact of 20 years absence of earthworms and three levels of N fertilizer on grass land soil environment. Agriculture, eco-systems and environment, 36 (1-2) : 75-85.

Colborn, T. and Clement, C. (Ed.) 1992. Chemically induced alteration in sexual functional development : The wildlife/human connection. *Pub. Princeton Sci. Pub. Princeton J.*, pp. 67-216.

Croll, B.J., 1991. Pesticides in surface waters and ground waters. *J. Inst. Water and Environ. Mgt.* 5 (4) : 389-395.

Cross, J.M. and Hose, J.F. 1988. Evidence of impaired reproduction in white croacker from contaminated areas of *S. California. Marine Environ. Res.*, 24 : 185-188.

Crossland, N.O., 1984. Fate and biological effect of methyl parathvan in outdoor ponds and laboratory aquaria. *Exotoxicol. Environ. Safety.* 8 : 482-495.

Crossland, N.O. and Bennett, D., 1984. Fate and biological effects of methyl parathion in outdoor ponds and laboratory agnaria, 1 : fate. *Eco-toxicol. Environ, safety,* 8 : 471-481.

Crutchfield, S.R., 1989. Agricultures effect on water quality. In : Agriculture - food policy review. US Agril. Ppolicies in challenging world Pub : USDA No. 6200 pp. 369-382.

Curley, A., Coeland, M.F. and Kimbrough, R.K. 1969. Chlorinated hydrocarbon insecticides in organs of skill born and blood of new born babies. *Arch. Environ. Health.* 19 : 628-632.

Curley, A., Burse, V.W., Jennings, E.L., Tomatis, L. and Akazaki, K., 1973. Chlorinated hydrocarbon pesticides and related compounds in adipose tissue from people of Japan Nature, 242 : 338-340.

Curtney, K.D., 1977. Prenatal effects of herbicides : evaluation of prenatal development index. *Arch. Environ. Contam. and Toxicol.,* 6 : 33-46.

Dahama, A.K. 1977. Management of agriculture pollution. In : Organic for sustainable agriculture. Pub. Agr. Botanica, Bikanes, pp. 82-83.

Daji, J.A. 1988. Soil structure. In : A Textbook of Soil Science Pub : Media Promoters and Publishers Bombay pp. 89-301.

Dale, W.E., Copeland, M.F. and Hayes, Jr. W.J., 1965. Bull. World Health Org., 33 : 471.

David, B.V., 1992a. Pesticide industry in India. In : Pest management and pesticides : Indian Scenario (Ed. B.V. David) Pub.: Namrutha Pub. Madras, pp 225-230.

David, B.V., 1992b. Registration of pesticides and guidelines. In : Pest management and insecticides : Indian scenario (Ed. B.V. David) Pub.: Namrutha Pub. Madras pp. 276, 325.

Davies, J.E. 1973. Environmental pollution by pesticides, Pub. : CA Plenum Prss, pp. 313.

Davis, G.M. and Lewis, I., 1956. Outbreak of food poisoning from bread made of chemically contaminated flour. *Brit. Med. J.,* 11 : 393-398.

De guise, S., Lagace, F. and Beland, P., 1994. True hermophrodism in St. Lawrance Begula whale. *J. Wild Dis.,* 30 : 287-290.

Dhaliwal, G.S. 1990. Pesticide contamination of milk and products. In : Food contamination from environmental sources. Ed : Nriagu, J.O. and Simmow, M.S. Pub. John Wiley and Sons, Inc., New York, pp. 359-389.

Dhaliwal, G.S. and Kalra, R.C. 1978. DDT residues in butter and infant formula in India. *Pestic Monit J.,* 12 : 91-93.

Dhaliwal, G.S. and Pathak, M.D., 1993. Pesticides in developing world : A boon in bane. In : Pesticides: their ecological impact in developing countries, pp. 8-24.

Dikshit, T.S.S., Kumar, S.N., Tandon, G.S., Raizada, R.B. and Ray, P.K. 1989. Pesticide residues in edible oils and oilseeds. *Bull. Environ. Contam. Toxicol.* 42 : 50-56.

Dodey, J.L. and Kammer, A.G., 1953. Chlordane Intoxication. *J. Am. Med. Assoc.,* 153 : 723.

Dolgova, L.G., Pavlyukova, M.F. and Grishko, V.N., 1988. Influence of NO_2, NH_3 and their mixture on microflora. *Biologicheskie Nanki.,* 12 : 92-96.

Domagalski, J.L. and Dubrovsky, N.M. 1992. Pesticide residues in ground water of San Joaquin valley California. *J. Hydrology,* 130(1-4) : 299-338.

Donahue, R.L., Miller, R.W. and Shikulana, J.C. 1990. Fertilizer and optimum yields. In : Soils - an introduction to soils and plant growth Pub : Prentice Hall of India, pp. 282-285.

Donna, A., Crosignani, P., Robutti, F., Beta, P.G., Bocca, R., Mariani, N., Ferriario, F., Fissi, R. and Berrino, F., 1989. Triazine herbicides and ovarian epithelial neoplasm. *Scand J. Work Environ. Health,* 15 : 47-53.

Dudhani, A.T. and Sengupta, S. 1991. Status of banned and bannable pesticides in India. Pub. Voluntary health association of India, New Delhi, pp. 65-148.

Dutka, B.J., Mcinnis, R., Jurkovic, A., Liu, D. and Castillo, G., 1996. Water sediment ecotoxicity studies in temuco and Rapel River basiu of chile. *Ecol. Modelling,* 90 (1) : 27-42.

Dutta, I.N., Attri, B.S. and Ratanlal, 1977. Pestology, 1 (2) : 7-10.

Edwards, C.A. 1966. Insecticides residues in soils. *Res. Rev.*, 13 : 83-132.

Edwards, C.A., 1986. Agro-chemicals as environmental pollutants. In: Control of Pesticides and Residues in Food : A guide and directory (Ed. B. Van Hofsten and G. Elaspern). Pub. : Swedish Science Press, Upsala, Sweden, pp. 1-19.

Ekstrom, G., Hemming, H. and Palmborg, M. 1996. Swedish pesticide risk reduction, 1981-1995: Food residues, health hazards and reported poisonings. *Rev. Environ. Contam. and Toxicol* 147 : 119-140.

EPA, Environmental Protection Agency, 1981. Office of pesticde programs : Summary of reported pesticide incidents involving EBD, EPA Rep., 431, pp. 1-20.

Epstein, S.S. 1976. Carcinogenicity of heptachlor and chlordane. *Sci. Total Environ.* 6 : 103.

Epperson, C.S., Elson, R. and Cashe, W., 1976. Biological effects and persistence of methyl parathion in clear lake California. *Environ. Entomol.*, 5 : 1116-1120.

http://faostat.fao.org/site/422/DesktopDefault.aspx?PageID=422#ancor

FAO, Food and Agriculture Organization, 2002, FAO Fertilizer year book, Vol. 52, pp. 17-35.

FAO, Food and Agriculture Organization, 2003, FAO year book production, pp. 50-52.

Fendick, E.A., Mather-Mihaiach, E., Houck, K.A., St. Clair, M.B., Fausta, J.B., Rockwell, C.H. and Owen, M. 1990. Ecological toxicology and human health effects of heptachlor. *Rev. Environ. Contam and Toxicol.* 111 : 61-119.

Fleischer, S. and Stibe, L. 1989. Agriculture kills marine fish in 1980s. Who is responsible for fish kills in 2000s. *Ambio.*, 18 (6) : 347-350.

Fox, G.A., Gilman, A.P., Peakall, D.B. and Andreka, F.W., 1978. Behavioural abnormalities in nestling lake Ontario gulls. *J. Wild life Mgt.*, 42 : 477-483.

Furie, B. and Trubouritz, S. 1976. Insecticides and blood dyscrasios : Chlordane exposure and self limited refractory megaloblastic anemia, *JAMA* 235 : 1720-1722.

Gass, R. 1991. Atrazine - Production : Gesundheit Sri Siko bei Chemiearbeitern (Poster). Internationale Gesellschaft fur Epidemiolgie. Europaische Regional tagung, Basel, Surtzerland.

Gaur, A.C. and Bardia, M.C. 1970. The effect of lindane and dieldrin on soil microbial population. *Indian J. Microbiol.*, 10 : 33.

Giwercman, A., Carlson, E., Keiding, M. and Skakkabaek, M.E. 1993. Evidence of increasing incidence of abnormalities in human testis : A review. *Environ. Health Prospect* (Suppl.), 101 (2) : 65-71.

Goes, E.h., Savage, E.P., Gibbons, G., Aaronson, M., Ford, S.A. and Wheeler, H.W., 1980. Suspected foodborne carbamate pesticide intoxications associated with ingestion of hydroponic cucumber. *Am. J. Epidemiol.*, 111 : 254-259.

Gracia-Banuelos, M.L., Meza - Montenegro, M. 1991. Principales vias de contaminacion por plaugicidas en nenatos lactantes residents en pueblo yagui, Sonara, Mexico, IT SOM-DIEP, 1 (2) : 33-42.

Green, E.S., Miller, D.J., Bruscato, F.N. and Holt, R.L. 1980. Investigation of pesticide residues in human adipose tissue in northeastern Lousian area. *J. Agri. Food. Chem.*, 28 (1) : 76-78.

Gupta, S.C. and Kalra, R.L., 1983. Insecticide residues in market farm gate samples of vegetables from Ludhiana, Punjab. *Indian J. Ecol.* 10 : 20-24.

Gupta, S.K., ergese, S., Chatarjee, S.K. and Kashyap, S.K. 1982. Organochlorine insecticide residues in cooked meal samples in India. *Pesticides* 16 : 8-9.

Gupta, Y.P. 1989. Pesticide poisoning. *Sci. Rep.*, 26 (9) : 491-492.

Guth, J.A. 1994. Monocrotophos - environmental fate and toxicity. *Rev. Environ. Contam. And Toxicity*, 139 : 75-127.

Haines, R.G. 1971. Ingestion of aldicarb by human volunteers : A controlled study of the effect of aldicarb on man unpublished report with addendum, Feb. 11, Union Carbide Corp., pp. 32.

Hall, J.K., Murray M.R. and Harturig, M.L. 1989. Herbicide leaching and distribution in untilled and tilled soil. *J. Environ. Quality,* 18(4):439-443.

Hallberg, G.R. 1989. Mitrate in ground water in US. In: Nitrogen management and ground water protection Ed. R.F. Follett, pp. 35-74.

Handa, B.K. 1986. Pollution of ground waters by nitrates in India BHU-JAL News, Quarterly J. Central Ground Water Board, 1 (3) : 16-19.

Handa, B.K. 1987. Nitrate content of ground water in India. *Fert. News,* 32 : 11-29.

Handa, B.K., Kumar, A. and Bharadwaj, R.K. 1991. Studies on Pal Lako, Srinagar J & K : Eutrophication status. BHD-JAL News, Quarterly J. Central Ground Water Board, 6 (4) : 26-35.

Handa, B.K. 1975. High nitrate and potassium ion concentrations as indicators for ground water pollutions in India. Abstract, International Symposium on Geochemistry of natural waters, Burlington, Outario, Canada, pp. 65-98.

Handa, B.K. 1991. Phosphate content of ground water in some parts of India. *Fertilizer News*, 36 (10) : 11-17.

Hardell, L., Erikson, M., Lenner, P. and Lundgern, E., 1981. Malignant Lymphoma and exposure to chemicals especially organic solvents, Chlorophenols, phenoxy acids - A case control study. *British. J. Cancer,* 43 : 169-176.

Hardin, B.D., Bond, G.P., Sikow, M.R., Andrew, A.D., Beliles, R.D. and Niemein R.W. 1991. Testing selected workplace chemicals for teratogenic potential. *Scand. J. Work Environ. Health.* pp. 766-775.

Harris, C.F. and Warren, G.F. 1964. Adsorption and desorption of herbicides in soil weeds, 12 : 120-126.

Heinrichs, E.A., Gastal, H.A. deo and Galileo, M.H.H. 1979. Incidence of natural control agents of veloel bean caterpillar and response to its predator to insecticide treatments in Brazillian soybean fields. *Pesqui Agro-pecu Brasil*, 14 : 79-87.

Hoar, S.K., Blair, A., Holmes, F.F., Bysen, C. and Robel, R.J. 1985. Herbicides and colon cancer, *Lanceli* : 1277-1278.

Infante, P.F., Epstein, S.S. and Newton, W.A. Jr., 1978. Blood dyscrasis and childhood tumore and exposure to chlordane and heptachlor. *Scand. J. Work Environ. Health,* 4 : 137-150.

IRDC, 1980. A teratology study in rabbits. Study No. 401-060 submitted to USEPA. International Research and Development corporation. Rep. No. CDC : 244-369.

Ishamuru, Y., Minato, K., Katayama, Y. and Aoki, A., 1987. Behaviour of HCH in small wooded watershed. *Bull of Kyoto Univ. Forests* No. 59 : 91-98.

Jacobson, J.L. and Jacobson, S., 1993. Four year followup study of children born to consumers of lake Michigan fish. *J. Gt. Lakes Res.,* 19 : 776-783.

Jadhav, G.D., 1986. Final technical report. Department of Environment (19/58/79 Env.) Govt. of India, new Delhi.

Jadhav, G.D. 1989. DDT and HCH residues in vegetable and fruit samples from Marathwada region. In: Abst. Natl. Symp. Impact and management of pollutants on crop productivity held at Haryana. Agril. University on 16-18 Feb. 1989, pp. 27.

Johannessen, Tore and Dahl, E., 1996. Declines in O_2 concentration along Norwgian skagerrak coast 1927-1993 a signal of eco-system changes due to eutrophication. *Limnology and oceanography*, 41 (4) : 766-778.

Johnson, W.W. and Finley, M.T. 1980. Handbook of acute toxicity of chemical to fish and aquatic invertebrates, USDI, Fish Wildlfe service. *Res. Pub.* 137.

Jones, D.D., Hannals, D.J., Buckland, H.J., Summer, C., Ludwig, J.P. and Giesy, J.P. 1995. Persistent synthetic chlorinated hydrocarbon in albatross tissue samples from midway Atoll cited in Rev. Environ. Cont. and Toxicology, 146 : 96-101.

Jorgensen, S.E., Nielson, S.C. and Marcussen, B., 1989. Removal of nitrates from ground water through the use of artificial wetlands. *Int. J. Water Res. And Dev.* 5 (3) : 209-214.

Kae, S.P., Murthy, N.B.K. and Raghu, K. 1989. Effect of carbofuran, Carboryl and their etabolites on rhizobium and azotobacter. *Bull. Environ. Cont. and Toxic.* 42 : 769-772.

Kalra, R.C. 1977. Annual report of project on studies on pesticide residues and monitoring pesticidal pollution.

Kalra, R.L., Sharma, M.L., Battu, R.S. and Gupta, S.C. 1983. Residues of DDT and HCH in butter and ghee in India, 1978-1981. *Environ. Pollut (Series B),* 6 : 195-206.

Kalra, R.L. 1989. Pesticide residues in food in India - An overview. In : pesticide residue in food consumer education and research centre, Ahemadabad, 79-98.

Kalra, R.L. and Chawla, R.P. 1983. Studies on pesticide residues and monitoring of pesticide pollution-Final technical report (Ludhiana : PAU), pp. 230.

Kansal, B.D., Grewal, M.S. and Dhaliwal, G.S. 1992. Status of water pollution and its control : Changing scenario of our environment Pub : PAU Ed. G.S. Dhaliwal B.S. Hansra, M.J. Yerath, pp. 147-156.

Karim, Z. 1992. Phosphatic fertilizers and environment. In: Proc. of regional FADINAP seminar on fertilizer and environment held at Thailand on 7-11 Sept., 1992, pp. 103-111.

Kathphal, T.S., Singh, G. and Yadav, G.S., Dhakar, J.S. and Singh, A., 1992. Monitoring of milk and milk products for DDT and CH contamination. *Pestic. Res. J.,* 4(2) : 123-131.

Keeney, D.R. 1989. Sources of nitrates to groundwater. In : N-management and ground water protection Ed. R.F. Follett, pp. 23-34.

Kennish, M.J. and Ruppel, B.E. 1996. DDT contamination in selected estuarine and coastal marine fish and shell fish of new Jersey. *Arch. Environ. Contam and Toxicol.,* 31 (2) : 256-262.

Klemmer, K.W., Budy, A.M., Takahasdhi, W. and Haley, T.J. 1977. Human tissue distribution of cyclodiene pesticides in Hawaii 1964-1973. *Clin. Toxicol.,* 11 (1) : 71-82.

Kloos, H., 1996. 1, 2 dibromo-3 chloro propane (DBCP) and EDB in well water in Fresno/Clovis metropolitan area of California. *Archives of Environ. Health,* 51 (4) : 291-299.

Kochman, M.M. 1928. Possible industrial poisonings with EDB : Munich Ed. Wochenscber, pp. 1334-1336.

Krishnamurthy, C.R. and Kutty, D. 1981. Agro-chemicals in maternal milk blood and cord blood : A source of toxicants for prenates and neonates. *Environ. Res.* 24 : 24-32.

Krishnamurthy, M. and Reddy, G.P.N. 1978. Residues of chlorinated hydrocarbons in market vegetables at Hyderabad. In : Pesticides residues in the environment in India. Ed : Edwards, C.A. Veeresh, G.K. Krueger, H.R., UAS, Bangalore, pp. 146-149.

Komarova, L.I. 1983. Problem of pesticide burden and their bearing on development of some pathological processes. In : Hygiene, toxicology of pesticides and polymers. *Kiev.* 13 : 177-182.

Krishnamurthy, S.V. and Reghupathy, A. 1990. Monitoring of HCH and DDT-R in groundnut and seasamum oils. *Pesticides Res. J.* 2(2): 145-150.

Kumar, Y. and Agnihotri, M.P. 1991. Persistence of deltamethrin in soil. *Pestic Res. J.* 3 (1) : 87-92.

Kurtz, F.W. 1983. Chemical exposure monitoring. *Residue Rev.,* 85 : 283.

Lal, R., Narayanarao, V.V.S. and Krishnamurthy, V., 1989. Pesticide residues in food : An overview of studies undertaken in India. In : Pesticide residues in food consumer education and Research Centre, Ahmedabad, 99-113.

Laster, M.L. and Brazzel, J.R. 1968. A comparison of predator population in cotton under different control programmes in Mississippi. *J. Econ. Entomol.* 61 : 714-719.

Laxminarayana, V. 1980. Sareening for organo-chlorine insecticide residues in samples of food stuffs drawn from Hyderabad. Secunderabad markets. Paper presented at the meeting of ICAR Export Committee, Ludhiana, April, 14-15, 1980.

Laxminarayana, V. and Krishnamenon 1975. Screening of Hyderabad market samples of food stuffs for organo-chlorine residues. *Indian J. Pl. Prot.,* 3 : 4-19.

Laxminarayana, V. and Menon, P. 1972. Exposure of general population in Hyderabad to DDT. *J. Food Sci. Technol.* 9 : 82-85.

Leet, Terry, Acquvella, J., Lynch, C., Margaret, A., Weiss, N.S., Vaughan, Th. and Checkoway, H. 1996. Cancer incidence among alachlor manufacturing workers. *American J. Ind. Med.,* 30 (3) : 300-306.

Letz, G.A., Pond, W.M. and Osterich, J.D. 1984. Two fatalities after acute occupational exposure to ADB. *JAMA* 252 : 2428-2431.

Lewis, R.E. and Davis, J.F. 1970. Soils-Handbook. *Kentucky Agri. Expt. Station Misc.,* 383, p. 28.

Lingers, P.D., Ridgway, R.L., Cowan, C.B., Davis, J.W. and Watkins, W.C. 1968. Biological control of bollworm and tobacco budworm by arthropod predators affected by insecticides. *J. Econ. Entomol.* 61 : 521-525.

Lola, M.V., Akhmet'eva, N.P., Grigor'ev, V.T. and Lakhtyuk, R.A. 1988. Content of biogenic substances in soil and their effect on eutrophication water body. *Water Resources* 15 (6) : 587-595.

Longabucco, P. and Rafferty, M.R. 1989. Delivery of non-point source phosphorus from cultivated muckland to lake Ontario. *J. Environ. Quality,* 18 (2) : 157-163.

Loosli, R. 1995. Epidemiology of Atrazine. *Rev. Environ. Contam. & Toxicol.* 143 : 47-56.

Lugei, F.M., Goursand, J. and Casalis, J. 1974. La pollution des lait humains francais pa les residues de pesticides organo-chlorives. *L. Alim Vie,* 62 : 40-69.

Luquet, F.M., Gaursand, J. and Casalis, J. 1974. La pollution des lait humains francais par les residues de pesticides organochores. *L. Alim Vie,* 62 : 40-69.

Ma, W.C. and Bodt, J. 1993. Difference in toxicity of chlorpyriphos to six species of earthworm in standard soil test. *Bull. Environ. Contam. & Toxicol.* 57 : 864-70.

Mackay, D. 1982. Correlation of bio-concentration factors. *Environ. Sci. & Techn.*, 16 : 274-276.

Mascher, F. and Marth, E. 1991. Influence of nitrate contamination of superficial ground water through agricultural cultivation in Styria (Australia), *Zentralblatt fur Hygiene und Umweltmedizin,* 192 (3) : 220-229.

Martinez-Toledo, M.V., Solmeron, V. and Gonazatezlopez, J. 1991. Effect of simazine on biological activity of *Azotobacter chrococcum. Soil Sci.,* 151 (6) : 459-467.

Martinez-Taledo, M.V., Salmeron, V. and Gonazalezlopez, J. 1992. Effect of organophosphorus insecticides, prephenphos on agricultral soil microflora. *Chemosphere,* 24 (1) : 71-80.

Martenson, A.M., 1992. Effects of agro-chemicals and heavy metals on fast growing rhizobia and other symbiosis with small seeded legumes. *Bioloy and Biochemistry,* 24 (5) : 435-445.

Matukhno, Yu, D. 1988. Nitrogen and phosphorus in river water. *Agrochimia* 10 : 82-88.

Mayer, F.L. Jr. and Ellerieck, M.R. 1986. Manual of acute toxicity : Interpretation and database for All chemicals and 66 species of fresh water animals. *Fish and Wildlife Service Pub.,* No. 160.

Medved, L.I. 1971. Success and problems of pesticide in hygiene. Use of hygiene pesticide toxicology and clinical pecture of poisonings. *Kiev.* 9 : 5.

Mehta, B.C., Singh, R.V., Srivastava, S.K. and Das, S., 1990. Impact of fertilizer use on ground water quality in parts of Ganjam district, Orissa, BHU-JAL News. *Quarterly J. Control Ground Water Board,* 5 (2) : 44-48.

Mirvish, S.S. 1977. N-nitroso compounds, nitrite and nitrate : Possible implications for Causation of human cancer. *Prog. Water Technol.,* 8: 195-207.

Moody, D.W. 1990. Ground water contamination in US. *J. Soil and Water Conservation*, 45 (2) : 170-179.

Moye, H.A. and Miles, C.J. 1988. Aldicarb contamination of ground water. *Rev. Environ. Contam. & Toxicol.* 105 : 99-140.

Mussalo-Rauhamaa, H., Pyysalo, H. and Antevro, K. 1988. Relation between the contents of organo chlorine compounds in Finnish human milk and characteristics of mothers. *J. Toxicol. Environ. Hlth.,* 25 : 1-19.

Muthukrishnan, P., Kandasamy, D., Balasubramanian, M. and Oblisami, G. 1981. Effect of organic amendments on degradation. In : Progress in soil biology and ecology in India. Ed. : G.K. Veeresh, *UAS Tech. Series,* 37 : 305.

Nagabhushan, R., Gyananath, G. and Sarojini, R. 1983. Toxicity of 3 organophosphatic compound and their effect on fresh water prawn. *Indian J. Comp. Anim. Physiol.* 1 : 71-75.

Nagvi, S.M.G. 1973. Toxicity of 23 inseciticides to tubicifid worm from Mississippi delta. *J. Econ. Entomol.* 60 : 70-74.

Narasimharao, 1990. AICRP on pesticide residue : Bulletin - Pesticide contamintion, APAU, Rajendranagar, pp. 6-15.

NAS, 1981. National Academy of Sciences. Nitrates : An Environmental Assessment. *National Academy Press,* Washington, pp. 65-106.

Nielson, K., Kaempe, B. and Jewen-Holm, J. 1965. Fatal poisoning in many by 2, 4-D Determination of agent in forensic materials. *Acta Pharmacol. Toxicol.* 22 : 224-234.

Oblisami, G., Ramaswamy, N.M., Purushothaman, D. and Rangaswamy, G. 1973. Preliminary studies on rhizobial nodulation of induced mutants of blackgram. *Madras Agric. J.* 60 : 25-26.

Ohmura, K. and Kurokawa, H. 1990. Change in water quality of small rivers flowing through agricultural districts. *Bull. Kokkaido Prefacture Agril. Expt. Station* No. 61, p. 31-39.

Oki, D.S. and Giambellula, T.W. 1989. Ground water contamination by nematicides. Pub : Tokya Publisheres, Tokyo, pp. 75-125.

Palawski, Buckler, D.R. and Meyer, F.L. 1983. Survival and condition of rainbow trout after acute exposure to methyl parathium, triphenyl phosphate and DEE. *Bull. Environ. Contam. Toxicol.* 30 : 614-620.

Pardovsky, K., Rosival, L. and Mesazarosova, A. 1977. Transplacental permeation of pesticides during normal pregnancy lesk Gynelil, 42 : 405-410.

Peterson, G.A. and Fyre, W.W. 1989. Fertilizer nitrogen management, In : N-management and ground water protection Ed. : R.F. Follett, pp 183-219.

Pimantel, D. 1987. Pesticides : Energy use in chemical agriculture, In : Towards a second green revolution Ed. : G.B. Marini-Bettelo.

Pishoka, N.P. and Timchenko, G.A. 1979. Effect of some organo phosphorus insecticides on bacterial spores of *Bacillus thuringensis* group. *Microbiologicheskii Zhumed*, 41 : 492-497.

Poland, A., Smith, D., Kuntzman, R., Jacobson, M. and Feldman, M., 1970. *Am. J. Obstret. Gynec.,* 106 : 939.

Prabhakarsetty, T.K. and Hosmani, M.M. 1980. Effect of herbicide residues on soil microflora. In : Pesticide residues in environment in India, UAS Tech. Series, 32 : 465-471.

Pustaji, A. 1989. Nutrient overloading of soil and protection of water from contamination in Hungary, Vestnik Sel Skokhozyai Stevannoi, Nauki, Moscow, No. 7, 143-147.

Racke, K.D. 1993. Environmental fate of chlorpyriphos. *Rev. Environ. Contam and Toxicol.* 131 : 1-154.

Raghu, K., Murthy, N.B.K. and Kumaraswamy, K. 1975. Fate and persistence of thiram in plants and soil. In : Origin and fate of chemical residues for food, agriculture and fisheries, IAEA, Vienna, pp. 137-148.

Raghupathy, A. and Kuttalam, S. 1992. Monitoring HCH and DDT residues in certain avian eggs in Tamil Nadu. *Pestioc. Res. J.,* 4 (2) : 149-153.

Ramachandran, M., Sharma, M.I.D., Sharma, S.C., Mathu, P.S., Aravindakshan, A. and Edward, G.J. 1974. *J. Com. Dis.,* 6 : 256.

Ramanand, K. and Sethunathan, M. 1990. Effect of pesticide contamination on degradation of carbofuran by soil bacterium. In : Transactions of 14th International Congress of Soil Sci. held at Kyoto on 12-18th Aug. 1990. Vol. III, pp. 190-195.

Ramesh, A., Tanabe, S., Subramanian, A.H., Mohan, D., Venugopalan, V.K. and Tatsukawa, R. 1990. Persistent organo chlorine residues in green mussels from coastal water of south India. *Mar. Pollut. Bull.* 21 : 587-590.

Rawesh, I.A., Gaaboub, I.A., El-Gayan, F.M. and El-Shazli, A.Y., 1975. Standard curves for Nuvoaron, malathion, Sevin, DDT, Kelthane tested against mosquito and microarustean Dephnia magna. *Toxicology* 4 : 133-144.

Reddy, H.R. and Anil Kumar, T.B. 1975. Toxicity of insecticides to sclerotium rolfsii, Pesticides, 9 : 48.

Reijnders, P.J.H., 1986. Reproductive failure in common seals feeding on fish from polluted coastal waters. Nature, 324 : 456-57.

Rice, 1977. *Can. J. Soil Sci.* 57 : 197.

Roach, S.H. and Hopkin, A.R. 1981. Reduction in arthropod predator population in cotton fields treated with insecticides for Heliothis spp. control. *J. Econ. Entomol.,* 74 : 454-457.

Rogan, W.J., Gladen, B.C., Mekinney, J.D., Carrearas, P., Thullen, J., Tingelstad, J. and Jully, M. 1986. PCBs in human milk : Effects on maternal factors and previous lactations. *Am. J. Public Health,* 76 : 172-177.

Rorcevic, N., Parkov, S., Galetin Smith, R., Vukovic, T., Vajinovic, M. and Dordjevic, M. 1987. Serum concentrations of organochlorine compounds during pregnancy and the new boru. *Environ. Contam. and Toxicol.* 38 : 117-128.

Ross, P.S., Deswart, R.L., Visser, I.K.G., Veddu, L.J., Murk, W., Bowen, W.D. and Osterhaus, A.D.M.E. 1994. Relative incompetence of new boru harbech seal *Phoca vitulina. Vety. Immunol. Immunoopath,* 42 : 33-348.

Rouz, F., Treich, I., Brum, C., Desoize, B. and Fournier, E. 1979. Effect of lindane on human lymphocyte responses to phythaemaggluthin. *Biochem. Pharmacol.,* 28 (16) : 2419-2429.

Russel, E.W. 1975. M-cycle in the soil. In : Soil conditions and plant growth ELBS and Longman. Pub. pp. 333-361.

Russelle, M.P. and Hargrove, W.L. 1989. Cropping systems : Ecology and management. In : N-management and ground water protection Ed. : R.F. follett Amsterdam, pp. 277-317.

Sabharwal, A.K. and Belsare, D.K. 1986. Persistence of methyl parathion in carp rearing pond. *Bull. Environ. Contam. And Toxicol.,* 37 : 705-709.

Sahid, I., Hamzah, A. and Aris, P.M. 1992. Effects of paraquat and alachlor on soil micro-organisms in peat soil. *Pertanika,* 15 (2) : 121-125.

Sanborn, J.B. 1979. SD9129/ Daphnia magna strains, Shell Dev. Comp., 6-4-78.

Sandman, ERIC, Loss, M.A. and Van Dyk, L.P. 1988. Microbial degradation of 2, 4-D in soils. *Rev. Environ. Contam. And Toxicol.* 101 : 2-45.

Sathpathy, J.M., 1974. Effect of soil treatment with granular insecticides on soil organisms. *Indian J. Entomol.,* 36 : 139-141.

Sarin, M.M., Krishnaswami, S., Trivedi, J.R. and Sharma, K.K. 1992. Major ion chemistry of Ganga source water : Weathering in high altitude Himalaya. *Proc. Indian Acad. Sci.* 101 (1) : 89-98.

Scheunert, I., 1992. Transformation of pesticides in soil chemistry of plant protection, 8 : 23-75.

Schrader, S.M., Ratelife, J.M., Turner, T.W. and Hornurg, R.W., 1987. Use of new field methods of samen analysis in the study of occasional hazards to pre-production the example of EDB. *J. Occup. Med.,* 29 : 963-966.

Sexton, W.F. 1966. Report on aldicarb, DSEPA pesticide petitions No. 9F0798 section C. USEPA office of pesticide program, Washington, D.C.

Shah, P.G., Patel, B.K., Raj. M.F. and Pathanwadria, R.D. 1992. Monitoring of HCH and DDT inghee and butter samples in Gujrat. *Pestic Res. J* 4 (2) : 117-123.

Shank, R.C. and Magee, P.N. 1981. Toxicity and carcinogenicity of N-nitroso compounds. In : Mycotoxins and N-nitroso compounds : environmental risks vol. V, Ed. Shark R.D. Pub. CRC Press, Boca Paton, pp. 185-217.

Sharma, R.C., Bhaskaran, M. and Bhinde, M.K. 1979. A simplified chemical method of DDT estimation in the body fat. *Indian J. Expt. Biol.* 17 : 1367-1370.

Sheela, S. and Vasnatharajan, V.N. 1977. Pesticide resistance of fensulfothion in soil. *J. Environ. Sci. Health,* 13 : 15-35.

Shevals, Y. and Waldman, M., 1989. Advances in management of fertilizer application and the resulting effect on the pollutant load in soil. In : Inorganic contaminants in vadose zone Ed. BarVosef, B. and Burrow, N.J. and Goldshmid, J. Berlin, pp. 241-265.

Shirasu, 1980. Microbial mutagenicity study, Inst. Environ. Toxicol. Kodira, Japan.

Siddiqui, M.K.J., Sazena, M.C., Bhargava, A.K., Seth, J.D., Krishnamurthy, C.R. and Kutty, D. 1981. Agro-chemicals in maternal milk blood and cord blood : A source of toxicants for prenates and neonates. *Environ. Res.* 24 : 24-32.

Sidramappa, R., Rajaram, K.P. and Sethunadhavan, N., 1973. Degradation of parathion by bacteria isolated from flooded soil. *Applied Microbiol.,* 26 : 816-849.

Singer, R., Moses, M., Valciukas, J., Lilis, R. and Selikoff, I.J. 1982. Nerve conduction velocity studies of workers employed in manufacture of phenoxy herbicides. *Environ. Res.* 29 : 297-311.

Singh, A. and Chawla, R.P. 1979. Organo-chlorine residues in fruit samples. *Indian J. Pl. Prot.* 6 : 95-96.

Skrivan, J. 1971. Methemoglobin in pregnancy. *Acta Univ. Carol. Med.,* 17 : 12.

Smith. J.H., Mabey, W.R., Bohonos, N., Holt, B.R., Lee, S.S., Chu, T.W., Bomberger, D.C. and Mill, T. 1978. Environmental pathways of selected chemicals in fresh water systems. *EPA 600/7-78-074* USEPA, Washington, pp. 248-332.

Smith, M.C., Thomas, D.L., Dottcher, A.B. and Compbell, K.C. 1990. Measurement of pesticide transport to shallow ground water. *Trans ASAE* 33 : 1573-1582.

Smolders, A.J.P., Roelofs, J.G.N. and Hartog, C.D. 1995. Internal eutrophication of aquatic eco-system mechanism and possible Acta Botanica Gallica, 142 (6) : 707-717.

Snawder, J.E. and Chambers, J.E. 1989. Toxic and developmental effects of organo-phosphorus insecticides in embryo of South African Clowed frog. *J. Environ. Sci. Health.* 24 : 205-218.

Sodal, D.P., Holm, O. and Vestol. J.A. 1989. Agricultural policies and environmental protection : the possibilities of a reduction in pollution from agricultural land. *Landbrukso konomisk formu,* 6 (2) : 20-23.

Spindler, M., 1983. DDT : health aspects in relation to man and risk/benefit assessment based thereupon. *Res. Rev.,* 90 : 10-14.

Spynu, E.I. 1989. Predicting pesticide residues to reduce crop contamination. *Rev. Environ. Contam. And Toxicol.* 109 : 90-91.

Srinivasulu, S. and Rangaswamy, G. 1973. Studies on the effects of three granular organophosphorus insecticides on soil microflora. *Indian J. Microbiol.,* 13 : 89-95.

Srivastava, S., Siddiqui, M.K.J. and Seth, T.M. 1983. Organo-chlorine pesticide residues in groundnut oil. *J. Food Sci. and Technol.* 20 : 25-27.

Stewart, D.K.R. and Fox, C.J.S. 1971. Persistence of organo-chlorine insecticides and their metabolites in Nova-Scotian soils. *J. Econ. Entomol.,* 64 : 367-371.

Stover, J.A. and Guitjens, J.C. 1990. Aldicarls in vadose zone : review. *J. Irr and Drainage Engg.,* 116 (1) : 36-49.

Subramanian, A.N., Tanabe, S., Taksukawa, R., Salio, S. and Miyazaki, N. 1987. Reduction in the testesterone levels by PCBs and DDE in Doll's Tortoises in North Pacific *Mar. Pollut. Bull.* 18 : 643-650.

Subramanian, B.R., Tanabe, S., Hidaka, H. and Tatsukaua, R. 1983. DDT and PCB isomers and congeners in Antarctic fish. *Arch. Environ. Contam and Toxicol.* 12 : 621-626.

Sullivan, J.H. 1980. Pesticide residues in imported species : A survey for chlorinated hydrocarbones. *J. Agric. Food Chem.* 28 : 1031-1034.

Swaminathan, R. and Prasad, N.M. 1978. Screening the efficient rhizobial isolates for tolerance to certain pesticides. *Pesticides,* 12 : 133.

Swaminathan, R. and Sullia, S.B. 1969. Influence of pesticide malathion on groundnut microflora. *Curr. Sci.* 38 : 282.

Szczepanski, Ch. V. 1990. Today's research for tomorrow's market or how to hit a moving target. In : Recent advances in chemistry of insect cowrol-II. Ed. L-Crombie species publication No. 79, Royal society of chemistry, Cambridge, pp. 1-16.

Takahishi, W., Saidin, D., Takei, G. and Wong, L. 1981. Organochlorine pesticide residues in human milk in Hawaii, 1979-80. *Bull. Environ. Contam. and Toxicol.* 27 : 506-511.

Talati, J.B., Shah, P.G., Raj, M.F., Patel, B.K. and Judal, G.S. 1991. Organochlorine pesticide residues in ground and cotton seed oils. *J. Mah. Agric. Univ.,* 16 (1) : 125-126.

Tanabe, S., Milura, S. and Tatsukawa, F., 1986. Variation in organochlorine residues with age and sex in antarctic minke whale. Proc. 8th Symp. On polar Biology 44 : 174-181.

Ter Haar, G. 1980. An investigation of possible sterility and health effects from exposure to EDB : Bahbury Rep., 5 : 167-188.

Tewari, S.N. and Sharma, T.C. 1978. Study of distribution of chloinated organic pesticides in different autopsy materials of human poisoning cases using TLC and UV specirophotometric techniques Chem. Era 14 : 215-218.

Thakra, S.K., Dewan, R.S. and Gulati, K.C. 1969. Buildup of DDT in human fat, oils and oilseeds as possible source. *Pesticides,*3 (2) : 13-14.

Thorpe, E. and Walker, A.l., 1973. The toxicology of dieldrins II : Comparative long term oral toxicity studies in mice with dieldrin, DDT, phenobarbionate, beta-BHC, gamma-BHC. *Food Cosmet, Toxicol.,* 11 : 433-442.

Tilson, A.A., Jacobson, J.L. and Rogan, W.J. 1990. Polychlorinated biphyenyls and the developing nervous system : Cross species comparisons. *Neurotoxicol. Teratol.,* 12 : 239-248.

Tisdale, J.L., Nelson, W.L. and Beaton, J.D., 1985. Soil fertility and fertilizers.Pub. Macmillan Co. pp. 175-491.

Tobin, J.S. 1970. Carbofuran : A new carbaate insecticide. *J. Occup. Med.,* 12 : 16-19.

UNEP, 1991. Fresh water pollution, UNEP GEMS Environmental library No. 6, Nairobi, pp. 241-242.

USDHHS, 1978. US department of health and human services occupational health guidelines for endrin Pub. No. 81-123. Pub. National institute of occup. Safety and Health, Cincinnats.

USEPA, 1985. Drinking water criteria document for hexachlorobenzene ELAOCIN-424, Pub. USEPA, Environmental criteria and assessment office, Cininnatr.

USEPA, 1985a. Health effects criteria document for nitrate and nitrite. USEPA Criteria and standard division, Office of drinking water, Washington.

USEPA, 1986. Guidelines for carcinogenic risk assessment. Fed Reg., 51 (185) : 33992-34004.

USEPA, 1987. Drinking Water criteria document for heptochlor, heptachlor epoxide and chlrodane, Rep. No. ECAO-CIN-406, USEPA Environment criteria and assessment office, Cincinnah.

Varanasi, U., Stein, J.E., Tilbury, K.L., Brown, D.W., Meador, J.P., Sloan, C.A., Clark, R.C. and Chan, S.C. 1994. Chemical contaminants of grey whates stranded along west coast of M. America. *Sci. total Environ.* 145 : 20.

Verma, G.S. 1992. Raic acificiation in India caused by nitrates. *Atmosfera*, 5 (4) : 233-239.

Walton, G., 1951. Survey of literature relating to infant methemoglobinemia due to nitrate contaminated water. *Am. J. Publ. Hlth.* 41 : 980-996.

Wasserman, M., Nogueira, D.P. and Tomatisd, L. 1972. Storage of organo-chlorine insecticides in people of Sao Paulo Brazil. *Ind. Med.*, 41 : 22-29.

Wasserman, M., Tomatix, L. and Wasserman, D. 1974. Epidemiology of organo-chlorine insecticides in adipose tissues of Isrelis. *Pestic Monitor. J.* 80 : 1-15.

Webster, E.P. and Shaw, D.R. 1996. Off site runoff losses of metolachlor and metribuzin applied to differing soybean production system. *Weed Technol.* 10 (3) : 556-564.

WHO, 1987. Evaluation and inference profile of joint FAO/WHO meetings on pesticide residues in food. 1961-1986.

WHO, 1988. WHO recognition of classification of pesticide by hazards. Guidelines to classification 1988-1989, WHO/UBC, 88/953.

Wiese, I.H., 1976. Pesticides and South African population. *S. Afr. Med. J.* 50 : 1801.

Wolcott. A.R., Foth, H.D., Davis, J.F. and Shikulana, J.C. 1965. Nitrogen corriers : I Soil effects. *Soil Sci Soc. America. Proc.* 29 : 405-410.

World Bank, 2011, data.worldbank.org/indicator/AG.PRD.FOOD.XD

Wong, L.C.K., Wiuston, J.M., Hong, C.B. and Hgensen, J. 1979. Study of carcinogenicity and toxicity of 1, 2-dibromoethate in rat. *Toxicol. Appl. Pharmacol.*, 63 : 155-165.

Chapter - 3

FUNDAMENTALS OF ORGANIC FARMING

3.1 BACKGROUND

Essentially, the method of organic farming is not a new found technology at all. Some principles of organic farming were followed ever since the cultivation of crops was practiced in most ancient civilizations. Use of digested or undigested manures, raw animal dung, forest litter, dead bodies of animals, fisheries wastes, oil cakes, adoption of mixed farming systems and many other practices were found to be followed in both Indian and Babylonian ancient agriculture. The origin of practices such as shifting cultivation, crop rotations can be traced to enrichment of soil fertility, which is the Lower leaf/se aim of organic farming. Most such practices were aimed at conservation of resources such as plant nutrients, by making use of naturally available materials. In most regions of deciduous forests, natural senescence of leaves and their subsequent decomposition is found to be beneficial for plant growth and soil health. The growth of many plants and health of soil are at their best, with little interference by man, as is even now witnessed in most forest ecosystems. Nature has provided an in-built system to support the soil health and plant growth. It is the interference by man, which has created the problems of imbalance in the eco-systems. The human interference by ways of intensified crop husbandry for increased food production has resulted into paucity of nutrients in soil, necessitating the artificial supply of plant nutrients. As the selected crops are grown on intensified scale in agriculture, naturally available balance between competitive forces among plant communities is lost showing the manifestations in the form of shortage of nutrients as well as susceptibility to invasions by pests/diseases. They were otherwise controlled by natural forces. Present agriculture is an evolution of such repeated human interferences. The pressure on cultivated land continuously mounted by increased global population, necessitated the artificial supply of nutrients to increase the agricultural production. To match with exponential growth of population and resultant spurt in the food requirement, the agricultural production had to reach taller targets. This forced the use of chemical sources for supply of plant nutrients instead of naturally available materials to achieve faster and larger increase in food production. The use of chemical sources of nutrients especially from 1950's resulted into spectacular increase in agricultural production. But, in the process, the importance of natural sources of nutrients was relegated. The organic farming, in its new found scope, encompasses the all round attempt on the use of these natural sources of plant nutrients and natural methods of control of pests. The importance of natural sources of plant nutrients and natural methods of pest control are the main theme of organic farming.

3.2 DEFINITION AND MEANING

Although organic farming was practiced since the period of ancient civilizations, the term organic farming was introduced by Lord Northbourne in 1940. He had a vision of sustainable, ecologically stable, self contained and biologically balanced farming in defining the organic farming. The origin of name organic lies in signifying the biological basis agriculture, because the term organic is derived from organisms, signifying the entire biota on earth.

Number of definitions are coined to define organic farming. The most accepted definitions based on the holistic meaning of organic agriculture are as follows.

IFOAM (International Forum for Organic Agriculture Movement), the most trusted and accredited global body in the field of organic farming defined the organic agriculture in 2006 after the detailed deliberations of officially formed task force for the purpose.

- **Organic agriculture is the production system that sustains the health of soil, ecosystem and people, by relying on ecological process, biodiversity and natural cycles and adapted to local condition than the use of inputs with adverse effects (IFOAM, 2008).**
- USDA's National Organic Standard Board defined organic agriculture as ecological production, management system that promotes and enhances biodiversity, biological cycles and soil biological activity by adopting management practices that restores, maintains and enhances ecological harmony as well as minimised use of off farm inputs (NOSB, 1995)

The organic farming, as commonly perceived by most people, does not entail with mere use of inputs prepared out of wastes/natural products. The over simplified meaning of organic farming in terms of use of manures, composts, biofertilizers and bio pesticides does not really cover the holistic meaning of organic farming. The meaning of comprehensive definition of organic farming revolves around promotion and maintenance of biological activity and biological cycles to achieve an ecologically stable agriculture production. The supply of nutrients and control of pests are the natural corollaries of biologically stable system. Hence, understanding the meaning of organic farming in a limited sense of providing nutrients and control of pests restricts the broad views of organic farming.

The definitions such as **"use of naturally available resources for crop production to supply the nutrients and to control the pests"** and **"form of agriculture excluding the use of synthetic fertilizers and pesticides"** mainly thrusts on the natural resource base than on stabilization of natural processes. In fact, the objectives envisaged in such definitions are naturally achieved if the biological stability of the system is aimed at.

3.3 NOMENCLATURE

In its new found form, the organic farming is practiced in various dimensions. In the last 10 to 12 years, the practice of organic farming has been expressed in different terminologies in different parts of world. Some of them are modificiations of age old concepts, while some of them stress on the newer concepts and unconventional sources or organic sources.

Essentially, all the terminologies presently in vogue aim at use of naturally available sources for supply of nutrients and for control of pests for increased agricultureal production. All the expressions about organic farming are the variations of these two themes, with a stress on one or more sources or one or more principles of organic farming. Organic farming signifies the use of any material (of organic origin) to supply plant nutrients, control of pests and to improve of soil health for better agricultural production. It also includes age old and new found practices to assist the natural processes in an agricultural ecosystem so that plant growth is brought in tune with the harmony of the eco-system. Such processes include decomposition, biological N fixation, nitrification and ammonification etc. These processes, which were interfered severely by the so called chemical farming, are also the bases of sustainability of the system.

Thus, the organic farming signifies the use of wide varieties of natural materials and practices. The materials may include animal dung, urine, digested/undigested manures, oil cakes, plant residues, farm wastes, bodies of dead animals, fish meal etc. They may also include many unconventional sources such as city wastes, sewage, sludge, industrial wastes, wastes of slaughter houses, human exareta and urine, marine wastes etc. The practices forming a part of organic farming include, green manuring green leaf manuring microbial inoculation, several cropping practices (such as trap cropping, crop rotations, inter-cropping etc.) for better pest control and many other practices which aim at improvement of soil health. Such practices may also include the use of many bio-agents, predators and natural enemies for natural control of pests as well as the use of earthworms and many other soil flora/fauna as a means of building up the soil fertility to bring it to the tune of natural harmony in an agricultural eco-system.

Many new terms are popularly used to represent different types of organic farming. Some of them are :

1. Natural farming

It signifies the method of farming, where cultivation of a crop is left to the natural forces in a given place. It means cultivation of crop with least human interference. Thus, seeding and harvesting are the only operations involved in such farming. It is comparable to primitive agriculture, where no cultivation is done, no weeding is carried out, no manure/fertilizer is applied and pests are not artificially controlled. Every thing is left to nature.

The inspiration for such an extreme form of farming is drawn from the perception that nature can take care of crop, left to itself. Fukavokaa was instrumental in demonstrating the importance of natural farming to the modern world. In this Japanese farm, he was able to demonstrate that it is possible to manage the crop with least interference, with no chemical fertilizer/pesticide and simultaneously build up the soil fertility. Mokicho Okoda called his farming, based on the principle that inherent power of the soil can take care of all needs, as nature farming. This later led to establishment of Mokicho Okada Association (MOA) which was responsible to establish various MOA farms in many parts of world to demonstrate benefits of nature farming. This may be an useful system to develop soil microflora and soil fertility in long run. But it is certainly not an encouraging method of farming to achieve higher agricultural production, as the crop yields are bound to suffer very heavily in such a system. It is not a practice suitable for all types of agro-climate situations, as many drier eco-systems can not support the crop growth as well as build up of micro flora to a desirable level, if left to nature.

2. Zero-chemical farming

It is a method of farming, where the use of synthetic chemicals is avoided in toto. The cultivation of crop relies totally on naturally available materials. In its verbatim meaning, zero-chemical farming implies all the aspects of organic farming. The nutrients are supplied through all available organic sources. The pests of a crop are controlled by all possible non-chemical means. The human interferences in the cultivation of a crop is so tuned that it is synchronized with the harmony of local eco-system so that crop is able to establish by making use of natural processes. As compared to natural farming, zero-chemical farming is more efficient. Because, the supply of nutrients and control of pests is not left to nature. They are achieved artificially without using any chemicals.

3. Bio Dynamic farming

Biodynamic agriculture owes its origin to the spiritual insights and perceptions of Dr. Rudolf Steines, an philosopher and scientist. He emphasised on use of forces with living nature. (Gold, 2007). His prescriptions include the practices a farmer can perform by use of one or more natural process. The biodynamic farming practices include herbal preparations against pests, methods to enhance decomposition in manures and composts. He also advocated a practice of preparation of energy rich manure concentrate by use of cosmic rays, which was believed to enhance the soil fertility. It can be considered as a variant of organic farming - partially covering the use of naturally available material. Adoption of biodynamic farming may not cover all the aspects of organic farming.

4. Biological farming

It is a system of crop production, in which the producer attempts to improve the biological balance between the flora and fauna, so that a stabilsed situation of agro-eco-system situation helps in natural control of pests or in supply of nutrients. Although the biological and organic farming share common objectives of strengthening the biological activities in soil, many ecological aspects and managemental aspects of organic farming are not in the ambit of biological farming. Hence, this also covers partial objectives of organic farming (Gold, 2007).

5. Humus farming

It is a type of farming, wherein the crop is expected to grow by utilizing the humus formed out of constant decomposition of biomass.

This type of farming requires a typically tropical climate with warm temperature, high humidity and more insulation. The crop mainly depends upon humus for all its requirements. No artificial fertilizers and chemicals are used. Even, application of organically decomposed manure is not necessary. Although this type of farming is ideal to be propagated, the congenial conditions are found only in equatorial and other rain forest regions of the world.

6. Compost farming

In this system of farming, copious amount of compost (prepared out of city wastes or agricultural wastes) or conventionally prepared farm yard manure is applied to meet the nutrient requirement of crops. However, there is scope to use chemicals for control of pests or for amending the soil in this kind of farming. Heavy reliance is made on digested manure as the main source of nutrients in this type of farming.

7. Sewage farming

This type of farming is an off shoot of vast liquid wastes generated out of populous cities, where large nutrient rich liquid wastes are generated at a limited space. In most rural areas, the population being sparse, the sewage generated is too small to be utilized for farming activities. Sewage farming aims to utilize the nutrient rich wastes in productive manner. The sewage being a human waste, is also a rich in large number of microbes and on its application to agricultural lands it is expected to enhance the microbial activity. Its inherent nutrient composition and the ability to enhance the microbial activity may jointly help to achieve the objectives of organic farming. But, sewage farming may not achieve all the objectives of organic farming, as sewage may not achieve the biological balance of the all the components of an agro-eco-system. In addition to, it may contain large harmful chemicals like heavy metals (cadmium and selenium), which may ultimately damage the soil fertility or may be retained in the food produced out of sewage. Sewage farming is also not possible in every urban situation, as its success around per-urban areas mainly depends upon the effective sewage collection mechanisms (by internal sewer system) and sewage treatment facilities. Soild part of sewage, often called as sludge, is also used to replace manure in some variations of such farming. The age old method in Indian Villages to use open ground toilets in the field around the villages was a crude method to use the solid wastes. In any case, sewage farming may be considered as a part of organic farming, mostly practiced around cities.

8. Integrated farming

Integrated farming is a type of farming, where animal based enterprises awre integrated with crop husbandry activities in such a way that each one is supportive and dependent on other. It can be an improvised mixed farming, where input out put relations between animal based activities are rationalised to match with crop husbandry activities. The focus is on maintaining the healthy integrations between them so that in the long run both crop and animals are benefitted by mutual dependence. Such a farming system is also matched with ecological components of a given site. Integrated Farming System (IFS) justifies most objectives of organic farming, as a healthier biodiversity is achieved and enhanced microbial activity is attained to ultimately improve the productive ability of soil. However, the pest control by healthier predator-pest balance may or more not be achieved by integrated farming, unless special efforts are made to deliberately manage the biological control systems.

9. Homa farming

It is a type of farming, where a traditionally performed homa by uttering specific mantras is expected to relieve the site of performance from all evil sources of causing the pest attack. It is a more a belief based farming than science based phenomenon. Although many success stories of homa farming are reported, scientific veracity of such claim is hardly tested.

3.4 PRINCIPLES OF ORGANIC FARMING

At conceptual level, the organic farming aims at achieving the regeneration and continuance of natural processes of plant growth in a given eco-system by making the eco-system as a self-sustainable system. The organic farming aims at making crop husbandry as self-supporting and self-sustainable activity in a given eco-system rather than its dependence on outside inputs. In this process, organic sources that are available in a given eco-system would be used for supplying plant nutrients and control of pests and diseases along with various cultural practices. There may not be any possibility of evolving a common list of practices and common set of organic sources for all the crops and all the regions. The actual variations of components of organic farming may vary from place to place, crop to crop or even from time to time. But, the basic concept of making the whole system self-supporting for continued production will remain in all forms of organic farming. Organic farming is indeed not restricted to single practice of supplying organic manures or green manuring alone or even controlling pests by one of the non-chemical method alone. It is an integrated evolution of all possible organic sources and practices synthesized for a particular eco-system with the sole aim of strengthening crop husbandry. All the processes involved in soil-plant relations should be synchronized with natural processes such as decomposition, erosion, percolation, biological N fixation, nitrification and many microbial activities to achieve the desirable level of productivity.

The basic principles of organic farming practices can be enumerated as follows :

1. A crop should be able to grow and yield successfully utilizing the nutrients supplied out of a given soil under ideal condition created for buildup of fertility by enhanced microbial activities.
2. Pests and diseases of a crop should be essentially controlled by natural enemies, predators, bio-control agents or by use of natural products/bio-extracts.
3. All possible organic sources available in nature could be used in digested/semi-digested condition to supply the plant nutrients. This may result into adoption of practices relating to manure/compost preparation outside agricultural field or sometimes *in situ* decomposition in an agricultural field.
4. Natural resistance of some crops to pests and diseases should be conveniently exploited for the benefit of crop production. No practice, which would suppress the natural resistance, is encouraged in organic farming.
5. The process of biological nitrogen fixation should be encouraged in all possible ways during the process of crop husbandry. Alteration of cropping systems, inoculation of suitable bacterial culture and management of substrate for better bacterial activity in the soil may help in better nitrogen fixing activities in the soil.
6. The human interference in the crop husbandry activities should never obstruct the natural processes of an eco-system and should never upset the natural balance available between components of an eco-system. On the contrary, such activities should augment and be supportive of natural processes.
7. Biomass management in a given situation to encourage self-supportive natural system of providing nutrients in conjunction with enhanced biological activity should be a prioritized consideration in organic farming.

8. Nature's friends such as earthworm and spiders should be recognized and their activity should be encouraged for the benefit of soil fertility, crop growth and pest control.

Although, the system of organic farming is based on these broad principles, actual practice of organic farming may involve practicing a few of them at a time, as ideal conditions as not available to follow all these principles in all situations. Depending upon the local resource base and local constraints, the practice of organic farming may have to be adopted based on some relevant principles. This would result into local practice of organic farming relevant to a particular eco-system and particular cropping system. Some examples of this nature can be cited.

1. Introduction of bacterial culture for pulse crop in cereal-legume system may be combined with application of digested manure to crop, without any component practice of biomass decomposition at site.
2. Intercropping cotton with pulse to encourage carabid beetles and spiders may be practiced in areas where shoot borer/ stem borer/ boll worms of cotton are in endemic dimensions.
3. Large generation of biomass in paddy-paddy systems may be useful to evolve a practice of biomass incorporation for *in situ* decomposition which may replace the use of digested manure. This may be combined with preparation of vermicompost outside paddy fields and its application in place of digested manure.
4. In a situation, where large aquatic weeds could be collected, preparation of manure by such weeds for later incorporation may be an easy, cost effective and relevant practice.
5. In sugarcane growing areas, where sugar factories generate large quantities of presumed, large scale preparation of nutrient rich manures using pressumed as basic material may be a useful practice. It can replace the other digested manures.

Thus, the organic farming principles are followed in different crops and in different proportions to evolve a locally suitable practice for achieving higher crop production. A combination of different practices without chemical component may have to be developed for a particular eco-system and sometimes even for a particular cropping system. Each practice or a group of practices so evolved is based on one or more principles listed above.

The main basis of evolving any organic farming practice, however, should toggle around the concept of developing a self supporting and sustainable eco-system to support the crop husbandry by natural processes.

3.5 SCOPE OF ORGANIC FARMING

The organic farming is not a technology by itself. But it consists of set of practices evolved around several principles based on widely varying character or resources in different eco-systems. The totality of each eco-system and mechanism of delicate balance available between components of each eco-system would ultimately decide which principle would weigh more or less and ultimately which practice should be predominant in organic farming.

In this background, the assessment of scope of organic farming should adopt widely varying techniques for different eco-systems and for different principles of organic farming. Further, the assessment should also include the harmony achieved between new set of organic farming practice and components of eco-system.

The scope of organic farming cannot be assessed on an uniform scale at global level. The criteria of assessment and in turn the extent of scope of organic farming vary according to soil, climate, cropping system, vegetation, irrigation, allied activities (animal husbandry, sheep rearing, poultry, piggery, fisheries, etc.) as well as harmony achieved between different components of each eco-system. Many examples can be cited to justify this.

1. Scope for preparing farm yard manure by a farmer with large number of livestock is certainly brighter than that of a farmer maintaining a few cattle.
2. An irrigated farmer is able to easily adopt green manuring practice or the practice of returning the crop residues for enrichment of soil, as the availability of moisture can facilities the decomposition. But a dry land farmer is not able to do so easily.
3. Temperate climate does not favour faster decompositions of organic residues as compared to tropical climate due to cold temperatures prevailing in temperate climate for most part of year.
4. A farm nearer urban agglomerations has advantage of obtaining decomposed compost or poultry manure at cheaper rates compared to interior rural areas.

5. Regions of better rainfall distribution support two crops and result in sufficient biomass for incorporation. As compared to this, poorer distribution of rainfall may offer little scope for biomass incorption.

Scope of adopting organic farming exists in every farming situation with varying degree. Within the meaning of broad concepts and varying principles of organic farming, it can be adopted for every crop and in every situation. Each farming situation offers unique scope of adopting some principles of organic farming and evolving a set of practices that may be specific to such situation. But a clear analysis of all the features and components of each situation should be undertaken to identify constraints and resources to evolve such practices. Some situations (dry eco-systems) may vividly offer lesser scope for adopting the principles of *in situ* decomposition. But, they would be ideal situations for using pre-decomposed manure. It is only required to plan, evolve and implement the practices in the right perspective.

As compared to chemical farming, the organic farming has no much scope, if dramatic results are expected in a short time. In some situations, the organic farming practices may not be cost-effective in short run as compared to use of fertilizers which are supplied at subsidized rates. However, the assessment of scope of organic farming on such consideration is indeed undermining the serious damages caused by chemical farming to our most eco-systems. The need of the hour is not to relegate the importance of organic farming by such considerations but to regenerate and strengthen the self-supportive production system, which was cheaply damaged in a short span of 30 to 40 years. Several scientific studies have been reported to show that organic farming system can increase the crop yields and stabilize it to the level of chemical farming in long run, although in the initial 1 to 2 years, yields may be reduced to some extent. The long term perspective of practicing organic farming can lead to a better scope in most agro-eco-systems than short term analysis of yield and economic gain from a cropping system. Organic farming does not only aim at higher crop yield or returns. But, it aims at developing long term self-sustainable practices which not only result into increase crop yield and returns but also sustain the whole production system to achieve perpetual productivity. In this perspective, the organic farming has excellent scope and should be encouraged not for short term gains but for long term returns.

Although the scope of adopting organic farming exists in every farming situation, the evolution and adoption of practices involves a great deal of planning and physical work on the part of farmer as compared to chemical farming. Simple and short-cut solution such as use of fertilizers to supply plant nutrients is certainly an easy job as compared to making use of all available organic source in a farm to convert them into useful plant nutrients. Similarly, chemical control of a pest or disease may involve a great deal of understanding the pest dynamics and its place *vis-a-vis* other biological and abiotic components in a given eco-system. But to control it by organic farming techniques, it needs greater planning and understanding the life style of pest, its food habits, its breeding habits and its place in an eco-system.

Chemical farming can never be a compatible component in any agro-eco-system and it has to be followed as an extraneous component forced upon an eco-system. Application of any manufactured agro-chemical would not match with the natural processes and such chemicals serve as impending factors for many natural processes of an eco-system. In organic farming, the practices are so developed that they are compatible with most of natural processes of the eco-system and thus accepted by nature without any impairment of the harmony evolved by an eco-system. For these reasons, organic farming practices are regarded as self-supportive and sustainable in the long run. The scope of adopting organic farming practices should not be curtailed for the reasons of some difficulties in evolving and adopting them. The scope of organic farming has to be perceived in the long run vision of developing self supporting sustainable production system than short run minor impediments.

3.6 ADVANTAGES AND LIMITATIONS OF ORGANIC FARMING

Qualitative and quantitative assessment of organic farming practices in various agro-eco-systems involved in most field and horticultural crops reveal that advantages outweigh and limitations. Major advantages of organic farming are :

1. As both nutrient needs and pest control are met by non-chemical methods, the entire eco-system, for which a package of organic farming is developed, remains free from all sorts of hazards by chemicals. The soil fertility, soil fauna, soil micro flora and crop growth would be retained in perfect harmony leading to sustainable systems. Deleterious effects of chemical on human health, animal health, crop growth, natural predators, microbial population and their activity, availability of nutrients are completely obviated.

2. Pollution by agro-chemicals at various levels of manufacturing, handling, packing, application by various inhalation, ingestion, dermal routes is fully avoided. Non-point contamination of water and agricultural products as well as the health complications due to such contamination are also avoided.
3. The soil regenerates its original status of a dynamic living system with harmonious interactions of its biological, physical and chemical components. As the soil organic matter is improved, the porosity, infiltration and structure of soil are drastically improved. This helps in better moisture conservation and would reduce the erosion losses substantially.
4. In long term analysis, organic farming practices are more cost effective, as sustainable yield levels could be achieved at no extra cost of chemical brought from outside. As the systems can retain their self-supportive nature in long run, assured returns are possible in most crops without any cost of fertilizer and other agro-chemicals.
5. Organic farming can result into a fairly pollution free products from agriculture and in turn, helps the society to be pollution free. Human and animal population are not exposed to the short term and long term ill-effects of consuming agro-chemicals through organic food.
6. Large subsidies on fertilizers and other agro-chemicals now draining the national exchequer and foreign exchange reserves are curtailed. According to one estimate, the subsidies offered on fertilizers amounting to Rs. 35,000 crores annualy would be sufficient to provide nation wide intensive training programmes on benefits of adoption of organic farming system.

Some major limitations of organic farming are :

1. Uniform practices of organic farming can not be developed for all situations and for all crops. Organic materials used in organic farming also differ vastly in various places. Packages of organic farming practice involving different components will have to be developed separately for each eco-system.
2. Dramatic and large benefits in terms of yields and returns cannot be expected in short time from organic farming practices. The increase in the yields and returns, commensurate with the resource structure of each eco-system, can be gradually increased, stabilized and sustained over a longer period of three to eight years.

3. Many organic manures are bulky in nature with low concentrations of nutrients. Their carriage to distant places may pose problems of increased cost and labour. This will be a major limitation, where in situ decomposition of bio-mass due to moisture constraints is not possible and pre-digested manure has to be inevitably used.
4. Most high yielding varieties/hybrids of cultivated crops, developed in the past 3-4 decades are tuned to their high responsiveness to application of fertilizers. Their nutrient demands are so high that, meeting their requirement by organic sources alone is economically non-feasible in all the situations. Some of them may not respond positively to organic farming practices immediately. New varieties/hybrids may have to be identified, with their responsiveness to organic manures, as the main criteria for selection.
5. Although many non-chemical methods of pest control are already developed, control of the weeds by non-chemical methods may have to be solely done by cultivation methods. This may pose some problems in countries, where extensive agriculture is widely practiced with large agricultural holdings managed by small manpower.
6. Many more efforts are needed to identify natural enemies, predators of some pests. Wherever, natural enemies and predators of pests are entirely absent in an eco-system, the pests have to be managed by cultural and breeding methods, which may involve more time and labour.

Most of these limitations can pose threat to the adoption of organic farming partially or fully. But, the concerted research efforts combined with strong commitment by farmers can overcome these limitations. It is certainly not an easy task to regenerate the self-supportive characters of an agro-eco-system for sustained productivity. But, it is not beyond the possibility of human efforts to re-establish production system based on long terms considerations. In addition, practice of organic farming is a prioritized imperative than an option. It is just not possible to desist from practicing the organic farming due to limitations. Every limitation has to be tackled scientifically and a suitable solutions can be found.

3.7 IMPACT OF ORGANIC FARMING ON PRODUCTION AND PRODUCTIVITY

Large scale use of fertilizers and other agro-chemicals approximately coincided with alarming growth of population. Hence, it is frequently held by advocates of chemical farming that, global food crisis is greatly mitigated by leap in agricultural production, which is the result of large scale use of chemicals including fertilizers. They also emphatically hold a view that any attempt to switch over to organic farming, which is tantamount to reduced or zero use of chemicals, would lead to global crisis in food production. As it is a well known fact that induction of organic farming on a large scale would suppress the yields of many crops in the initial years, their apprehensions are indeed valid. It is already proved experimentally that yield of paddy, wheat and many other food crops are reduced in the initial 2-3 years after supplying the nutrients only through organic sources.

In this background, two pertinent questions are necessary to be considered to test whether it is viable to adopt organic farming on large scale.

1. Availability of organic sources of nutrients.
2. Reduced yields.

Availability of organic sources of nutrients

It is obvious that the huge nutrient requirement of these crops is not met by organic sources and as a result the crop yields will have to suffer. To meet their whole nutrient demand, large quantities of organic manures are needed. Some of the reports have indicated that yields from organic farming can be as much as those obtained by chemical farming, if the whole requirement of nutrients is met out. This leads to a question whether it is possible to generate and supply scuh large quantities of organic meanures for entire cultivated area so that yield levels are maintained?

If the organic farming is to be practiced only with digested manure as source of nutrients, then it is amply clear that it is not a practical proposal to supply the required quantity of organic manures to vast cultivated area. But, organic farming is certainly not a mere practice of applying digested organic manure. It is far more than this, It aims at making a production system sustainable and self-supportive by giving the ability to the soil to generate and supply the required quantity of nutrients. This aim is achieved by developing a set of practices to decompose all available biomass in an eco-system, besides, depending upon digested sources of manures. This may include incorporation of crop residues, harvesting wastes, weeds, cattle wastes and any other biomass in an eco-system besides depending upon digested sources of manures. If an eco-system produces lesser biomass, it may be necessary to supplement the requirement through digested sources of nutrients.

Reduced yields

It needs no imagination to perceive that at global or national level, the reduced crop yields by following organic farming can upset the whole economy besides pushing up the problem of widespread hunger. Attempt to switch into organic farming should not lead to such chaos. But,

(a) The yield reduction due to organic farming is not reported in number of crops, which include almost all horticultural crops, many pulse/oilseed crops, wherein both traditional and improved varieties have been able to maintain their yield levels inspite of following organic farming practices (Arecanut, coconut, mango, sapota, guava, papaya, pomegranate, limes, applie, austard apple, leafy vegetables, all flower crops, pulses such as horsegram, blackgram, lentil, kidney bean and oilseeds such as seasum, niger, safflower etc.)

(b) In all the remaining crops including the cereals, the yield reduction may be predictable when :

- The entire nutrient requirement is not met by organic sources alone either due to prohibitive cost or due to impracticability of transp ɔrting bulky manure.
- Nutrient demand by a variety/hybrid is too high to be met by organic sources alone (eg. some sugarcane varieties requiring 250-350 kg N/ha).

Both the above issues are worth consideration, when the merits of organic farming are discussed. When, the yield reduction is not a criteria in number of crops and organic farming is already followed in many such crops, there should be no reason why chemical farming should be imposed. Instead, the practices of organic farming may be further reviewed and strengthened to acquire full range of benefits from the present system (eg. methods of manure preparation, timely and scientific methods of application, in situ incorportion of green manures etc.).

Similarly, the yield reduction recorded in many crops has not maintained a consistent trend. In most of such crops, the reduction in yield due to organic farming is temporary. It is reported to improve and made stable after a lapse of 3 to 5 years. The sustained yield improvement due to organic farming practices is assisted by:

1. Thorough assessment of presently followed practices such as methods of manure preparation, time and method of its application.
2. Full exploitation of available sources in a system to utilize all biomass generated for production (eg. weeds, crop residues, stubbles, harvesting wastes, green leaf manuring, green manuring) and make the whole system sustainable in respect of its ability to supply nutrients.

Where the nutrient demand is too high to be met by organic sources alone (eg. hybrid cotton, cereal, sugarcane), the integrated nutrient management involving chemical and organic sources may be inevitably followed.

Hence, the apprehensions of reduction in crop yield and resultant shortage of agricultural produce should not be the impediment to implement the organic farming practices. A steady transformation from chemical farming to organic farming would not only ensure that country would not face food shortages but would also make our production system sustainable for perpetual production. Besides, organic farming would save our eco-system from the onslaught from indiscriminate use of chemicals. That is, our production system comprising of soil and its biotic/abiotic components would be saved from being turned into unproductive sterile system.

3.8 REFERENCES

Gold, Mary, V., 2007. Sustainable Agriculture : definitions and terms, USDA publication for Alternative Farming Systems Information Centre, pp. 1-13.

IFOAM, 2008. Definition of organic agriculture http://www.ifoam.org/growing-organ.

NOSB, 1995. NOSB Definition of Organic. http://www.ota.com/Standards/nosb/definition. html.

Chapter - 4

HISTORY AND EVOLUTION OF ORGANIC FARMING

History of organic farming can be truly traced in the history of agriculture itself. The ancient agriculture in different civilizations was organic in nature, simply because agriculture was practiced in the cradle of nature since its inception. The use of chemicals has only started in the last 60-70 years. Many of ancient agricultural practices are even now followed in all the countries. They have obviously found place in the newly conceived modern organic farming with or without some modifications. Hence, the history of modern organic farming and its genesis from the ancient times must be traced in different civilizations and their agricultural practices, documented elaborately in the annals of literature.

In the early era of scarce populations and vast natural resources, practicing agriculture was restricted to few fertile pockets in most civilizations making way to the system of shifting cultivation- which has been in vogue even in modern days (in the style of **slash and burn agriculture or jhum cultivation**). The organic farming practices of then followed agriculture were adequate to produce enough food to the thin population. But all these practices of different civilizations have not found light in the form of published literature due to limitations of documentation. Hence, many of such practices are deeply buried in the realms of history and not traced easily. Nevertheless, some practices across the globe are documented. The early literature of India and China is replete with experiences and practices of agriculture. For example, ancient crop rotation practices of India (3500 BC), composting and mulching practices in China (2000 BC) and Sumerian wind break techniques (1500 BC) are found in the books of history of civilizations.

It will be indeed appropriate to deal with all the ancient civilizations of world in order to trace the history and evolution of organic farming. Due to limitations of space and scope of this book as well as due to non-availability of historical documents of all the civilizations, the history and evolutions is traced based on Indian and American/European evidences. It may be noted here that American National Library took up the task of tracing this history as a part of documentation on sustainable agriculture. The ancient publications were traced from first century BC till 1979 and presented. They are summarised in However, this attempt excluded the historical evidences of important Indian, Chinese and Mesopotamian civilizations possibly due to constraints traceability of documents/scripts/ proofs of reports of both pre-Christian and Christian era, as many these documents are even now in vernacular/local languages other than English. An attempt is made in this chapter to trace the history and evolution of organic farming from available Indian historical evidences in comparison with the history traced from the evidences in Europe/USA.

Before the history of organic farming is traced, it must be sufficiently clear that organic farming does not comprise of a single practice, but connotes holistic and whole farm agriculture with a range of practices- which can be classified in two groups, based on fundamental principles of organic farming

(a) Practices related to crop nutrition and crop management

(b) Practices related to crop protection

Judicious and locally acceptable combination of practices from both the sets practiced simultaneously on a sound ecological adaptation will lead to organic farming. Both these sets include many routine agricultural practices followed as regular crop management practices, but help to achieve the cause of organic principles (crop protection and crop nutrition). For example, practice of sowing in the appropriate season helping to avoid a disease or method of ploughing improving the fertility status, although they are routine practices during the cultivation of crops. They help to achieve the natural and organic way of controlling disease or increasing the soil fertility. The perception of organic farming includes all such practices from both the sets, besides the popular organic practices, that directly supply nutrients (manuring and green manuring). For such reasons, the history of organic farming must be traced in the history of these practices, now considered as a part of modern organic farming. These practices could be categorised into following groups for the sake of convenience of tracing their history as the roots of organic farming.

1. Cropping systems
2. Appropriate time of sowing
3. Germination of seeds

4. Preparation and use of manures
5. Liquid organic manures
6. Control of pests and diseases
7. Biological control of pests
8. Bio fertilizers

An attempt is made to trace the history of organic farming under each category of practices in the chronological order briefly with references.

4.1 ORGANIC FARMING IN INDIAN HISTORY

4.1.1 CROPPING SYSTEMS

It is now well known that various cropping systems like sequential/rotation cropping, intercropping, mixed cropping, relay cropping and other multiple cropping systems are part of modern agriculture. They have been adopted to increase the agricultural production, besides number of other operational and economic advantages in terms of security against crop failures and diversification of farm produce. But, some of the cropping systems distinctly serve the cause of organic farming and therefore considered as part of organic farming. Ex., rotational cropping offers advantages in terms nitrogen fixation or improved nutrient use efficiency (by using the residual nutrients). Sequential system involving leguminous crops helps succeeding crops by their ability of biological nitrogen fixation. All these are part of organic farming, because they focus on natural ways of nutrient/pest management. It is interesting to study when the cropping system practice was started. Although the scientific advantages of cropping systems were recently understood, the farmers have followed the practices of crop rotation and intercropping since historical and even mythological age.

Crop rotation is the earliest recorded organic practice. Rotating one crop with other in sequence is now scientifically understood to offer advantages in terms of convenient pest avoidance, better and efficient use of soil moisture, besides nitrogen fixation by leguminous crops. All these advantages were perhaps not clearly understood in ancient times, but crop rotation was in vogue from very early age. Some mythological and historical evidences in respect of crop rotation are as follows:

- The earliest reference of rotating barley or wheat in winter with rice in summer by neolithic settlements at Saraikota as early as 3100-2800 BC in Swat (now in Pakistan) (Mehra 2004)

- Legumes like peas and chickpea were included in crop rotation involving cereals in the Harappan phase of Chacolithic period at Nausharo in late 3000 BC (Mehra 2004). This practice, continued for many centuries must have paved the way for now popular cereal-legume sequence
- Inclusion of peas/chickpea in the sequential systems was reported from Atranjikhera in Balandshar in late Indus-Saraswathi civilization (2500-1500 BC) (Saraswat, 1992)
- Popular crop rotation of blackgram-rice was practiced prior to Sangam period (200 BC), as mentioned in poems of Tolkappian (Jeyaraman, 2004)

These evidences suggest that crop rotation must have originated in India. It must have been practiced in India for centuries with various modifications, to include different types of crops – often guided by the experience. Many historical evidences even suggest that crop rotation involving legumes could sustain the soil fertility perpetually. Having noticed its beneficial effects in terms of pest control and nitrogen fixation, it is now regarded as an important practice of modern organic farming.

Similarly, long before the present day practice of intercropping was evolved, several centuries ago, mixed cropping was followed in India. Perhaps mixed cropping must have gradually been metamorphosed into intercropping, as the traces of mixed cropping are even now noticed in some regions/some crops (ex: niger + groundnut). Mixing two or more crops offers distinct advantages like efficient utilization of soil moisture, insurance against crop failures as well as simultaneous production of adequate fodder and food. The modern scientific knowledge has added in recent years some more advantages such as efficient use of available nutrients, as well as effective control of pests (as trap crop). Both intercropping and mixed cropping are now regarded as practice of modern organic farming, as the nutrients are made available by the leguminous crop and mutually complementary of crops can help the growth and yield by synergistic effects (ex; Sugarcane+ soybean). The modern concept of in situ green manuring with green manure crop like sesbania as intercrop is the extrapolation of original concept of intercropping. Some historical evidences in respect of mixed cropping and trap cropping are as follows:

- Earliest reference about mixed cropping was made by Tholkappier in his Tholkappian poem during 200 BC in

Tamilnadu. It contained references of growing fox tail millet with lablab (Jeyaraman, 2004).

- Planting shatapushpa (*Antheum sowa* Kurtz) and kubberakshi (*Caesalpina crista* L.) in between smaller trees to get rid of insect attack was recommended by Chakrapani Mishra in Vishwavallabha written 1577 AD (Nene, 2004a). This conceptualisation must have led in the recent days to adopt what is now called as **"trap cropping"**. It is clear that growing other plants in between crop to avoid insect attack must have been practiced in various crops and regions of India, before the terminology "trap cropping" was coined by the modern entomologists.

Now, crop rotation, mixed cropping and intercropping are part of any form of agriculture. But, they are more natural practices of organic farming, as organic farming connotes the effective use of all the organisms (the term organic is derived from organism) in an eco-system, that are integral part of crop production or crop protection directly or indirectly. Hence, successful organic farming will have to include successful cropping system practices, as the impact of cropping systems are predominant on both crop production and crop protection domains.

4.1.2 APPROPRIATE DATE OF SOWING

The growth and yield of a crop depends upon when it is sown. Sometimes, time of sowing even decides the crop duration too, when it interferes in leaf expansion and flowering. Sowing date decides the exposure of the crop to different cycles of day or night length/ wetting/ drying/freezing/thawing- which have great influence on crop growth Often, time of sowing decides its vulnerability to pests and diseases or the ease with which they can be controlled. Sowing at appropriate time is, therefore, an important practice to control the pests and diseases in the most natural way Because, it is possible to avoid the occurrence or the severity of pests and diseases (for this reason, it is considered as cultural and non-chemical way of plant protection) and to increase the crop production. The delayed sowing or early sowing may attract more pests and diseases, and it is well known principle in organic farming that it is more prudent to avoid the pests/diseases than to control them after allowing their attack.

There are number of historical and mythological evidences to prove that sowing date was considered as an important strategy in agriculture. The early farmers might have understood the importance of adopting appropriate date of sowing as natural means of avoiding the pests and diseases. The following evidences can be examined to substantiate this:

- Earliest reference about season specificity to grow a crop in a particular season was found in Neolithic settlements at Saraikota of Swat during 3100-2800 BC (Mehra 2004). They grew barley and wheat in winter as well as rice in summer.
- Seasons for wet crops (called kedara), winter crops (called haimana) and summer crops (called graishmika) were clearly recognised by 300 BC as documented in Koutilya's Artha Shastra (Nene 2004a).
- Similarly, specificity of growing pea and chickpea during winter was recognised and adopted in agriculture at Atranjikhere of Bulandshar district during 2000-1500 BC in late Indus-Saraswathi phase of Indian civilization (Saraswat, 1992).
- Koutilya's Artha Shastra even classified the crops based on their sowing dates:
  1. Vrihi (rice), priyangu (*Setaria*), tila (sesamum) and varaka (*Vigna aconitifolia*) were the crops recognised for early sowing.
  2. Mudga (*Vigna radiata*), Masha (*Vigna mungo*) and Shabhya (*Lablab*) were recognised for mid-season sowing.
  3. Kusumbha (safflower), Masura (*Lens esculentis*), Kulutha (*Macrolyta*), Yava (barley), Godhuma (Wheat), Kalaya (Field pea) Atasi (Linseed) and Sarshapa (Mustard) were recognised for late season sowing.

Although such clear evidences of cropping systems and sowing date were found pertaining to period up to first century BC, further historical evidences do not mention about these practices. But, the fact that even now cropping systems and date of sowing are part of our agriculture can lead us to deduction that they must have been followed in later period also.

4.1.3 GERMINATION OF SEEDS

A healthy and vigorous crop can be established only when a healthy seedling is germinated by a healthy seeds. This important fact (which is the essence of seed technology) was known to early practioners/advisors of agriculture in India. They included number of practices to make sure that seedlings were established. Some documented evidences of such practices leading to good germination are:

- Kautilya's Artha Shastra of 400 BC is the earliest documented evidence to show that seeds dipped in water with a piece of gold and sown with a specific mantra ensured that good germination is achieved (Use of gold piece and chanting of mantra was a part of holy belief; but pre-soaking of seeds ensuring the promotion of pre-germination reactions is even now known to ensure speedier and vigorous germination).
- In the same source, report about encapsulation of cotton seeds by dung leading to better germination is made. It is now known that dung treatment to seeds helps in better imbibition of water even under partial moisture stress and thus lead to good germination.
- Vrikshayurveda by Surapala in 1000 AD (Nene, 2004a) made an elaborate conceptualisation of seed treatment in broader perspective for healthy and vigorous seedlings. The range of treatments included water treatment, treatment with plant extracts or treatment with dung+ honey etc.

Replicating these practices, the modern Indian organic farming has evolved the practice of "beejamrutha", where the seeds are treated before germination in a specific mixture of dung, urine and other local materials.

4.1.4 PREPARATION AND USE OF MANURES

Most organic practitioners of today's agriculture are familiar with different types of manures, either pre-digested or undigested. Especially among them, farm yard manure deserves the attention as its popularity is so much that for many organic farmers - practising organic farming is tantamount to use of farm yard manure alone. For more than century, the only pre-digested manure known to the world was farm yard manure. It is really interesting to trace the history of preparation and use of different types of manures, including green manures.

4.1.4.1 Farm yard manure

Farm yard manure is basically manure prepared out of decomposition of crop residues and wastes with animal dung, where the dung acts as starter bacterial culture and accelerates the decomposition. It is expected to be pre-digested or well decomposed manure, although at farmers' level sometimes partially/semi decomposed status is acceptable. Like all manures it is rich in organic matter and serves as base for mineralisation of nutrients, the rate of mineralisation dependent upon the level of decomposition. The nutrients available in organic form are slowly mineralised to available form-hence all the nutrients are not released at a time.

But, how the use of farm yard manure was started? When did present form of farm yard manure commence? To answer these questions, some historical evidences are necessary to be examined:

- Although first mention about the use of animal manure is made by Thiruvallavar in his poem Thirukural in 70 BC along with green manuring and sheep penning (Jeyaraman, 2004), no report is available in further historical evidences to prove that dung based decomposed manure was being used in the early ages.
- But evidences like Vishva vallabha written by Chakrapani Mishra in 1577 AD (Nene, 2004a) and Persian compilation by Dara Sikoh (son of Jahajahan) during 1650 AD (Akbar, 2000) in his book Nuksha Dar Fanni Falahat (meaning-art of agriculture) suggest that animal dung was directly being applied to the crops. Even the reference of use of dung directly for crop production is found in Rigveda and ancient epic Ramayana. So, probably till late 17th century, dung based decomposed manure was not in existence, while dung was directly used. Indian mythology considers cow as mother (Gomatha) and perhaps all the earlier literature mentioned the use of cow dung as manure although it is not clear whether it meant the dung of animals other than cow. Dara Sikoh mentioned the use of dung of pigs, besides dung of other animals.
- In the broader connotation of manure as any thing added to soil to grow a crop, addition of ash from yagna kunda (specific place, where the fire was created to offer the prayers to lord of fire called Agni) practiced in vedic period was perhaps starting point of the use of manures (Katyal and Rattan, 2005). It was called vedic farming or rishi-krishi (agriculture by saints). As such yagnas used dried cow dung cakes and as it was noticed then that ash of the cow dung cake benefitted the crops, direct use of cow dung must have been originated.
- Dara Sikoh (1650AD) was the first person to recommend the use of inorganic salt peter (sodium nitrate) (Akbar, 2000) to increase the crop production. Then, he called it as manure. This development must have been the stimulus later for manufacture and use of fertilizers. He also recommended the use of animal dung, night soil and even human urine.

- It is indeed obscure when the decomposed manure was prepared and used in the form of farm yard manure. From Moghul period till early 1900s, no mention is found about the use of decomposed manure. It is not found even in the report of Lord Wellesley in 1805 submitted to the Court of Directors of East India Company on the study of agriculture in India. The survey report by Francis Buchanan in 1801 covering Madras and Malbar regions is also silent on farm yard manure. The report of Famine Commission of 1880 by J.A. Voelcker also does not mention about the use of farm yard manure (Katyal and Ratan, 2005).
- It can be deduced from available literature that the system of decomposing crop residues as farm yard manure using cattle dung must have started sometimes late in late 1800 in crude form. It must have started by accidental observation that residues decompose quickly when in contact with animal dung (when they are heaped together). For many years, heap method (dumping the wastes and crop residues with dung as heap) must have been used, although it achieves partial decomposition. Later, Albert Howard during his stay in India between 1900 and 1940, the crude method of preparing the dung based manure was refined as Indore method of composting (in 1929) the farm yard manure in underground pits (Bhattachrayra and Chakraborthy, 2005). He is now recognised as father of modern organic farming.
- During 1934, Bangalore method of anaerobic decomposition by Acharya, later in 1938, Coimbatore method (pit-cum heap method), and during 1980, NADEP method of aerobic decomposition were standardised. Pioneering work at IARI, New Delhi on both aerobic and anaerobic processes of decomposition and bacteria responsible for such decomposition added scientific basis to manure making. A method similar to NADEP called Japanese vat method was later added as late as 1995.
- Presently, the farmers follow different methods of FYM preparation, depending upon their convenience. Heap method, though unscientific, is very popular in farming community – as it does not call for extra care and does not involve more labour. Some lazy farmers adopt the pit method as a dumping yard for all the wastes generated in the farm and house and treat it as manure making method. However, such practice does not lead to good decomposition because,

seldom is the turning done; nor layer wise deposition takes place. Most of Indian farmers follow pit method (Indore method) – which can be made to work both under aerobic and semi-aerobic conditions. The practitioners of organic farming also practice latest fully aerobic methods like NADEP or Japanese vat methods.

4.1.4.2 Vermi-compost

Among other types of manures now being used in organic agriculture, vermi-compost deserves a historical investigation. The earliest historical evidence of earthworm inhabited in soil is Kashyapiya Krishi Sukti (800 AD), in which saint Kashyapa [He lived in Kosala Coudh in central Uttar Pradesh] described the presence of earthworm as desirable character of good soil (Katyal and Ratan, 2005). Although earth worms must have been part of ancient agriculture, no other evidence is found in available literature about preparation of vermi-compost with manmade efforts. But, over generations, it is well experienced that earth worms can create loose friable soil, due to its constant movement in the soil, besides increasing soil aeration and infiltration rate. Only in the last 25-30 years, a technique to prepare manure called vermi-compost was developed and popularised. Now it is also well known that vermi-compost is richer than FYM in nutrients and can significantly improve the physical characters of the soil.

4.1.4.3 Green manures

- Like many other organic farming practices, even green manuring (green leaf manuring also) has its roots in ancient civilization. Although pulse crops were included in crop rotation and in mixed crops from 200 BC, utilizing the green foliage of pulse crop to manure other crops was not reported in ancient and medieval literature except a mention about green manures by Thiruvalluvar (70 BC) (Jeyaraman, 2004). Francis Buchanan, who surveyed south India in 1801 recorded that leaves and shoots of many species of plants were used as manure. Such species included *Robinia mitis, Aspedias gigantean, euphorbium tirucalli and Eruthroxylon sideroxyloides*. The usefulness of *Vicia faba* (called as baqla then) was recorded in Persian manuscript Irshadu'z Zaraat (Katyal and Rattan, 2005), wherein its leaves, branches and roots were used to improve the quality soil.

- Before direct application of green parts of plants, Indian farmers must have observed that in crop rotations, the crop residues dried roots, stem and leaves of crops like *Macrotyloma uniflorum* and *Crotolaria juncea* left in the field used to enrich the soil. This was observed and recorded by Mollison (1901) [cited by Katyal and Rattan, 2005]. He appreciated the wisdom of Indian farmers to grow the crops even without manures but by including the leguminous crops in rotation. M.S. Shivaraman made pioneering efforts to spread the concept of manuring with Gryricidia during first half of 20th century to obtain bumper rice yields in south India without any other manure (Katyal and Rattan, 2005). Many other green tree crops of leguminous family were identified for their green manuring potentiality in the modern agriculture commencing from 1950s.
- All other types of organic manures like poultry manure, city compost etc and concept of enrichment of manures by fortification were added in the last 30-40 years mainly to harness the productive potentiality of vast wastes, besides as strategy to dispose them.

4.1.4.4 Liquid organic manures

Long before the present day liquid manures were made available in the market, a liquid manure called "kunapajala" was discovered and used for centuries during ancient civilization. The documented references of "kunapajala" are found in following ancient works (Nene, 2004b):

- "Vrikshayurveda" by Surapala in 1000 AD in eastern India. He claimed that it was practiced by earlier sages and he only repeated its use after verification.
- "Lokopakara" by Chavundaraya II in 1025 AD in south India (Karnataka state).
- "Sarangdhara paddhati" by Sarangdhara during 1283 – 1301 AD (Chapter Upavana Vinoda meaning arboriculture) in the present Uttar Pradesh.
- "Vishwavallabha" by Chakrapani Mishra (1577 AD).

[Surapala was a sage and a scientist of Medieval period, Chavundaraya II was a poet-scholar in western Chalukya king at Kalyan (in Bidar) in the court of Jayasimha II. Sarang-dhara was also scholar in the court of Hammira of Sakam-baridesa (modern

Bundelkhand). Similarly, Chakrapani Mishra was a scholar in the court of Maharaja Pratap of Mewar in Rajasthan].

All these scholars, who reported about the use of kunapajala in the span of 600 years between 1000 AD to 1577 AD, described it as a liquid organic manure to enhance the plant growth, flowering and fruiting of tree crops. They described the method of its preparation as follows:

- Excreta, marrow, flesh, blood and other remains of dead animals (initially boar was recommended and later the remains of any animal/bird was recommended) mixed with water and stored underground along with husk (paddy/ wheat?). Then the contents were cooked with sesame oil, honey and soaked black gram.
- Put the contents in pot for a fortnight (incubation in the modern sense) in warm place.

The Sanskrit word "kunapa" means stinking smell of dead body and "jala" means water. During the course of preparing kunapajala, most of the nutrients and useful substances of dead body would be transferred to solution and thus the liquid manure was prepared.

No scientific investigation on its nutrient content, microbial load, and its benefits are documented. But, a Maharastrian mathematician turned sage called Ayangarya reported excellent benefits by the use of kunapajala in mango and coconut. He even developed "herbal kunapa" (called sasya kunapa) using sour mango fruits and soapnut, and reported that sasya kunapa gave excellent results in chilli. He even reported the preparation of "manujala", which is fermented vegetable waste with human urine (Ayangarya, 2004). A special kunapa called "indsafari" was prepared by him by fermenting star fish in cow's urine. Indsafari was reported to be every effective to control tea mosquito (*Helopeltis*) (Ayangarya, 2005) Similarly, he reported the preparation of "kukkuta kunapa" and "mushaka kunapa" respectively by the use of dead bodies of chicken and rats (Ayangarya, 2006). He used all these kunapa jala @ 1 per cent concentration for spray to improve the growth of perennial tree crops and to control their pests in north eastern India. Unlike earlier workers, he did not use it for soil application. Recently, Narayanan (2006) of Dindigul district of Tamil Nadu effectively used "mushaka kunapa" by recording the details of methods to prepare it (2 weeks fermentation of rat pieces in mixture of 5 kg cow dung, 3 litres of cow urine, 500 g of sugar and 250 g of black gram. Later, 1 litre of cow milk and 100 ml honey was added).

This ancient wisdom of India may serve the requirements of both crop production (in terms of readily available nutrients or growth promoting substances) and crop protection (in terms of pesticidal values to control both insect and non-insect pests). Till now, no scientific study is reported to corroborate this historical evidence with scientific reasoning. Nor any attempt is made to recommend its use in different crops. But, as pointed out by Neff *et al.* (2003), distinct advantages of dissolved organic nitrogen in kunapa jala can facilitate the quick availability to crops than traditionally applied organic manure, which has to be slowly mineralised. If "kunapa jala" can directly supply liquid organic nitrogen, it sounds logical that "kunapa jala" should be able to benefit the crops (dead body of any animal is rich in both protein and non protein nitrogen and extraction of nitrogen from animal body as liquid manure is an efficient way of recycling naturally found nitrogen).

The use of dead animals as source of nutrients to increase the fertility of soil must have been very old observation, as renowned and mythological epic "Ramayana" mentioned 'all dead things – rotting corpse and stinking garbage returned to earth are transformed into wholesome things that nourish the life. Such is the alchemy of mother earth'. Even today, with or without knowledge, in many rural areas, the dead animals are buried in gardens. The beneficial effect of cow's urine in herbal kunapa is now recently testified by the findings of CIMAP, Lucknow (Central Institute of Medicinal and Aromatic Plants). They have distilled a chemical from cow's urine which has bio-enhancing characters (Nene, 2004b). Khanuja *et al.* (2003) have been able to identify a 'cow urine distillate fraction' that enhances the activity of antimicrobial agents (US patent No 6410059) which substiantiates the utility of cow's urine for bio-control of plant diseases.

Two other liquid manures worth discussion are 'Panchagavya' and 'Jeevamrutha'. Panchgavya is a mixture of five ingredients (Panch = five in Sanskrit) viz., cow dung, cow urine, milk, curd and ghee/ honey). Panchagavya was used since Vedic period as traditional practice to purify the body and soul (it is mentioned in 'Garudapurana' and 'Varaha purana'). Purification of body was necessary on the occasions of births and deaths in the family or even when prohibited food was eaten. But, the use of 'Panchagavya' was first time reported by Natarajan (2003) as growth promoter of plants. Many variations in the preparation have been made by him, as compared to original prescription, to add sugarcane juice, coconut water, banana etc. He even advocated its fermentation aerobically for a week. In the last 5-6 years, large numbers of reports are made to highlight the diversified benefits of the use of 'Panchagavya'.

Jeevamrutha is also a mixture of cow's urine, dung, pulse flour and jaggery incubated for a period of 5-10 days. Its preparation was recommended more popularised in Maharashtra, Karnataka and Andhra Pradesh by Palekar (2005), as part of nature farming. His prescription that application of jeevamrutha can take care of all the needs of the crop (both protection and nutrition) may not be scientifically acceptable, but jeevamrutha does contain proliferated mass of beneficial microbial population besides small quantity of nutrients. It appears from the historical evidence that jeevamrutha is a modification of 'sasyagavya', which itself is adaptation from ancient technology on 'kunapa jala'. Because, the ingredients used and method of incubation followed in preparation of kunapa, sasygavya and jeevamrutha are similar.

This author is currently handling a research project on standardisation of preparation and application of jeevamrutha. It has been found that jeevamrutha can be promoted as rich microbial liquid culture to enhance microbial activity in soil to ultimately improve the fertility status, as both nitrogen fixing bacteria and phosphate solubilising bacteria are found to be 5-7 times increased, as compared to original count in cow's urine and dung. It incidentally contained a small quantity of nitrogen (0.1 to 0.5%), phosphorous (0.02-0.04%) and potassium (0.2-0.4%) besides micronutrients like zinc, iron, manganese and calcium. Its regular application (5-7 times in a cropping period) can increase the organic carbon status of soil. The incubation of dung and urine (both containing many useful bacteria) in the presence of added nitrogen (supplied by pulse flour) and utilizable simple carbohydrate (sucrose as jaggery) in preparation of jeevamrutha is ideal for large scale proliferation of bacteria. Its application increased the bacterial load in soil and therefore, its regular application (5-7 times during cropping period) is responsible for improved fertility by enhanced biological activity. The macro nutrients, secondary and micronutrients contained in jeevamrutha may also help the crop, as these nutrients are in readily utilizable form, rather than complex forms, which need mineralization – as found in case of most manures.

4.2 CONTROL OF PESTS, DISEASES AND OTHER DISORDERS

Perhaps, our knowledge of agriculture is so heavily influenced by western literature that we were made to believe that early Indians had no knowledge of plant protection and protecting the plants from disorders surfaced only after western thinkers put forth their ideas of plant protection in the form of plant pathology and entomology. Such a perception of agricultural scientists, heavily dosed by western influence, led the policies in post-independence period and percolated in to present day recommendations of pesticides as the only solution against every pest. Of late, sustainable issues have crept in agriculture and now it is realised that the pesticides do not offer sustainable plant protection. But, in the archives of agriculture, large numbers of historical evidences are found to control a pest/ disease/ non- pathogenic disorder by number of non-chemical ways. An attempt is made to present the Indian evidences of plant protection measures followed in early days.

Long before the appointment of E.J. Butler, Imperial mycologist to study and recommend the measures to control Indian crop diseases in second half of 19th century, the knowledge of plant protection was well developed in India, but seldom documented in English. Not only insect pests, non-insect pests, diseases and non-pathogenic disorders were identified, but also number of simple and practical control measures were recommended and practiced during ancient and medieval period. Many of them were mechanical, cultural, ritualistic as well as offensive (to kill the pests) measures developed based on environmental and ecological principles. Most control measures included locally available eco-friendly materials derived from nature and therefore, in tune with nature's cycles. In the last 15-20 years a realisation is now made in scientific circles that such measures are more sustainable and after reinventing such eco-friendly approaches, they are now included them in Integrated Pest Management Programme as well as programme on Organic Pest Control.

Both biotic and abiotic plant disorders were identified in India in the ancient times. Biotic disorders included both insect and non-insect pests as well as diseases, while abiotic disorders were mostly due to environmental stress like drought, frost, heat and wind etc. Surapala's Vrikshayurveda (1000 AD) by Surapala classified the disorders as internal and external. The internal disorders were supposedly caused by the imbalance of internal mechanisms due to variations of vata, pitta and kapha. This concept, in fact, is an extrapolation of "tridosha" concept (tri= three; dosha= aberration) of Ayurveda shastra originally applicable to human beings. Literally, Ayurveda (Ayu= life; veda= knowledge) extends to the plant life also and not restricted to human health alone, as commonly perceived by all. Surapala's proposition of "tridosha" was preceded by centuries of knowledge of crop protection, as evidenced in following works.

- Earliest reference of plant protection in Rigveda (8000 BC)
- References of disorders and their control in Atharvaveda (1000 BC)
- Kautilya's Artha shastra (300 BC)
- Amarasimha's Amarakosha (200 BC)
- Patanjali's Mahabhashya (200 BC)
- Krishi- Parasara by saint Parasara (100 BC) (considered as oldest textbook of agriculture in world)
- Sangam literature of Tamilnadu (200 BC- 100 AD)
- Vrahamihira's Brhat Samhita (600 AD)
- Kashyapiya Krishi Sukti by saint Kashyapa (800-900 AD)

Surapala's external disorders included both biotic and abiotic causes including insects.

An attempt is made in forthcoming pages to elaborate the identification of the disorders as well as their control measures followed in ancient and medieval period in India.

4.2.1 Internal disorders

Although the concept of "tridosha" proposed by Surapala in Vrikshayurveda cannot be explained in terms of knowledge of modern plant physiology, it must be recognised that even now, number of plant disorders are ascribed to impairment physiological processes. They might have been termed by Surapala as internal disorders. He even identified each dosha to represent a specific function. So, vata representing physiological processes involved in movements (Plants lack locomotory organs like animals, but still many movements in a plants are observed. These movements include water movement and movement of sap); pitta representing reactions and biosynthesis of various ingredients in plant body and kapha representing balance between growth and development (may also include the imbalance between vegetative and reproductive growth). Harmony between these mechanisms ensures that plant can prevent or resist the disease and disharmony makes them prone to diseases. Considering the level of knowledge of plant physiology and plant pathology of his days, Surapala needs to be commended for his attempt to identify the reasons for disorders and classify them, besides classifying the disorders themselves.

Now it is well known the physiological mechanisms of plant life and human life are completely different and disorders of plant life might not occur due to vata, pitta and kapha – typical components of ayurveda. But, ignoring the terminology, the internal mechanisms of hormones, translocation, sink-source relations as well as water relations do exist even in latest knowledge of plant physiology. The balance between these internal mechanisms is still an open and nascent subject, with no much explanation. Are the observations of Surapala an eye opener to modern challenges of resistance/ vulnerability of a crop to a disease as a function of such balance between different physiological mechanisms?

Most of physiological disorders identified today are not coupled with corrective suggestions (except the use of growth regulators in some cases). Mere identification of physiological disorders will not offer practical solutions to the on-going problems of the growers. But, Surapala had his own sets of solutions to control these internal disorders (Nene, 2004a). Some of them are narrated here:

- Sprinkling of kunapa (a liquid extract of dead animal body incubated for 10-15 days) against imbalances of vata.
- Fumigation by burning animal fat, ghee, hemp, horse hair, cow's horn against imbalances of vata (must have been alternative to Kunapa).
- Sprinkling of decoction of panchamula against kapha disorders (panchamula = roots of five plants namely *Clerodendrum phlomidis, Aegle marmelos, Stereospermum suaveolens, Gmelina arborea* and *Oroxylum indicum*).
- Drenching the tree base by decoction of milk, honey, Yastimadhu (*Glycyrrhiza glabra* L) and *Madhuca indica* against pitta disorders.
- Drenching of tree base by decoction of triphala (triphala= dried fruits of three plants namely, *Terminalia chebula, Terminalia bellirica* and *Emblica officinalis*) against pitta dosha.

Many such recommendations were also made and presumably practiced more than 500 years after Surapala lived as evidenced by writings by Chakrapani Mishra in 1577 AD in the chapter called Taruchikitsa (meaning treatment of trees) in his 'Vishva vallabha'.

- Bitter juices of plants with watery decoction of oily flesh or powdery cow dung against the disorders caused by vata.
- Sprinkling water mixed with nirgundika (*Vitex nigunda* L), aragvadha (*Cassia fistula* L) leaves of Panchashakha and ambu (*Pavonia orodata*) or by smearing their roots with this

mixture could set right the disorders caused by vata (symptoms like poor growth, faded leaves, slender trunk etc).

- Feeding such weak trees with kulmashanka (*Cassia asus* L), goat dung, bark of amra (*Mangifera indica* L) and aswagandha (*Withania somnifera* L) could also mitigate the effects of vata.
- Vata related disorders suffering from loss of turgidity and knots on aerial parts can be corrected by smearing a mixture of cow dung, bhallataka (*Semecarpus anacardium* L) and marrow or simply by sprinkling kunapa jala.
- Sprinkling decoction of rasna (*Venda tesellata*) aswagandha (*Withania somnifera* L), pavanari, naga (*Mesua ferrea* L) and kana (*Piper longum*) or water mixed with shatapushpika (*Antheum sowa* Kunz) can correct disorders caused by vata
- Use of oil, cool and sweet juices and sprinkling of cool water with cow's milk, sita (*Pueraria tuberosa* L) and vida (probably medicinal salt) can cure bile (pitta) related disorders. The trees affected by such disorders of imbalance in internal fluids exhibit slender trunk with pale whitish leaves and dry branches.
- Similarly, treatment with ushira (*Vitivera zizaniodides* N), musta (*Cyperus rotundus* L), honey, milk and sarpis (*Sansevieria rozburghivia* S) can mitigate the disorders caused by bile.
- A tree suffering from phlegm (kapha) disorder has oily leaves and bark, pale undersized flowers and slimy/tasteless fruits, coupled with slow shoot development. Such trees could be cured by the use of decoction of bark of plaska, arjuna (*Terminalia arjuna* L), udumbara and saptaparna (*Alstowa scholars* L) and nimba (*Azardiractita indica* L)
- Decoctions of vasas (*Adhotada vasica* N), musta and rohita (*Tecomelia undulate*) are also useful to mitigate the effects of phlegm.
- To mitigate the bad effects of excess watering (probably coupled with poor drainage), incision with deep sharp axe to allow the sap to flow and closing the wound by paste of bidanga or application of ashes of bark of plaksha and arjuna were recommended.
- To overcome the effects of indigestion (symptoms: no leaf development; loss of turgidity; ants infecting leaves) triphala

[fruits of amla (*Emblica officinalis*), bibhitaka (*Terminalia bellarica* and haritaki (*Terminalia chebula*)] or kuberakshi (*Caesalpinia crista*) could be used to treat the roots.

These reports open up a new untapped field of investigation in modern agriculture – dealing with physiological disorders caused by imbalances of internal mechanisms, a field not yet dealt by modern plant physiologists or agronomists. The applicability of 'tridosha' concept to plants may be investigated, with the knowledge of modern plant physiology. These earlier report also lead us to an important conclusion that many disorders could be corrected by local preparations using various herbs. The medicinal values of large number of plants were identified centuries back and unfortunately now they are not being even regarded as possible control measures for many non- pathogenic complications.

The scientific veracity of these techniques can be certainly questioned and applicability of these techniques can be doubted. But, the techniques of Surapala and Chakrapani Mishra cannot be dumped without testing. Even after testing, these measures provide clues to a new direction of finding new organic techniques to overcome imbalances in plant body. Many disorders, not caused by any pathogens are even now tried to explain in terms of modern knowledge of plant physiology. But, the concept of imbalance between the internal mechanisms has been never a subject of consideration to identify the disorder. The concept of balance between internal mechanisms could be tried by the agricultural scientists in annual and perennial crops to explain the internal disorders and corrected.

4.2.2 Abiotic disorders

Vihiramihira (600 AD) was the first person to identify the abiotic disorders such as those caused by cold, frost, speedy wind and scorching sun (Bhat, 1981), which were later classified as external disorders by Surapala (1000 AD) and Sarangdhara (1300 AD) (Majumdar, 1935). The symptomatic description of the disorders were

- Scorching heat: pale leaves, yellowing, older leaves dry out, roots attacked by insects (possibly secondary attack)
- Frost: Drying of leaves
- Stormy winds: Breaking of branches, uprooting/twisting of trees
- Fire, lightening: Drying up of trees
- Mechanical wounds: Drying of branches, blackening of wounds

- Excessive water: Foul smell (due to rotting?), reduced fragrance of fruits, stunting plant and reduced leaf size

Note: All these reports were recorded in perennial fruit trees. It is not very clear why such observations were not recorded in annual crops, although annual crops were grown.

Many control measures were also suggested to check abiotic disorders using natural, organic and locally available materials. Perhaps this was the starting point to identify the non-chemical organic ways to deal with plant disorders. Some of them suggested in Surapala's Vrikshayurveda are cited (Sadhale, 1996) here:

- Mechanical wounds were sought to be healed up by pastes prepared from plant extracts.
- Frost damages were sought to be healed up by covering the base of plants with residues.
- Sprinkling of kunapa jala with milk to mitigate the damages by frost of scorching heat
- Trees struck with lightening could be made to sprout again, when anointed with mixture of vidari (*Pueraria tuberose*), sugar and nagajivha (red arsenic) with sesame.
- Trees dried up due to fire can be smeared with mud from a region, where lotus is naturally grown.
- Broken trees can be smeared by paste of bark of plaska (*Ficus lacoe*) and udumbara (*Ficus glomerata* R) mixed with honey, ghee, wine and milk on broken part and the broken part be firmly fastened to the plant.
- Dressing the wounds by the paste made of the barks of banyan + mustard seeds + ghee or paste made of Vidanga (*Embelia ribes*) in thick mud
- Flowering and fruit set can be boosted in unproductive trees with
  (a) Irrigating with kunapa jala and milk or
  (b) Treating the branch or tip of branch (where flowers are expected) with the mixture of sesame, barley, kulattha (*Macrotyloma uniflorum*) green gram and black gram

4.2.3 Biotic disorders

The modern agricultural science has focused more on the biotic disorders, which include insect pests, pathogens as well as non-insect pests. The former two have led to two separate branches of agricultural science called entomology and plant pathology, the latter (non insects pests) often causing heavy damages are even now generally neglected. Even today, we do not have effective solutions to the problems of wild boar, jackal, elephant damage, bird damages etc.

The Indian historical evidences suggest that all of these were subjects of investigation, more than 2000 years back. These evidences could be dealt separately in two issues viz., **Identification of Disorders and Protection/Treatment Practices**.

Identification of biotic disorders

The number of evidences available on identification of biotic disorders in the Indian history is surprisingly high, to the disbelief of modern plant pathologists and entomologists. Our early thinkers, experimenters, scholars and even sages had an elaborate knowledge of plant disorders caused by other living beings. Due to the lack of technology of microscope, many of these microbes were not seen by them. But, their ingenuity in distinguishing the disorders based on the differences in external symptoms and their experimental spirit in finding newer methods to control the disorders must be honourably applauded, considering the type of facilities they had and the level of scientific knowledge they had at that time.

Considering the vast number of evidences on identification of biotic disorders, it is more convenient to acquaint with them in tabular form. The following table evidently indicates that the knowledge on identifying the biotic disorders was of high order in ancient and medieval India.

S N	Name of disorder	Reported in	Causal organism now identified	Reference
1	Bird as pest	Rigveda (3500 BC)	Parakeets, sparrow, crows	Nene and Sadhale, 1997
		Kautilya's Artha Shastra (400 BC)	Parakeets, sparrow, crows	Shamasastry, 1926
		Sangam literature of Tamilnadu (100 BC)	Parakeets, sparrow, crows	Balambal,1993
2	Crow as pest	Atharva veda (1000 BC)	Crows	Devi Chand, 1997
3	Rat as pest	Kautilya's Artha Shastra (400 BC)	Rats attacking store grains & fields	Shamasastry, 1961
		Atharva veda (1000 BC)	______Do______	Devi Chand, 1997

4	Other animals as pests	Atharva veda (1000 BC)	Wild boar	Devi Chand, 1997
		Patanjali's Mahabhasya	Deers	Krishnamrthy, 1993
		Krishi- Parasara of Parasaramuni (100BC)	Goats & buffaloes	Sadhale, 1999
5	Hopping insects Attacking the crop	Krishi-Parasara of Parasara muni	Locusts & termites	Sadhale, 1999
6	Gundhi bug in rice	Krishi- Parasara	Leptocorisa varicornis F.	Sadhale, 1999
7	Pandaramundi (white earhead) in rice = stem borer	Krishi- Parasara	Tryporyza incertulas Walkers.	Sadhale,1999
8	Store grain pests	Amara kosha (100BC)	-	Jha, 1999
9	Phanerogamic semiparasite (mentioned for their medicinal values)	Susruta (400 BC)	Loranthus longiflorus Disr (vandaka)	Nene, 2003
			Cuscuta reflexa Roxb. (dodder)	Pandey and Chunekar, 1999
10	Striga in sorghum	10th century document	Striga (called kura-nti/ kshetranashini)	Watt, 1889-93
11	Mildew of paddy Blight of sugarcane	Buddhist document kallavagga (100BC)	-	Raychoudhary,1964 Rhys Davids,1881
12	Fever of plants	Manasollasa of Someshwaradeva (1126-1138 AD)	Described as incurable diseases	Sadhale and Nene, 2004
13	Various diseases of grain crops	Manasollasa of Someshwaradeva (1126-1138 AD)	Churnaka for smut Lala for ergot Madhuka for blight Granthika for rootknot Jyothis for chlorosis Rupaka for powdery mildew	Sadhale and Nene, 2004

Practices for protection / control of plant diseases

Although earliest attempts to control the plant diseases included the measures like utterances of specific mantras, whose technical efficacy cannot be confirmed, instances of use of both organic and inorganic materials for control of plant disease are aplenty in the Indian history. Few selected citations are presented here to substantiate that early Indians had a detailed knowledge of plant diseases and their control.

- Oldest document on the use of organic materials to control crop disorders is Kautilya's Artha Shastra (300 BC) (Nene, 2003). He recommended plastering the cut ends of sugarcane setts with a mixture of honey, ghee, fat of hogs and cow dung. Treating cotton seeds with dung for bio-control of diseases was also suggested by him. Kautilya's suggestions also included pit sterilization before planting trees by burning residues as well as exposure of seeds to dew in night and drying them on subsequent day to activate mycelia and kill them by sunlight.
- Varahamihira suggested the use of milk, ghee and cow dung for dressing seeds and smoking them by burning animal flesh or turmeric before sowing (Nene, 2003).
- Surapala (1000 BC) suggested several practices for control of plant diseases. They included use of kunapa, smoke of animal fat, panchamula etc. All of them have biocidal properties. For example, honey, mustard and liquorice have antimicrobial properties; cow dung and cattle urine have antiseptic properties. Cowdung can promote bio-control of many pests.
- 17th century document of Dara Shikoh (Akbar, 2000) recommended smearing cow dung to cuttings of fig before planting, garlic extract, common salt for insect control.
- 19th century document from Rajastan (Javalia, 1999) suggested measures like foliar application of oil (sesame) to protect from frost, sprinkling of curd with asafoetida to prevent powdery mildew and use of asafoetida and curd with Embelia ribes to control anthracnose in oranges.
- Someshwara Deva (1126 AD) (Shamastry, 1926) was first to suggest treatment of seeds with ash.
- Dipping seeds in salt solution was a practice in 19th century (Gupte and Raje, 1896).

These indicate vividly that plant protection procedures as well as identification of pests were not new to ancient Indians. The technological limitations of spreading this knowledge restricted its spread. Many of the materials recommended by Surapala in Vrikshayurveda are now found to have pesticidal values.

4.3 BIOLOGICAL CONTROL OF PESTS

One of the natural ways to control a pest is its biological control. In fact, the modern science of crop protection recognises 'pest management' as a strategy of protecting the crops, rather than old theory of 'control of pests' The field of biological control of pests heavily relies on the management of pests based on the principle of naturally existing food chains – involving the pest and its enemies. In the biological control, shortly referred frequently as bio control, pests are managed in such a way that their population is over dominated by the population of its natural enemy. In the field of biological control, both predation and parasitism were employed since beginning to bring about the pest control.

The history of bio control of pests is traced to Egyptian records of 4,000 years ago, where domestic cats were depicted as useful in rodent control. Insect Predation was recognized at an early date, but the significance of entomophagy and its use was not recognised till late 17th century, except for a few early human populations in Asia, where a sophisticated agriculture had developed. The predation was first recognised as an effective bio-control method to control the pests. The Chinese citrus growers placed nests of predaceous ants, *Oncophylla smaradina*, in trees where the ants fed on foliage-feeding insects. Bamboo bridges were constructed to assist the ants in their movements from tree to tree. Date growers in Yemen went to North Africa to collect colonies of predaceous ants which they colonized in date groves to control various pests. But, somehow the later to use predation as tool for pest control were missing in the annals of world history, including the Asian history till as late as 18th century. But, more works on parasitism were reported during 17th century.

The first record of insect parasitoidism is attributed to the Italian, Aldrovandi (1602) who observed the cocoons of *Apanteles glomeratus* being attached to larvae of *Pieris rapae* (the imported cabbageworm). But, he mistook the cocoons for insect eggs. Printed illustrations of parasitoids were found by Goedart (1662), who described parasitoid as "small flies" emerging from butterfly pupae. Antoni van Leeuwenhoek (1702) attempted to describe the phenomenon of parasitoidism in insects. He drew a female parasitoid ovipositing in aphid hosts. Vallisnieri (1706) [$] first correctly interpreted this host-parasitoid association and probably became the first to report the existence of parasitoids. Cestoni (1706) reported other parasitoids from eggs of cruciferous insects. He called aphids as "cabbage sheep," and their parasitoids, "wolf mosquitoes." Darwin (1800) discussed the useful role of parasitoids and predators in regulating insect pests. An ever-increasing number of references to entomophagous and entomogenous organisms appeared in the literature of 18th century, largely in the form of papers dealing with parasitoid biology. Diseases of silkworms were recognized early in the 18th Century. De Reamur (1726) described and illustrated *Cordyceps* fungus infecting a noctuid larva.

Although the instances of predation and parasitoids were reported till the end of first half of 18th century, the efforts to use them as bio control method were not crystallised till 1762, when first successful importation of an organism from one country to another for biological control took place with the introduction of the myna bird from India to the island of Mauritius, for locust control [$]. Further development of modern biological control awaited the recognition of the fact that insect pest problems were of greater importance.

4.3.1 Biological Control Efforts in the Early 19th Century

A number of innovative efforts were reported during the first half of the 19th Century that attempted to use the beneficial entomophagous insects. Darwin (1800) recommended protecting and encouraging syrphid flies and ichneumonid wasps because they destroyed considerable numbers of cabbage-feeding caterpillars. Kirby & Spence (1815) showed that predaceous coccinellids controlled aphids. Hartig (1827) recommended the construction of large rearing cages for parasitized caterpillars, with the ultimate aim of mass release. Ratzeberg (*ca.* 1828) called particular attention to the value of parasitic insects with publication of a large volume on the parasitoids of forest insects in Germany. He did not believe that parasitic control could be augmented by humans. Bassi (1834) first demonstrated that a microorganism, *Beauvaria bassiana*, caused the muscardine disease of silkworms. Kollär (1837) writing an article for farmers, foresters and gardeners pointed out the importance of entomophagous insects in nature's economy. He studied parasitoid biology and was the first to report the existence of egg parasitoids. Boisgiraud (1843) reported that he used the predaceous carabid beetle, *Calasoma sycophanta*, to successfully control gypsy moth larvae on poplars growing near his home in rural France. He also reported that he had destroyed earwigs in his garden by introducing predaceous staphylinid beetles.

4.3.2 Biological Control in the Late 19th Century

Asa Fitch (1855) was the first entomologist to seriously consider the transfer of beneficial insects from one country to another for the control of an agricultural pest. Fitch suggested that the European parasitoids of the wheat midge, *Sitydiplosis mesellana*, be sent into the eastern United States.

Walsh(1866) supported Fitch's work and he became the first worker in the United States to suggest that insects be employed in weed control. He proposed that insects feeding on toadflax, *Linaria vulgaris*, be imported from Europe to control yellow toad flax plants. The first actual case of biological control of weeds was, however in Asia, where around 1865 the cochineal insect *Dactylopius ceylonicus* was introduced from southern India into Ceylon for prickly pear cactus control (*Opuntia vulgaris*). Originally, *Dactylopius* had been imported to India from Argentina in 1795, in the mistaken belief that it was the cochineal insect of commerce, *D. cacti*.

Riley (1870) has been named the father of modern biological control. He shipped parasitoids of the plum curculio from Kirkwood, Missouri to other parts of that state. In 1873 he became the first person to successfully transfer a predator from one country to another with the shipment of the American predatory mite, *Tyroglyphus phylloxerae* to France for use against the destructive grapevine phylloxera. The results were not particularly successful, however. In 1883, Riley directed the first successful intercontinental transfer of an insect parasitoid, *Apanteles glomeratus*, from England to the United States for control of the imported cabbageworm. He was Chief Entomologist of the U.S. Department of Agriculture. In 1872, Riley began his interest in the cottony-cushion scale, *Icerya* spp. which was considered the most important citrus pest in California. The successful biological control effort against the cottony-cushion scale spirited many biological control attempts in many countries.

In 1923, with the transfer of biological control work to the Citrus Experiment Station and Graduate School of Subtropical Agriculture of the University of California, the later works on biological control were concentrated at California. With the importation of *Chrysolina* beetles from Australia for Klamath weed control, the beginning of biological weed control in California was made in 1944.

Steinhaus (1946) established the first laboratory and curriculum in insect pathology at the University of California, Berkeley. Later he transferred it to the newly opened Irvine campus of the University and attempted to further insect pathology there. His untimely death in 1968 precluded this goal. The Division of Biological Control became the Department of Biological Control at University of California in 1954. The biological control unit at the new campus of the University of California in Riverside, was by 1962, the most advanced research centre for that discipline in the world at that time.

Today, biological control is a specialised science, with several sub-disciplines such as biotechnology, insect pathology, insect microbiology, weed pathology etc, As many as 35 countries have attempted systematic research and commercial exploitation of bio control techniques for the effective control of crop pests. Success of biological control at the field levels, depends upon how well these organisms are produced in mass scale for the ultimate distribution to the farmers. Hence, in most countries, a chain of bio-control of laboratories were established both under government and private participation to artificially breed and multiply the beneficial insects/pathogens, to use them on mass scale in the field to control the designated pests. Most advanced countries including US, UK and many European countries have used bio-control as a major strategy in the recent years. The biological control was also popularly used as an integral component of Integrated Pest Management (IPM) to control the crop pests more effectively. Only in recent years, the biological control has found a place as an exclusive strategy in organic farming, along with other organic methods of pest control.

4.3.3 Biological control in India

Relatively, biological control of pests is a new field for India, as no evidence is found in any Indian documents earlier to 1957 about the attempts being made to control the pests by biological means.

The Commonwealth Institute of Biological Control (CIBC) established its Indian Station in 1957 with the aid of Canadian Colombo plan funds, with temporary need based substations in various parts of the country. The advent of CIBC marked the beginning of organized and systematic biological control research in India. During this period, our knowledge of natural enemies of crop pests and weeds increased manifold. Several PL-480 projects were obtained for the survey of natural enemies of major pests like aphids, diaspine scales, adelgids, etc. Bio-systematic studies on the Indian fauna of important groups of insects such as trichogrammatids, coccinellids and Coccoidea were also carried out during this period. To a limited extent, work on forest entomology was also done. The CIBC brought out a catalogue of parasitoids and predators and published technical bulletins comprising papers on specific items of research and also technical communications reviewing the bio control work done in different regions of the world.

The All India Coordinated Research Project on Biological Control of Crop Pests and Weeds was launched in 1977 under the aegis of the Indian Council of Agricultural Research and funded by the

Department of Science and Technology for the first two years. The AICRP had 10 centres initially. Since 1979, the project has been included under the regular research activities of ICAR. The Project Coordinator's cell and the unit working on basic aspects were initially attached to the Indian Institute of Horticultural Research, Bangalore, and later shifted to the premises vacated by the CIBC Indian Station to serve as the main centre of the project in 1988. The main centre was named as Biological Control Centre and the entire programme functioned under the administrative/financial control of the National Centre for Integrated Pest Management (ICAR). In the eighth five-year plan, the project was elevated to an independent Project Directorate of Biological Control, with its headquarters in Bangalore.

At present, the Project Directorate of Biological Control, Bangalore, established in 1993, is the nodal agency in the country that organizes biological control research at the national level. It has 16 centres spread across the country for carrying out field studies on bio suppression of pests of crops. The Directorate at Bangalore has carried out basic research on the biosystematics of important groups of insect bio agents such as trichogrammatids, chrysopids, coccinellids, and tachinids. The reference collection maintained at PDBC has been catalogued in the form of a technical bulletin and also available in a retrievable, electronic format. The holdings contain 3441 species under 216 families and 16 orders and a small collection of insect type material is also being maintained. The huge collection of unsorted and unidentified insects maintained at PDBC is a veritable store house for taxonomic study and unfortunately, for want of expertise in various groups, these collections have not been touched for many years.

The Network Project on Insect Biosystematics is being operated at PDBC since September 2005 for survey, collection and systematic studies on insects with particular emphasis on insect bio agents belonging to the orders Hymenoptera, Neuroptera, Coleoptera, and Diptera. Under the project, intensive taxonomic studies are being carried out on trichogrammatids, coccinellids, Chalcidoidea, thrips, aphids, etc. Multimedia-aided and traditional identification tools have been generated for several insect parasitoids and predators. A website has been hosted on the Aphid fauna of Karnataka and their bio agents.

The National Biodiversity Authority (NBA) has recently recognized PDBC as one of the Designated National Repositories (DNR) for various kinds of biological specimens, both dead and live.

In order to exploit the agricultural arthropod resources from various agro-climatic zones the following thrust areas have been proposed under the Bureau's programme:

- Exploration, identification, characterization, conservation and utilization of biodiversity of beneficial insect resources.
- National and international exchange of beneficial insects.
- Development of digitized inventories, information storage and retrieval systems on insect resources.
- Consultancy and human resource development for exploitation of insect resources. Variations in pest populations from geographical locations and their parasitoids identified through molecular approaches. Parasitoids for abiotic stress and enhanced biological attributes developed.
- Utility of effective symbionts and toxins produced in insect cell line cultures and cloned products of bio pesticides with increased virulence and persistence.
- Establishment of repository for insect derived beneficial bacteria and testing for crop use.
- Exploitation of acarine microbial associates.
- Insect symbioses for pest and disease management in crop plants.

Under private sector, organisations like Bio-Control Research Laboratory (PCI unit) and others have also been doing a laudable service to the farming community by producing natural enemies in large quantities for effective implementation of bio-control programmes.

4.4 BIO FERTILIZERS

Biofertilizers can be defined as "microbial inoculants which contain live or latent cells of selected strains of nitrogen fixing, phosphate solubilizing microorganisms used for application to seed, soil or composting areas to accelerate certain microbial processes; thus augmenting the availability of nutrients in an easily assimilable forms to plants."

No early historical records are available to show that bio fertilizers were used in any form in India. The first study on legume Rhizobium symbiosis was conducted by N.V. Joshi and the first commercial production started as early as 1956. However, the bio fertilizers were popularised by proper policy initiatives only in ninth plan by the Ministry of Agriculture. The use of bio fertilizers was promoted with the setting up of the National Project on Development and Use of Biofertilizers (NPDB). The commercial history of bio fertilizers began with the launch of 'Nitragin' by Nobbe and Hiltner, a laboratory culture of Rhizobia in 1895, followed by the discovery of Azotobacter and then the blue green algae and a host of other micro-organisms.

In the last 20-25 years, the use of bio fertilizers has gained greater importance, apparently as a strategy to adopt integrated nutrient management technique. Large number of university sponsored and private funded laboratories are established to produce the biofertlizers for distribution. Although the free distribution of biofertilizer cultures was promoted initially, purchase of cultures is promoted in recent years. In the last 10 -15 years, cultures of following bio fertilizers are sold popularly:

1. Different host specific strains of Rhizobium for symbiotic N fixation.
2. Phosphate solubilising bacteria for solubilising fixed phosphorus.
3. Azospirillum for non-symbiotic N fixation.
4. Azotobacter for non-symbiotic N fixation.
5. Vascicular Arbiscular Mycorrhizae (popularly called as VAM) for increasing phosphorus availability.
6. Trichoderma culture for hastening the decomposition.
7. Pleurotus culture for decomposing hard lignified substrate.

The use of bio fertilizers is more naturally compatible with other organic practices, because of their common feature of using natural resource for nutrient availability. Hence, it is an organic farming practice effectively used for sustainable nutrient supply, when it is concurrently used with other related practices.

Two fundamental requirements responsible for success of bio fertilizers are:

- The high initial load of microbial population at the time of application of culture
- Favourable conditions in soil for proliferation

Both these conditions are naturally satisfied in organic farming than in any INM situation or pure chemical situations – because all other organic practices favour the improved microbial activity. The success of use of bio fertilizers, is therefore, naturally higher in organic situation.

4.5 REFERENCES

Akbar, R. (Tr.), 2000. Nuskha dar Fahni-falahat (The Art of Agriculture). Agri-History Bulletin 3 Asian Agri-History Foundations, Secunderabad, India.

Asa, Fitch, 1855. Sixth, seventh, eighth and ninth reports on the noxious, beneficial and other insects of the state of New York. Albany, New York. 259.

Ayanagarya, V.S. 2005. INDSAFARI – An organic pesticide for tea. Asian Agri- History **9**: 317-319.

Ayanagarya, V.S., 2004. Manujala: A liquid manure. Asian Agri- History **8**: 319-321.

Ayanagarya, V.S., 2006. Mushika kunapa. Asian Agri- History **10**:157-159.

Balambal, V., 1993. Agriculture in Sngam age In: Agriculture in ancient India Itihas Patrika Prakashana Thane, pp 26-39.

Balkundi, H.V. 2000. Ancient and Medicinal History of Indian Agriculture and its relevance to Sustainable Agriculture in the 21st century In: Ancient and Mediaval History of Indian Agriculture (eds. S.L Choudhary, G.S. Sharma and Y.L. Nene). Proceedings of Summer College of Agriculture, Udaipur 313 001, Rajasthan, India. Pp: 39-45.

Bassi, A. 1934. Del mal del segno, calcinaccio o moscardino, mallatia che affigge in bachi da seta e sul modo di liberarne le bigattaie anche le piu infestate. Part I: Theoria. Orcesi, Lodi. P. 1-9, 1-67.

Bhat, M.R. 1981. Varhamihira's Brhat samhita. Part 1. Motilal Banarasidas, Delhi 110 007, India. Pp. 527-535 (Reprint 1992).

Bhattacharya, P. and Chakraborty, G. 2005. Current Status of organic farming in India and other countries. Indian J. of Fertilizers 1(9): December 2005 pp 111-123.

Bhudharakrida (V-1). Government Branch Press, Mysore, India.

Darwin, E. 1800. Phytologia. Publ., London.

Devichand, 1997. The Atharvaveda (English) Pub: Mushiram Manoharalal Publishers Pvt. Ltd. pp 939.

Goedaert, J. 1662. Metamorphosis et Historia Naturalis Insectorum. Jacques Fierens, Middelburgh.

Jeyaraman, 2004. Agriculture in Sangam literature of Tamil Nadu In: Glimpses of Agricultural heritage of India Pub: Asian Agri-history Foundation Ed: Y.L. Nene pp: 49-51.

Jha, V., 1999. Amarakosha (Hindi) Pub: Motilal Banarasidas Delhi.

Katyal, J.C. and Rattan 2005. Soil management in ancient, medieval, and pre-modern India and its relevance in modern day sustainable Agriculture. Agricultural heritage of Asia: In Proceedings of Int. Conference 6-8 December 2004, Hyderabad pp -61-77.

Khanuja, S.P.S., Alok Kalra and Darirkar, M.P. 2003. Scientific studies on utilization of biological activities of cow urine and its arka (distillate) for agriculture and health. Paper presented in National seminar on cow in agriculture and Human health held on 16-12-2003, Kota, Rajastan Souvenir and abstracts pp 93.

Kirby, W. & W. Spence. 1815. An Introduction to Entomology. Longman, Brown, Green & Longmans, London. 285 p.

Kollär, Vincent. 1837. In London's Gardner's Magazine. 1840. [English translation].

Krishnamurthy, R., 1993. Agriculture in ancient India and its social aspects In : Agriculture in ancient India Itihas Patrika Prakashana Thane, pp 40-55.

Mehra, K. L. 2004. Agricultural Foundation of Indus- Saraswathi civilization In: Glimpses of Agricultural heritage of India Pub: Asian Agri-history Foundation Ed: Y.L. Nene pp:15-26.

Majumdar, G.P., 1935. Upavana-vinoda (a sanskrit treatise on arbori-horticulture) Pub: Indian Research Institute, Calcutta pp 128.

Mollison, J. 1901. A Textbook on Indian Agriculture. Vol 1. Government of Bombay, Bombay, India. PP. 9-105.

Narayanan, R.S., 2006, Application of gunapa jalam (kunapa jala) as liquid biofertlizer in organic farm Asian Agri-history 10: 161-164.

Natarajan, K. 2003. Panchakavya-A Manual. Other India Press, Mapusa 403 507. Goa, India.32 pp.

Neff, J.C., Chaplin III, F.S. and Vitousek , P.M. 2003. Breaks in the cycle: Dissolved organic nitrogen in terrestrial ecosystems. Frontiers in ecology and Environment 1: 205-211.

Nene, 2003. Crop disease management practices in ancient, Medieval and pre-modern India. Journal of Asian Agri- history 7 : 3 pp 185-201.

Nene, Y.L. and Sadhale, N., 1997, Agriculture and biology in Rigveda Asian Agri-history 1: 177-190.

Nene, Y.L., 2004a. Ancient classics on Indian Agriculture In: Glimpses of Agricultural heritage of India Pub: Asian Agri-history Foundation Ed: Y.L. Nene pp: 134-148.

Nene, Y.L., Kunapajala, 2004b. A liquid Organic Manure of Antiquity. In: Glimpses of Agricultural heritage of India Pub: Asian Agri-history Foundation Ed: Y.L. Nene pp: 66-78.

Pande, G.S. and Chunekar, K.C. 1999. Bhavaprakasha Nighantu of Bhavamishra (1500-1600 AD) Pub: Chaukhambha Bharati Academic Press Varanasi pp: 984.

Raychoudhary, S.P., 1964. Agriculture in ancient India ICAR Pub new Delhi pp 167.

Réaumur, M. De. 1726. Remarques sur la plante appellée a la Chine Hia Tsao Tom Tchom, our plant ever. Mem. Acad. Roy. Sci. (21 Aug 1726). P. 302-305.

Rhys Davis, 1881. Vinaya Texts part 3. Sacred books of East No 20 IV – XII of Kallavagga Oxford University Press Oxford.

Riley, C. V. 1870. Parasitic and predaceous insects in applied entomology. Insect Life **6**: 130-141.

Sadhale N. and Nene, Y.L., 2004. On elephants in Manasaollasa – 2 Diseases and treatment Asian Agri- History **8**: 115- 127.

Sadhale, N. (Tr.) 1996. Surapala's Vrikshayurveda. The Science of Plant Life-History Bulletin No. 1. Asian Agri- History Foundation, Secunderabad 500 009, India. 104 pp.

Sadhale, N. 1999. Krishi Parasara Agri-History bulletin No 2 Asia Agri-histroy Foundation Secunderabad pp 134.

Saraswat, 1992. Archeological remains in ancient cultural and socioeconomical dynamics of Indian subcontinent, Palaeobotanist **40**: 514-515.

Shamasastry, R. (Ed.) 1926. Abhilashitartha Chinttamani of Someshwara Deva.

Shamashastry, R. 1926. Abhilashitartha Chintamani by Someshwara Deva Bhudarakrida (V-I) Govt. Branch Press, Mysore.

Steinhaus, E. A. 1946. Insect Microbiology. Comstock Publ. Co., Inc., Ithaca, New York. 763 p.

Van Leeuwenhoek, A. 1702. Letter in Nr. 266 of the Philosophical Transaction 1700-1701, Vol. 22, p. 659-72. Smith & Walford, London.

Walsh, B. D. 1866. Practical Entomologist. June 1866. P. 15-19.

Watt, G. 1891. A Dictionary of the Economic Products of India. Vol. V. Cosmo Publications Delhi, India. 676 pp. (Reprint 1972).

Chapter - 5

POTENTIALS OF ORGANIC FARMING

The practice of organic farming is not tantamount to use of any one particular organic source of material or use of a single organic farming practice. Each type of organic farming (nature farming, ley farming, non-chemical farming etc.) can be followed by adopting variety of methods of organic farming. The principal methods of organic farming are broadly classified as :

- *In situ* decomposition methods
- Use of pre-digested/semi-digested manures
- Use of N-fixing and P-releasing microbial bio-agents
- Cultural methods (to supply nutrients/control pests)
- Use of predators and parasites
- Use of eco-friendly methods to control the pests and diseases

The ultimate aim of the organic farming can be achieved by several methods. Depending upon the resource base of an eco-system and time/cost factors, some methods may attain predominance in a given situation. All the methods may not be practicable in all the situations and for all the crops. Sometimes, a particular requirement such as pest control may get pre-dominance over other aspects of organic farming (for instance, control of boll worm in hybrid cotton is of utmost importance than other factors to increase the cotton yields). This may result into emphasis on particular method of organic farming to control of pests than methods to supply plant nutrients. Thus, each eco-system and each crop may require different set of methods of organic farming based on priority. Sometimes, a judicious mixture of many methods may have to be developed based on critical analysis of resources available at command, cost of materials, convenience of their use and favourable situation for decomposition.

Each method may ultimately end up in one or more practices involving various sources of organic materials. Many sources may have to be combined, sometimes, to practice a method (for example, preparation and use of farm yard manure may involve crop residues, poultry litter or industrial wastes). A single source may be used by more than one method (for example, crop residues /farm wastes can be directly incorporated in the soil or can be used to prepare manures). Taking all such possibilities, the practices of organic farming are evolved based on any convenient method involving all available sources in a given eco-system. The use of all possible materials and methods to supply the required plant nutrients and to control the pests and diseases are most important than means to achieve the objectives of organic farming.

Broadly, the methods of organic farming and corresponding sources of organic materials are grouped in Table 5.1.

Table 5.1. Organic farming methods showing corresponding sources

Methods	**Sources**
FOR NUTRIENT MANAGEMENT	
1. For in situ incorporation	
a) Green manures	Annual, perennial green manure crops
b) Crop residues	Crop residues after harvest of crops including the stubbles, roots, left over in soil (stem, leaf, roots) All harvesting wastes
c) Poultry manure	Poultry litter
d) Urban and rural wastes	Sewage, sludge, garbage, sullage and industrial wastes
e) Recycling the weed biomass	Recycling the weeds grown on the farm Incorporation of aquatic and terrestrial weeds in the whole eco-system with or without treatment
f) Recycling agro-based industrial wastes	Incorporation of sugar industry wastes like begasse, pressmd Incorporation of decomposalbe wastes from cotton, silk, wool, tea, tobacco, rice and other agro-based industries
g) Fish/marine wastes	Use of fish industry wastes and fish rejections
h) Use of oil industry products	Use of non-edible oil cakes Use of oil cakes with pesticidal properties
i) Sewage farming	Use of city sewage water directly for farming Use of industrial sewage water for farming with pre-treatment, if necessary
2. Use of pre-digested / semi-digested manures	
a) Farm yard manure	Preparation of digested farm yard manure using crop residues, harvested wastes, weeds, agro-

	industry wastes and cattle dung and urine
	Application of biogas slurry to prepare farm yard manure
b) composting	Preparation of compost using garbage from rural and urban wastes
	Preparation of compost by using city sewage/sludge/ other city wastes
c) Other livestock wastes	Dung and urine of sheep, goat, pigs and other domesticated animals
d) Biomass conversion of un-conventional sources of nutrients	Slaughter house wastes Human excreta and urine
3. Bio-fertilizers	
a) N-fixing agents	Use of rhizobium, azotobacter, azospirillum bacterial culture
b) N-containing vegetation	Use of BGA (blue green algae) and azolla
c) P-solubilizing microbes	Use of P solubilizing fungi/bacteria
d) Vermiculture	Use of vermicompost prepared elsewhere
	Use of earthworms in the field to convert the biomass *in situ*
e) N-fixing crops and trees	Leguminous crops and leguminous trees
4. Cultural methods	
	Crop rotation with pulses for N-fixation
	Intercropping with pulses for N-fixation
	Minimum tillage for nutrient conservation
	Agro-forestry methods (alley cropping)
	Strip cropping and vegetative/live bunds
	Mulching
	Cover crops
FOR CONTROL OF PEST AND DISEASE	
a) Predators of pests	Encouraging the growth of predators like spiders and beetles in the cropping environment
	Suitable intercropping practices
b) Natural enemies	Rearing and releasing the natural enemies of crop pests (Ex. Trichogramma against boll worms)
	Spraying virus cultures known to be enemies of pests (Ex. NPV against spodoptera)
c) Cultural methods	Intercropping with trap crops (Ex. Bhendi in cotton)
	Proper choice of crop rotation to expose of the soil insects by deep cultivation
	Proper moisture conservation methods

d) Commercial use of plant products/extracts	Use of neem cakes/pongamia cakes Use of neem kernel oil Use of neem based pesticides
e) Local innovative techniques (differ from area to area)	Use of asafoetida for ant control Use of neem leaves against store grain pests Beating drum for bird scaring Use of puffed rice to attact birds, which eat away the germs/pests of the crop
f) Resistance built by breeding or bio-technology/ genetic engineering	Breeding of varieties/hybrids with multiple resistance against pest and diseases

The above list is neither exhaustive nor exclusive. Some of them may be conventional, unexploited/under exploited. Some more are unconventional but may have large potentialities for being used as organic sources. Any farmer or agricultural scientist can add to above list any material or any practice, which can either supply nutrients or control the pests/ diseases of the crop by non-chemical means. A deep rooted perception to make use of all the available organic sources at a given site or a given eco-system to either control the pests/disease or to supply nutrients can lead to successful organic farming. Many more materials can be added, if one can perceive the basic essentiality of avoiding the use of chemicals in all possible ways. Some disease/ pest/ weed may be obnoxious and known non-chemical method may not be effective to control them economically. Such weeds could be used to prepare manures if they are cut before flowering.

In order to develop a desirable set of organic farming practices, several materials can be used singly or in a mixture. Each one of these materials may differ in their mode of action, economic value, source of availability and effectiveness. But, all belong to a common group of materials used in organic farming.

The narrow meaning of the term `organic farming' - that it is mere application of farm yard manure is an over simplified expression in the context of broader principles and methods of organic farming. It will be a gross under-estimation of the vast potentiality of the available organic sources of nutrients, if farm yard manure is perceived to be the only material to replace the fertilizers. But, use of traditional sources such as farm yard manure deserves scrutiny so that the valuable nutrients are not lost. Any new method to make use of available traditional source more effectively without loss of the nutrients is also to be regarded as a part of organic farming. Some sources may be completely new and deserve exploitation. Some of them are not even perceived to be sources of nutrients. The urgent need to make the organic farming successful calls for a combination of traditional and new sources, which may be more effective in certain situations. Thus, organic farming is not restricted to the use of one specific material or practice. It is a combination of several materials and practices - best evolved to suit to the resource base for a given agro-eco-system. It is in this broad perspective, the concept of organic farming needs to be understood and propagated as an effective system of sustainable agriculture in the present millennium.

As a matter of fact, the use of organic manures was emphasized and recommended along with the fertilizers ever since the fertilizers were recommended. This is reflected in the report of National Commission of Agriculture (Anon., 1976). It was realized in the planning circles of Government of India even in early seventies that the use of organic manures has to be simultaneously followed along with the fertilizers. The document clearly emphasises that continuous use of fertilizers alone would fail to retain good physical properties of the soil, besides, reducing the efficiency of applied fertilizers. As such, the loss of organic matter is faster in Indian soils due to rapid decomposition by higher temperature even in unfertilized soils. The natural loss of organic matter in tropical climate coupled with failure to build up the good physical properties as a result of use of fertilizers has reduced the fertility of our soils, which were otherwise fertile. When organic manures are not added along with fertilizer, the situation further aggravates.

It only needs a sense of comprehensive imagination to realize that each one of the above named resources is large enough to supply sizable organic manure, provided their utilization is properly planned. Each one of the above has a scope to be converted into a specified skilled profession/trade/industry, if the aim of supplying the organic manures from such source is realized and practiced in large scale. A new genre of agro-industries, aiming to use these materials to manufacture large quantities of organic manures can be visualized and implemented at national level. For this, various organized/unorganized sectors involved in the handling and the use of such materials need to be vitalized. Multi-disciplinary approaches have to be adopted to deal with such problems, as they involves the role of engineers, administrators, bankers, researchers, extension workers and of course farmers. Large scale manufacture of organic manures on industrial scale making use of urban/rural wastes has been attempted on successful scale in many western countries such as USA, Germany and France. Already, a beginning is made in our country too to make use of city wastes in places like Pune, Bangalore, Delhi and Calcutta.

Such large scale attempts need to be tried in all cities to make the organic farming successful. Some Indian sugar industries and distilleries have made successful attempts to manufacture the manures based on the use of their discarded by products such as pressmud. They have also tried to make use of their sewage for profitable farming. Such attempts are necessary to be replicated in all other sectors-including many government sectors (to handle the city garbage and sewage). All the sectors of the society have to get tuned to use the available resources to make the organic farming successful. The success of organic farming, in fact, need a kind of social change both in rural and urban areas.

The fertilizers are preferred by the farmers as compared to organic manures because :

- The fertilizers are concentrated and easy to handle and transport as compared to organic manures, which are bulky and contain lesser nutrients on percentage basis.
- The fertilizers are cheaper (per unit kg of nutrients) as compared to organic manures
- They can be readily purchased from the market without any effort on the part of the farmer and his family (in terms of labour)
- The response to the fertilizers is always very quick, while the organic manures need continuous application for several years to achieve consistent yield improvement.

In early seventies, the farmers were reluctant to use fertilizers. But, some of the above named advantages attracted the farmers and fertilizer usage became a common practice in our country. Now, most of Indian farmers are fully convinced that the crop production is successful only with the use of fertilizers. However, the use of fertilizers is wrongly perceived to totally replace the use of organic manures instead of supplementing them. The use of organic manures has been gradually losing the ground due to various reasons. The organic farming is aimed to reverse this trend, to regain the self-sustainable capacity of the soil. The deleterious effects of agro-chemicals on the whole agro-eco-system in general and on various components of the soil medium in particular warrant the disuse of the fertilizers altogether. This needs thorough reorientation of the farming community towards the use of organic sources.

Although fertilizers are highly concentrated source of nutrients, they supply the nutrients in large quantities often not commensurate with gradual requirements of plant growth. The supply of the nutrients often exceeds the inherent capacities of the soil to store the nutrients for future use. For these reasons, most of applied fertilizers are wasted resulting into low nutrient use efficiency.

The farming community in general, has been tuned to the use of easily available and quickly responding fertilizers. The general reluctance to adopt the methods of organic farming is due to reasons such as additional labour requirement (to prepare the manures), low nutrient content, bulky nature of the manures and slow/gradual release of nutrients as compared to fertilizers. It needs a great deal of efforts to collect and supply both digested and digestible organic mateirals on the farm to make the organic farming successful. Both the farmers and administrators should realize that the organic farming is essential no matter how much labour is involved in making it successful. Similarly, the whole set up of the policy makers, adminstrators, researchers and extension workers should reorient itself to make the organic farming successful.

5.1 GREEN MANURES

Green manuring refers to incorportion of live biomass into the soil in order to supply plant nutrients. As the decomposition of incorporated biomass is faster - if green and succulent tissues are used, the practice attained a name called `green manuring'. The crops grown to be incorporated are called green manure crops.

Ideally, a green manure crop should be a fast growing, non-woody and short duration crop so that the practice of green manuring is practically feasible in a given cropping system. The benefits arising out of green manuring practice are limited by factors such as :

(a) Quantity of green matter incorporated and actual nutrients contained in them.

(b) Time taken for decomposition

(c) Favourable factors available in soil for decomposition

(d) Stage of incorporation

(e) Release pattern of nutrients

The time taken for decomposition and stage of incorporation will ultimately decide whether a green manure crop can be useful in a cropping system or not. Hence, short duration crops of 50 to 60 days are selected as green manure crop under multiple cropping system. Invariably, all the green manure crops are incorporated prior to flowering to ensure easy and fast decomposition. Most green manure crops flower in 30 to 45 days period.

Quantity of green matter incorporated and actual nutrients contained in them depends upon the species selected, plant population, stage of cutting, the fertility of soil to grow green manure crop and the season.

Time taken for decomposition depends on quantity of biomass incorporated and favourable factors available in soil at the time of incorporation. Usually, succulent green matter decomposes fully in a span of 30 to 50 days under favourable soil moisture condition.

Green manuring is an important practice of supplying plant nutrients in organic farming. Large number of green manure crops are identified to suit to various situations. Some of them along with their potential green biomass yield and possible nutrient contribution are presented in Table 5.2.

The practice of green manuring usually involves either

Table 5.2. List of green manure crops with their nutrient composition

S. N.	Green manure crop	Green manure yield (t/ha)	Nutrient content (%)			Nutrient contribution (kg/ha)		
			N	P_2O_5	K_2O	N	P_2O_5	K_2O
	ANNUAL LEGUMES							
1.	*Sesbania rostrata*	24.90	0.70	0.30	0.50	174.30	74.70	124.50
2.	*Cassia mimosoides*	4.70	0.80	0.20	0.50	37.60	9.40	23.50
3.	*Crotolaria juncea* (Sunhemp)	21.20	0.50	0.10	0.40	106.00	21.20	84.80
4.	*Cyamopsis tetragonoloba*	20.00	0.90	0.30	0.60	180.00	60.00	120.00
5.	*Sesbania aculeata* (Dhanchia)	20.20	1.10	0.20	0.40	222.00	40.40	80.80
6.	*Sesbania speciosa*	7.80	0.70	0.10	0.60	54.60	80.00	46.80
7.	*Vigna unguiculata*	10.00	0.70	0.10	0.50	70.00	10.00	50.00
8.	*Vigna radiata*	7.70	0.60	0.10	0.50	46.20	7.70	38.50
9.	*Vigna trilobus*	5.30	0.60	0.10	0.40	31.80	5.30	21.20
10.	*Vigna sinensis*	15.00	0.50	0.40	0.60	75.00	60.00	90.00
11.	*Berseen*	15.50	0.40	0.50	0.50	62.00	77.50	77.50
	PERENNIAL LEGUMES							
1.	*Cassia hirsute*	2.50	0.60	0.40	0.60	15.00	10.00	15.00
2.	*Desmodium orvoides*	1.40	0.80	0.20	0.90	11.20	2.80	12.60
3.	*Glyricidia maculata*	3.00	0.80	0.10	0.70	24.00	3.00	21.00
4.	*Sesbania punctata*	3.70	0.90	0.20	0.50	33.30	7.40	18.50
5.	*Pongamia pinnata*	3.00	0.20	0.20	0.30	6.00	6.00	24.00

Abrol and Palaniappan (1988); Dahama (2001)

(a) Growing the green manure crops and ploughing the entire crop into the soil generally at the time of flowering, when it is most succulent and easily decomposable, so that it contributes sufficient organic nutrients. It is generally referred to as *in situ* green manuring. This involves considerable time and cost to grow the green manure crop in the same piece of land, where the crop is prepared to be grown.

(b) Bringing in green leaves as well as succulent stem/branches of plants grown elsewhere (generally grown on bunds or near agricultural lands) and incorporating them in the soil either with or without chopping them. This is referred to as green leaf manuring. This practice is very useful and suitable for bringing in nutrients at no extra costs. However, cost of transportation may be prohibitive to the whole practice itself in some cases, where such materials are brought from long distances.

Many green manure crops are known for their capacity to fix the atmosphere nitrogen (e.g., Sesbania, Sunhemp and cowpea). Green manures are expected to bring an extra quantity of nitrogen (if they are leguminous) while they are grown in the same piece of land and are able to leave substantital nitrogen in their root nodules underneath the soil. However, it is now confirmed that additon of nutrients through their decomposition is greater than extra contribution of N. Many crop response studies have indicated that the ultimate increase in the crop yield is not due to extra N expected to fix as very little nitrogen can be expected to be fixed before six weeks (Joffee, 1955), beyond which the incorporation is not advisable. This was later confirmed by many scientists (Singh, 1965).

Large number of reports are available in the literature about the beneficial impact of green manuring on various soil characters. Some of them are listed here (Table 5.3).

Table 5.3. Beneficial effects of green manuring on crop yields and soil properties

Crop	Green manure	Yield increase	Soil characters improved	Refereces
Rice	*Sesbania aculeata*	–	Soil aggregatesd increased by 62 per cent. Reduction in bulk density, incr-eased infiltration rate	Boparai *et al.*, 1992
Rice	*Sesbania aculeata*	20 per cent NPK, 100% over control	–	Setty and Channa Basavana, 1991

Rice	*Sesbania rostrata*	120 per cent over control	–	Kalidurai and Kannaiyan, 1991
Rice	*Crotolaria juicea*	–	Increased uptake of NPK from inorganic applied fertilizer	Balasubramaniyan *et al.*, 1991
–	*Vetch-oat mix*	–	Increased respiration, CEC and sum of exch. basic cations.	Hanes and Mucha, 1991
–	*Cowpea*	–	Labile Prelease improved	Hundal *et al.*, 1991
Rice	*Sesbania aculeata*	–	Increased number of bacteria, fungi and actinomycetes	Bhattacharya *et al.*, 1990
Capsicum	*Dolichos lablab*	Increased fruit yield	–	Cancio *et al.*, 1991
Rice	Glyricidia	–	Uptake of NPK improved	Vageesh *et al.*, 1990
–	Not specified	–	Humic acid and humic fractions increased by 2.2 and 7.1%. Free humus increased by 8.6%, reduced B.D., increased water stable aggregates and capillarity	Xiong and Cheng, 1991
Maize	*Glyricidia sepium, Sesbania sesan*	Increased grain yield by 2.5 per cent	–	Gutheridge, 1992
Soybean wheat	*Crotolaria juncea*	–	Increased availability of P	Magalhaes *et al.*, 1992

Benefits of green manuring may be many. But, many practical consideration limit their usage on wide area. Some of such limitations are:

(a) It can not be practiced in intensive agricultural systems, as green manure crop is not expected to occupy the land at the cost of profitable crop in such systems. In countries, where large holdings are prevalent and where there is no much pressure on the land to grow more number of crops, green manuring is a very useful practice to enrich the soil and successfully switch over to organic farming system. Despite the great pressure on Indian cultivated land, attempts are made in intensively cropped irrigated areas of our country to introduce a green manure crop in between double crop sequences in a year. Following are some examples.

Rice - Green manure - rice

Green manure - cotton

Rice- rice - green manure

(b) It cannot be a successful proposal in arid and semi-arid area, as limited rainy period is more gainfully used to grow profitable crops than green manure crops. Wherever the rainfall is ideal to grow two crops in succession, green manuring may be practiced. But, the unpredictable rainfall may limit the actual decomposition of incorporated green biomass. Practice of green manuring is more prevalent in rice, as the it can be incorporated and puddled in standing water to facilitate easy and faster decomposition.

(c) The cropping system involving green manure will generally end up in low short term gains - as cost of raising green manure crop is not realized immediately by corresponding increase in the production of main crop. Many farmers do not include green manure in their cropping system for this reason.

(d) In green leaf manuring, the succulent plant material has to be transported to the site of utilization. If the distance involved is more, the cost of transportation becomes prohibitive and in such situation green leaf manuring may be more unremunerative. Hence, practice of green leaf manuring is restricted to leaves/succulent stem of plants grown around the cropped region.

(e) In irrigated paddy, incorporation of green manure is expected to release methane in first 15 to 20 days. Planting the crop during this period may damage the crop.

(f) The release of nutrients from green manure incorporation in arable lands is slow and the benefit of green manuring may be available to succeding crops. Hence the immediate benefit to the crop - wich receives the green manure - may be to a limited extent. However, under puddled conditions, *in situ* green manuring has been found to release the nutrients within a couple of days after incorporation.

In the broder perspective of organic farming, green manuring still holds excellent promise in some situations such as :

(a) Where green manure crop can be simultaneously grown as intercrop along with the main crop without offering any competition to the main crop. Such systems can offer advantages of saving the extra time required to grow the green manure crop. Some such examples are:

- Sugarcane + french bean
- Cotton + cowpea, cotton + alfalfa
- Sugarcane + soybean
- Cotton + Sunnhemp

(b) Where 50 to 60 days period of water availability can not be used to grow any profitable crop, it is wise to grow green manure crop and incorporate after 4 to 5 weeks so that available water source is effectively utilized.

(c) Where the green leaf manure crop can be grown on bunds and/or in the vicinity of agricultural land and does not involve heavy cost of transportation.

(d) In the plantation crops with lower plant population per hectare (ex. Coconut, arecanut, cashew, mango, sapota, guava, apple etc.) the green manure crops can be grown around the basin and incorporated.

When green manuring is practiced, it may not be able to completely satisfy the nutrient demand of the main crop. It may be regarded as a concomitant practice along with other organic farming practices. Although no estimate has been made to indicate the area practicing green manuring, it may be projected that atleast 10 to 15 per cent of irrigated area follows this practice for better utilization of resources at the farm. Large part of this area is used to grow irrigated paddy crop.

5.2 FARM YARD MANURE

The term `farm yard manure' is a expression to signify any manure prepared in the backyard using the farm waste, cattle dung and urine. Since very early ages, its use is prevalent in our agriculture. In fact, only 50 years back farm yard manure was the only source of nutrients supplied to crops. Conventionally, FYM is prepared by spreading cattle dung, water and crop residues in a specially made pit at the backyard of every Indian household in the villages. Often, many house hold wastes are also added with farm wastes. The residues are allowed to decompose with cattle dung for a period of 6 to 8 months. The resultant manure is transported to cultivated land. The level of decomposition varies widely by number of factors. In practical rural situations, semi-decomposed and sometimes even undecomposed residues are also carried along with decomposed manure. Many variations of preparing FYM are found based on convenience, availability of organic materials, time available for decomposition.

Although, very little scientific attention was paid till 1970's to scientifically study the method of preparing FYM and possibility of improving the quality of decomposition, three principal methods of preparing FYM are standardized on scientific lines in the last 15 to 20 years.

1.	Open pit method	Allowing semi-aerobic decomposition of farm wastes in underground pit. This method generally requires turning over the decomposing material frequently
2.	Sealed pit method	Allowing anaerobic decomposition of farm wastes in underground pit. This requires sealing the surface of the pit with dung slurry
3.	Japanese method (vat method)	Allowing aerobic decomposition of farm wastes in an above ground structure of stones/bricks or thatched walls upto height of 1 meter

While the methods of spreading the cattle dung, farm wastes, cattle urine and water layer by layer remains similar for all the methods, the varying feature in these methods is the way in which decomposition takes place. In open pit method partial aerobic condition is provided as the top part of pit is allowed to be in contact with atmospheric air. A couple of turnings are given in this method which helps in achieving better decomposition. The time taken for decomposition is around 4 to 5 months.

In sealed pit method, anaerobic conditions are provided and the decomposition is delayed upto 6 to 8 months. But in vat method, complete aerobic decomposition is allowed and it takes 2 to 3 months.

Under anaerobic conditions, the decomposition results into manure rich in NH_4 - N but poor in NO_3 - N due to limited supply of oxygen. Population of nitrifying bacteria is less. Manure rich in NH_4 - N is desirable for agricultural fields as such nitrogen is readily available for exchange in the soil (cation exchange) particles and thus not lost for longer period.

Under aerobic conditions and semi-aerobic conditions more nitrification is allowed due to adequate supply of oxygen and large population of nitrifying bacteria. Manure generated by aerobic and semi-aerobic conditions is rich in NO_3 - N and poor in NH_4 - N. NO_3 - N is likely to be lost quickly by leaching and volatalization, although it can readily improve the available N status immediately after application.

Advantages and disadvantages of sealed pit method and vat method are tabulated in Table 5.4.

Table 5.4. Advantages and disadvantages of pit system and vat system

Advantages	**Disadvantages**
SEALED PIT SYSTEM (anaerobic and semi-aerobic)	
1. More NH_4-N is available in manure. It is less likely to be leached.	1. Time taken to complete decomposition is more (6 to 8 months)
2. No turning is needed. Labour requirement is less	2. The rain water can enter the pits and can wash way nutrients, as the bed has no impervious cement structure
3. The temperature often reach upto 65°C at the time of decomposition. This helps to kill the weed seeds	3. The underground water may create conditions not suitable for decomposition (when pit is in low lying situation)
VAT METHOD (aerobic)	
1. Time taken for decomposition is as less as 3 to 4 months	1. Manure contains more NO_3 - N fraction than NH_4-N due to adequate O_2 supply. NO_3-N is likely to be lost quickly
2. As the base of vat is stone or masonry structure, nutrients are not lost	2. Requires frequent turning involving additional labour cost
3. For faster decomposition earthworms could be released in vats	3. Initial cost is high- as farmer has to erect above ground structure using stones
	4. Temperature seldom reaches beyond 25 to 30°C. Hence, weed seeds are not killed.

In all the methods of manure preparation, a starter bacterial innoculum in the form of cattle dung is essential. Cattle dung contains dense bacterial population with mucus and intestinal worn out cells. The carbonaceous residues serve as substrate for their activity and as a result, the cellulose and hemi-cellulose substances are broken into simple substances. The crop residues and other form wastes get converted into granular pieces releasing large quantity of nutrients. As all the components of carbonaceous substrata are broken down, the decomposed manure contains both macro and micro nutrients. As the process of release of nutrients proceed even after the application of manures in the field, manurial benefits are derived over longer period of time in contrast to quick availability of large quantity of nutrients when fertilizers are applied.

Method of preparation of farm yard manure is standardized as follows

(a) The recommended pit size is 4′-5′ width x 3′-4′ depth and 15′-20′ length.

(b) FYM pits have to be located in a well drained area with no high water table.

(c) Suitable sealing materials to seal the pits (in anaerobic method) are dung slurry, polythene sheet.

(d) During rainy period, the rain water should be not allowed to stagnate around cr in the pit.

(e) Crop residues and dung are spread layer by layer with intermittent sprinkling of water at the time of filling pit.

(f) Vat system will have vat size of 1 m width, 1 m height and convenient length (5-7 m).

(g) The vats will invariably have stone/masonry structure at the bottom.

(h) Stone slabs are perpendicularly erected in such a way that space in between the wall provides sufficient aeration but not allow the manure to drop out.

(i) In va system, the turnings are necessary once in 10 to 15 days.

(j) The manure taken out of pit or var should be spread as early as possible and immediately harrowed in the field to utilize the maximum quantity of nutrients in manure.

Recently, some new practices are added in preparation of FYM. They are :

(a) Biogas slurry, which is rich is nutrients, is diverted into the pit along with other residues.

(b) Washings and urine from cattle byre are diverted directly to the pit.

(c) Ideally, the cattle urine can be collected in soaking medium (saw dust or straw) and later used for manuring.

(d) Superphosphate is added in between the layers of dung and crop residues to prepare P-rich manure (phospho compost).

(e) Micro-nutrient enrichment of FYM is tried by using various micro-nutrient fertilizers.

(f) Trichoderma (bacterial culture) is added to FYM pit in between layers of crop residues to improve/hasten the rate of decomposition.

The decomposability of plant residues and farm wastes essentially depends upon C : N ratio. It should be ideally between 1 : 15 - 1 : 20 for good decomposition. If substrate has high carbon status (resulting into a way C : N ratio of 1 : 50 / 1 : 60), the decomposition is slower, requiring additional sprinkling of urea between layers.

On an average, the FYM contains 0.5 per cent N, 0.1 per cent P_2O_5 and 0.5 per cent K_2O. But, some workers have reported total N in FYM as high as 1.65 per cent (Flynn *et al.*, 1999).

5.2.1 Response to applied FYM

The benefits derived by application of FYM are reported ever since the documentation on experimental results started. Farm yard manure application can result into chain of effects in soil besides increasing the crop yield. The response to applied FYM could be studied in five groups :

(a) Improved crop yields

(b) Improved physical properties of soil

(c) Improved nutrient availability

(d) Improved microbial activity

(e) Residual benefits to succeeding crop

5.2.1.1 Improved crop yields

Application of FYM is reported to have improved crop yields of almost all crops. An extensive documentation on the beneficial effects of FYM on crop yields is available from 1970s in the form of consolidated reports of All India Co-ordinated Agronomic Research Project. These reports are compiled based on data generated on diverse ecosystems of all the states of our country. They are summarized in Table 5.5 (Anon., 1969-70; Anon., 1970-71; Anon., 1971-72; Anon., 1972-73; Anon., 1973-74; Anon., 1974-75; Anon., 1975-76 and Anon., 1976-77).

Table 5.5. Response to applied FYM in different crops

Crop	No. of states	No. of experiments	Response (kg/ha)
Paddy	11	341	168
Wheat (irrigated)	8	210	202
Wheat (dry)	4	71	85
Sugarcane	8	258	7.3 ton
Cotton (irrigated)	4	71	56 (lint)
Cotton (dry)	5	294	16 (lint)
Potato	5	26	4124
Sweet potato	2	9	1920
Tapioca	1	3	455

Onion	1	2	4193
Garlic	1	1	508
Chillies	2	16	339
Bhendi	1	2	296
Tomato	1	3	520
Brinjal	2	5	427

In addition, large number of reports about beneficial effects of FYM on crop yields are available in the recent past. Recent reports are briefly summarized in Table 5.6.

Table 5.6. Response to applied FYM in different crops

S.N.	Crop	Beneficial effect on yield	Reference
1.	Rice-wheat	8.8 q/ha - increase over control	Singh *et al.*, 2000a
2.	Potato	7.2 q/ha - yield decline halted by FYM application as compared to inorganic source	Singh *et al.*, 2000b
3.	Groundnut	Increase of yield by 100 per cent as compared to control and 5 per cent as compared to 100 per cent NPK (23 yrs)	Vaishnav and Dixit, 1998
4.	Rice	Increase in grain yield by 0.98 t/ha compared to control	Singh *et al.*, 1999
5.	Oat	Increased yield by 6.3 tons DM/ha as compared to control	Tripathi *et al.*, 1991
6.	Cotton	Kapas yield increased upto 7.4 q/ha	Sagare and Guhe, 1991
7.	Sugarcane	Cane yield increased by 25.4 to 30.9 t/ha	Yadav and Prasad, 1992
8.	Phaseolus. soybean, wheat, potato	Increased root activity, increased root weight by application of FYM	Mitra, 1992
9.	Maize	Increased grain yield	Gaur and Mathur, 1979

Long term experiments on red sandy loam soils at Coimbatore between 1900 to 1972 have indicated that application of 12.55 t/ha of farm yard manure every year has increased the yields of jowar by 320 kg/ha, of ragi by 313 kg/ha of Indian millet by 256 kg/ha and cotton by 221 kg/ha respectively (Krishnamoorthy and Ravikumar, 1973).

Continuous application of farm yard manure at 10 t/ha was responsible to increase the yield of finger millet by 1447 kg/ha over absolute control and 152 kg/ha over continuous application of NPK alone (Anon., 1996). Out of 18 years of experimentation at CRIDA, application of 10 t/ha of farm yard manure increased crop yields over absolute control and improved the yield stability index from 0.12 to 0.42 in 15 years (Singh *et al.*, 1990).

5.2.1.2 Improved soil physical properties

Application of FYM is of greater significance for sustainability as it has great potentiality to improve the physical properties of soil besides supplying nutrients. Each of these physical properties has large practical implications in maintaining soil as a medium of production and great role in halting environmental degradation. Maintaining them and improving them in long run is essential part of sustaining the eco-system. Some physical properties of soils improved by application of FYM are cited in Table 5.7.

Table 5.7. Effect of FYM on physical properties of soil

Crop	Soil physical property improved	Reference
Rice-wheat	B.D. reduced to 1.32 as compared to 1.45 in control. Porosity increased by 5 per cent as compared to control, WHC increased by 5 per cent as compared to control	Singh *et al.*, 2000a
	Microporosity and total porosity, increased available water capacity and volumetric moisture content increased (Rothamsted)	Rose, 1990
Sugarcane	Organic matter status increased from 0.346 to 0.363 per cent	Yadav and Prasad, 1992
Coconut	Organic matter increased by 45 per cent and water holding capacity increased	Tennakoon, 1990
	Total carbaon, total N increased in coarse sand (>0.2 dia) fractions than fine sand (0.02-0.2 mm dia) fraction Micro-aggregates (>0.25 mm) increased	Aoyama and Tenimai, 1992
	Organic carbon increased from 1.8 to 2.2 per cent by FYM application	Goyal *et al.*, 1993
Rice-wheat	Cumulative infiltration rate increased by 5.66 to 7.91 cm/ m, increased CEC from 17.18 to 20.74 meq/100 g and decreased B.D. from 1.66 to 1.45 g/cc	Ganal and Singh, 1988
	Aggregation improved by 50.79 per cent, bulk density reduced from 1.5 to 1.42 g/c, porosity increased by 35 per cent	Rajput and Shastry, 1988
Potato	Increased soil aggregates, improved organic matter status	Omar, 1990
	FYM acted as strong buffering agent	Whalen *et al.*, 1999
Potato	Annual application of 5.6 Mg/ha improved the potato yield	Westernman *et al.*, 1999
Triticale	Soil organic matter, organic N, NO_3 - N, EC increased with FYM application	Hao *et al.*, 1999

5.2.1.3 Improved nutrient availability

The concept of improving the fertility of soil does not mean creating of flux of available nutrients, as is done by applying inorganic fertilizers. Sudden increase in available nutrient status in the soil may improve the uptake by plant roots immediately. But providing such a situation in soil results into :

(a) Loss of unutilized nutrients forming the basis of reduced nutrient use efficiency.

(b) Imbalance in availability of other nutrients (Ex. mutual antagonistic effects of N-K or Zn-P or synergestic effects of N-K).

(c) Fixation of unutilized nutrients into unavailable forms

(d) Decreased pH.

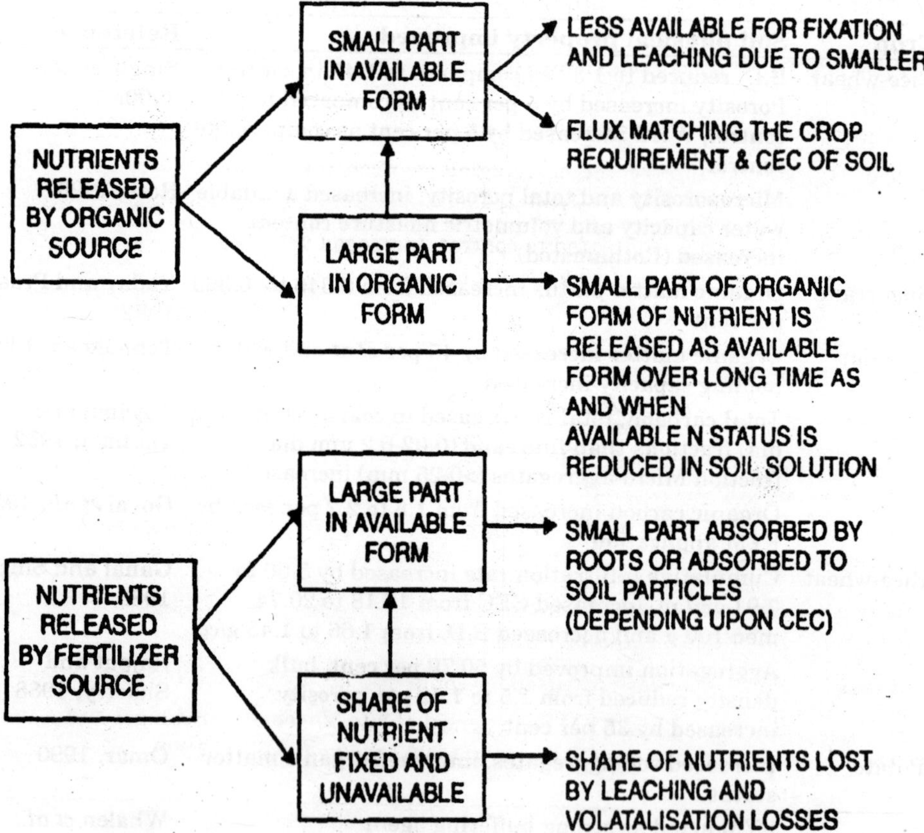

Fig. 5.1. Nutrient availability by organic/fertilizer source

Further, a flux of available nutrients at any one time of crop period does not match with crop requirement at all, as a growing crop can absorb the available nutrients gradually in small quantites throughout the crop growth. In this perspective, FYM or any organic manure is more suitable source of supplying available nutrients for a long period in a clear contrast with large amount of available nutrients being made available to crop by applying fertilizers.

A contrasting study on the release and availability patterns of nutrients by an organic source and that of nutrients by a fertilizer source is made to appreciate the benefits from organic source (Fig. 5.1).

Nutrient availability by applying FYM is improved, for a long period at a gradual rate. At any given point of time, available form of nutrient may be less. But, long period of making the nutrients in available form is tailor made for limited CEC of soil (for adsorption of cations) as well as to limited capacity of roots to absorb as they grow.

Large number of reports are available in the literature to indicate improved availability of macro and micro-nutrients by application of FYM. They are briefly summarized in Table 5.8.

Table 5.8. Nutrient availability by applied FYM

Crop	Soil physical property improved	Reference
Rice	N uptake increased by 22 kg/ha over control	Singh *et al.*, 1999a
Wheat	Increased uptake of N by 14 kg, P_2O_5 by 35 kg and 19 kg/ha as compared to control. Increased uptake of 7 kg P_2O_5 and 12 kg K-$_2$O/ ha over recommended inorganic fertilizer	Singh *et al.*, 1999b
Ragi	N uptake increased and increased av P status	Geethakumari & Shivashankar, 1991
	Addition of FYM increased transformation of P (from superphosphate and rock phosphate) to available P	Tripathi and Minhas, 1991
Oat	Uptak of N & P in oats was high when FYM was applied	Tripathi *et al.*, 1991
Wheat	Adverse impact of antagonistic interaction of Zn-P is mitigated by FYM application	Golakiya *et al.*, 1990
Beans	Zn uptake was increased by application of FYM	Allen and Jolley, 1999
–	Sites receiving FYM for a period of 3 to 30 years showed increased P levels upto 200 mg/kg in soil surface (0-15 cm)	Schlegel and Alam, 1999

5.2.1.4 Improved microbial activity

Farmyard manure is a product of microbial activity and contains large number of microbial population. Application of farm yard manure can increase the microbial activity in the soil both by activating the microbial action and by aiding the multiplication of microbial population.

Due to these properties, application of farm yard manure is in perfect tune with biological requirement of soil and helps to build the soil on sustainable basis. The capacity of soil to release, store and supply the plant nutrients is based on this microbial activity of soil.

The microbial activity in the soil is responsible for many processes like ammonification, nitrification, denitrification, nitrogen fixation, solubilising P as well as decomposition of carbonaceous materials. Success of decomposition of various decompsable materials like crop residues, dead roots, exudates from living roots, as well as dead micro-organisms depends on the soil microbial activity. The labile component within rhizosphere (or close to surface of roots) serve as ready source of energy rich C-compounds necessary to support high level of microbial activity (Clarholm, 1994).

Grazing of various bacteria by protozons and nematodes are important sources of NH_4 - N, P and other plant nutrients (Clarholm, 1985 and Ellito *et al.,* 1979). Protozoan predation improves release of organic carbon and nitrogen in presence of plant roots (Kuikman *et al.,* 1990). Predator-prey interactions among rhizosphere micro-organisms play key role in plant nutrient dynamics.

For all the above reasons, soil microbial activity is of utmost importance in release of plant nutrients and sustaining the fertility. Application of farm yard manure greatly catalyses the biological activity in soil. Therefore, has universal and time tested importance.

Large number of reports are available in the literature to show the enhanced microbial activity by application of farm yard manure. Increased organic carbon by application of FYM has been reported to help in increased population of bacteri, actinomycetes and fungi (Gaur *et al.,* 1990) in almost all types of soils (Table 5.10).

In addition to this, many research reports are available to show that application of FYM can increase microbial activity. Some of them are briefly summarized in Table 5.9.

Table 5.9. Effect of FYM on microbial activity in soil

Crop	Soil physical property improved	Reference
–	FYM treated soils had higher population of Azotobacter and Rhizobium	Tippannavar *et al.,* 1990
–	Increased number of bacteria, fungi and soil invertebrates (74 years of study)	Hickisch and Muller, 1990

–	FYM could restore positive N balance after mineralization and loss of N	Gill *et al.*, 1991
Coconut	FYM application increased microbial activity (Srilankan study)	Tennakoon, 1990
–	Soil microbial biomass increased (23 years of study at Hisar). Soil respiratory and enzymatic activity increased	Goyal *et al.*, 1993
–	Bacteria, fungi, invertebrates increased FYM application	Hickisch and Mulla, 1990
Potato, wheat	Increased microflora	Mittal, 1992
Bean	Manure application increased mycorrhizal colonization	Allen and Jolley, 1999

Table 5.10. Impact of organic carbon status on microbial population

Soils	Bacterial population (10^6/g)			Actinomycetes (10^6/g)			Fungal population (10^4/g)		
	1	2	3	1	2	3	1	2	3
Hilly soil (HP)	28.6	26.7	52.9	8.9	11.2	20.8	11.6	14.5	17.6
Alkali soils (Delhi)	3.4	8.8	29.6	0.7	2.6	16.3	4.6	6.3	5.2
Heavy clay (Patna)	11.7	16.5	17.6	2.3	3.1	3.5	12.4	13.5	13.7
Acid red loam (Kanke)	5.1	20.5	23.5	1.5	1.8	2.2	10.8	12.4	12.8
Siernozem (Hisar)	31.1	51.3	73.5	13.5	37.4	55.8	28.7	25.5	25.0
Alkali (Rajasthan)	0.4	0.5	1.1	16.0	15.5	30.4	27.2	39.0	48.2
Alluvial (Cuttack)	32.1	74.8	141.2	3.2	9.7	16.1	1.7	1.8	1.8
Red soil (Hebbal)	16.4	20.5	39.2	1.6	1.1	1.6	6.7	4.1	4.9
Medium black	3.9	2.5	2.5	3.3	4.3	3.6	13.3	1.7	13.3
Acid laterite	5.2	4.6	3.1	8.4	6.0	9.4	18.1	14.4	14.5

Source : Gaur *et al.*, 1990 Note : 1 - Control; 2 - 0-5% C soil; 3 - 2% C soil

5.2.1.5 Residual effect of farm yard manure

Like many other organic manures, farm yard manure leaves behind definite beneficial residual effects to the succeeding crop. The of total nutrients entering labile form in subsequent years beneficial residual effects are derived by :

(a) Part of total nutrients entering labile form in subsequent years.

(b) Improved physical properties help in soil enrichment as well as improved physical properties of soil which ultimately help in crop production in many ways.

Improved yield of succeding crop by FYM application to preceding crop was reported from as early as 1932 at Pusa in the experiments involving crop rotation. Fixed crop rotation experiments conducted by All India Co-ordinated Agronomic Research Project at various stations to study the residual effects of applied FYM on succeding crops are summarized in Table 5.11 (Anon., 1969-70; Anon., 1970-71; Anon., 1971-72; Anon., 1972-73; Anon., 1973-74; Anon., 1974-75; Anon., 1975-76 and Anon., 1976-77).

Table 5.11. Residual response of application of farm yard manure in various cropping system

States	Rotation	Kharif application only (response in kg/ha)	Rabi application only (Response in kg/ha)
U.P.	Rice-wheat		
M.P.	Rice	52-681	63-902
West Bengal	Wheat	158-590	152-887
A.P.	Rice-Rice		
Karnataka	Rice	151-1844	87-1044
Kerala	Rice	62-720	133-1521
Orissa			
Punjab	Maize-Wheat		
Jammu &	Maize	38-1358	188-2414
Kashmir	Wheat	437-2553	508-2571
M.P.	Jowar-Wheat		
Karnataka	Jowar	186-1821	267-1560
	Wheat	420-779	290-2130
Haryana	Bajra-Wheat		
Rajasthan	Bajra	114-288	43-197
Gujrat	Wheat	247-415	240-340
Tamil Nadu	Jowar-bajra		
	Jowar	297-1069	488-1366
	Bajra	358-526	657-1890

Many other workers have reported recently about beneficial residual effect of FYM application. Singh *et al.* (1999b) reported that N uptake was increased in succeeding year after application of FYM as compared to fertilizing through inorganic sources. Ramsingh *et al.* (1985) found that paddy yields were increased by 6 q/ha as a residual effect of FYM applied to previous crop. Nimje and Seth (1988) found the residual beneficial effect of FYM applied to soybean crop on succeeding crop of maize.

Although use of FYM offers a great potentiality to practice organic farming and several types of biomasses can be used to prepare FYM, some questions of feasibilities of using FYM on a large scale are frequently encountered. They are :

(a) Whether FYM can satisfy the huge nutrient demand of the crops?
(b) Whether it alone can satisfy the necessities of organic farming?
(c) Whether it is possible to supply sufficient FYM to the entire cropped area?

Whether FYM can satisfy the huge nutrient demands of the crop

High yielding varieties have very large nutrient demands. Apprehensions are always made whether organic sources can satisfy them. Before examining whether fertilizers have capability to satisfy the nutrient demands vis-a-vis organic sources, it is advantageous to know the manner in which nutrients are made available in soil when fertilizers/organic sources are applied.

Farm yard manure, on an average containing 0.5 per cent N, 0.1 per cent P_2O_5 and 0.5 per cent K_2O is not expected to replace the nutrient rich fertilizers. It is also not designed to satisfy the huge nutrient needs in one stroke. The nutrient demand of any crop for nutrients is also not sudden. The demand for nutrients is gradual. Fertilizers create sudden glut in the nutrient concentration in the vicinity of site of application immediately after the application and the crop is not able to absorb so much nutrients at that time. This invariably leads to loss of nutrients. This phenomenon applies, in general, to all the fertilizers, although minor variation do occur due to factors such as cation adsorption, fixation, microbial transformation etc. either to enhance or reduce the nutrient availability in narrow ranges. Whereas, by supplying the nutrients through FYM, the loss of nutrients are minimized simultaneously maintaining the supplies through out the cropping season. This is well represented in a study of the nutrient demands and supplies (by fertilizers and FYM) by hybrdi cotton in Fig. 5.2 (Joshi, 1999). Decomposing plant materials releasing

N at slower rates promotes a greater N uptake and thus a better synchrony between N availability and crop demand (Cobo *et al.*, 1999).

It is erroneous to simply presume that the nutrients supplied through the fertilizers are all made available to the crop. Major portion of the applied nutrients is prone to the losses. After 4 to 6 weeks after application, the crop has to depend upon the nutrients either fixed and later released to the soil solution (as in case of potassium and phosphorus) or adsorbed cation to the soil particles (as in case of ammonium or potassium ions). These subsidiary sources, will in turn not satisfy the needs of the crops either. In any case, it can not be argued that FYM will be able to satisfy the entire needs of the crop at all stages. But, the fertilizers also have no capacity to satisfy the nutrient needs at all stages of the crop. FYM, if applied over a period of time would improve the continuous supplies of the nutrients, as it helps in building up the natural capacity to self support the nutrient requirement by continuous microbial action. Moreover, the supplies through the FYM is mostly in the form of organic nutrients commensurate with the mineralizing capacity of the soil, whereas, the sudden supply of nutrients through the fertilizers, floods the soil medium by nutrient ions much beyond the capacity of microflora to immobilize into organic forms or capacity of the plants to absorb through various routes of mass flow or even interception by cation exchange capacity of soil. For this reason, the practice of applying the fertilizers is not in tune with the gradually increasing demand of the crops. The practice of splitting nitrogen dose is introduced considering these aspects. Despite of splitting the N dose, flooding the soil medium with nutrients and resultant losses cannot be fully avoided.

Whether FYM alone can satisfy the nutrient necessities of the crop

By virtue of its low nutrient composition, it can not be stated that application of FYM alone can satisfy all the nutrient needs of the crops, although some crops such as pulses, minor millets and some vegetables can be grown with the supply of FYM alone without hampering the yield levels. Application of FYM can go as a component organic farming practice along with the other practices so as to satisfy the total demand for nutrients. When the organic farming practices are combined to satisfy the nutrient demands, it must be borne in the mind that many organic farming practices (including the application of FYM) have residual effects. The practice of budgeting the nutrients on annual basis has no meaning in organic farming as the budgeting has to be done on the basis of total cycle, duly considering the residual effects of succeeding crops.

Whether FYM can be supplied to all the cropped area

It is general experience in rural areas that digested FYM is not available in adequate quantity. In this background, most pertinent question is whether digested FYM can be supplied to all cropped area - if organic farming is to be practiced. The cattle heads in India are increasing (Table 5.12) and biomass production/straw production is also increading in the last 20 to 30 years. Then, why should the country be short of digested FYM?

On the question of supplying the digested FYM to all the cropped areas, it should be clear that :

(a) The full potential of the materials available to prepare the FYM is not exploited, due to wastage of materials.

(b) There is general slackness on the part of the farmers to adopt the elaborate process of preparing the FYM, when concentrated fertilizers are readily available in the market at subsidied rates.

5.2.2 Essential materials to prepare FYM

Conventional sources : Crop residues, cattle dung, farm wastes, harvesting wastes, cattle urine and house hold wastes.

Unconventional sources : Weeds, agro-based industrial wastes, poultry waste, forest litter, other livestock wastes, marine/fisheries waste and animal carcasses.

Preparation and use of FYM was a part of our agriculture even well before the fertilizers were known to us. But, the crops and their traditional varieties grown earlier demanded lesser nutrients. Hence, even limited application of FYM was sufficient. In a changed scenario of high yielding varieties and more intensive agriculture, the practice of preparing the use of FYM needs to be in different mode.

Although large number of organic wastes can be used to prepare FYM, most of them remain unutilized/ under utilized. Apart from intensive use of conventional sources like crop residues, harvesting wastes and household wastes, large number of unconventional sources such as weeds, agro-based industrial wastes, poultry wastes, other animal wastes etc. need to be fully exploited. Some any traditional sources (like cattle urine) are also not intensively used.

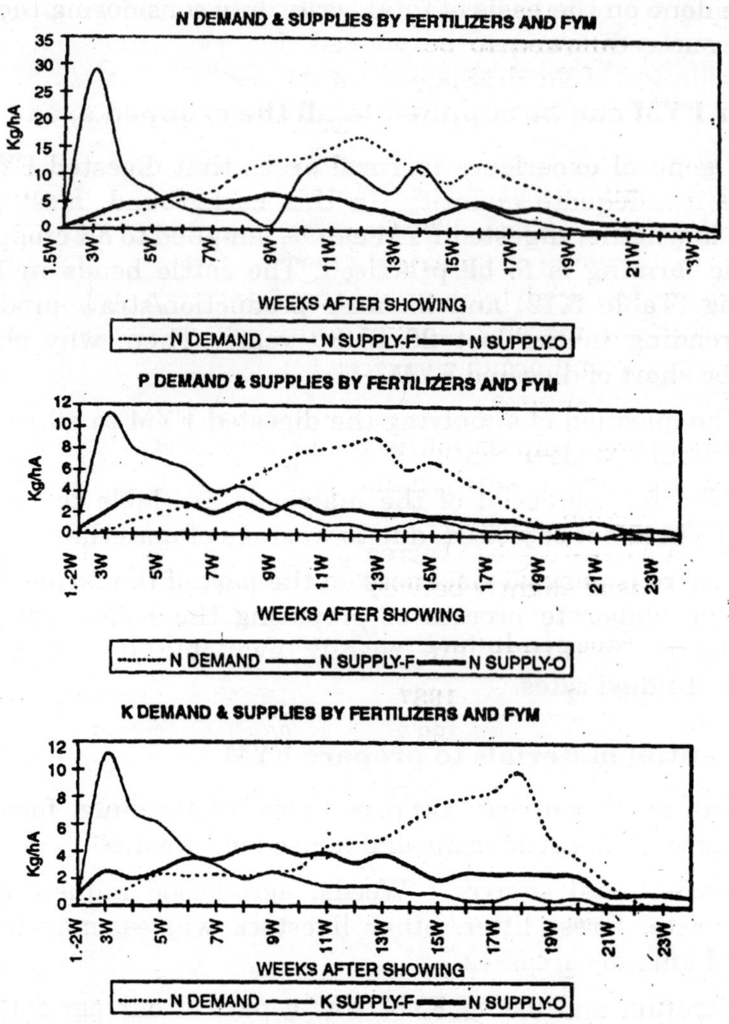

Fig. 5.2. Nutrient demand and supplies by fertilizers and FYM

5.2.3 Attitudinal change towards FYM

In order to make use of all possible organic sources, a basic change in the attitude of our farmers is needed. Change is needed in respect of following issues.

They have to realize that dung, crop residues, wastes as well as many unconventional sources of organic materials are, in fact a natural wealth and they need to be carefully used without waste. A large quantity of dung is not able to be collected by farmers because, more than 90 per cent Indian cattle graze outside sheds for more than 7-8 hours.

Valuable natural resources like cattle urine and washing wastes should never be allowed to be wasted. For this, they may have to modify the design of the drainage from their sheds.

It may be logically easy for the farmers to avoid labour consuming elaborate steps to prepare the manures and instead, purchase and concentrated fertilizers at cheaper rates. But, organic manure is not an equivalent to fertilizers. The benefits of organic manure are not restricted to supply of nutrients only. Organic manure is a comprehensive material to improve the soil environment in a more holistic way than merely supplying nutrients. Hence, comparison between fertilizer and organic sources can not be drawn for a limited purpose of nutrient supply.

General wrong impression is publicly perceived that number of cattle in our country is reducing and FYM production has come down due to such reduction. As a matter of fact, the number of cattle heads as well as other livestock has increased in our country (Table 5.12) till 2007, and decreased slightly between 2007 to 2012.

Table 5.12. Livestock in India (million)

	1982	1987	1992	2007	2012
Cattle	192.4	199.7	204.5	199	190
Buffalo	69.8	75.9	83.5	105	108
Sheep	48.7	45.7	50.7	68	61
Goat	95.2	110.2	115.2	140	135
Horses	0.9	0.8	0.8	0.6	0.6
Pigs	10.07	10.06	12.79	11.0	10.0
Other animals	2.32	2.22	2.42	2.43	2.41
Total	419.6	445.3	470.2	526.0	512.4

Source: 19th Livestock Census - 2012. All India Report

What is lacking in rural community is willingess to associate with FYM preparation and not the shortage of dung or crop residues. There is not other reason why FYM is not available in adequate quantity, when quantity of dung and crop residues have substantially increased in the last 25 to 30 years.

Consumption of crop residues by cattle is indeed not a deprivation of biomss for manure preparation. If the dung and urine from cattle are effectively collected, nearly 70 to 80 per cent of energy fed to as crop residues can be recovered back.

As much as 80 to 90 per cent of N, 60 to 70 per cent of P and 60 to 75 per cent of K contained in the crop residues fed to cattle are excreted by them in the form of dung and urine. In case of milch cows, these figures are in the range of 70 to 90 per cent, 50 to 59 per cent and 61 to 65 per cent, respectively (Gaur *et al.,* 1990). In this perspective, dung and urine from cattle are efficiently converted energy from crop residues. As Indian Cattle population is one of the highest as compared to other countries - it is ridiculous to allow such a huge resource of nutrients to go waste. An estimated potentiality of dung and urine availability from Indian cattle and other livestock as well as their nutrient value is estimated in Table 5.13.

Table 5.13. Potential dung and urine available from Indian livestock and their manurial value

Type of animal	Population (million)*	Daily production (kg)**		Nutrient content (%)**						Annual potential nutrient contribution (lakh t/year)		
		Dung	Urine	Dung			Urine			N	P_2O_5	K_2O
				N	P_2O_5	K_2O	N	P_2O_5	K_2O			
Sheep/goat	165.90	0.3	0.2	0.65	0.5	0.03	1.7	0.02	0.25	3.24	1.84	0.36
Pigs	12.79	2.0	2.0	0.6	0.5	0.2	0.4	0.10	0.50	0.93	0.56	0.65
Poultry	307.00	0.025	–	1.0	1.4	0.8	–	–	–	0.27	0.37	0.22
Cattle	204.50	5.0	3.3	0.5	0.3	0.9	1.2	–	0.75	48.23	11.20	52.07
Buffalo	83.50	6.20	3.80	0.8	0.4	0.4	1.8	0.3	0.2	34.07	11.03	9.88
Other livestock	3.22	3.4	2.9	0.4	0.2	0.1	1.1	0.4	0.3	0.54	0.30	0.16
Total	76691.00	–	–	–	–	–	–	–	–	87.28	25.30	63.34

*Anon., 1997 ** Gaur *et al.,* 1990

At present, not even half of the dung and 10 per cent of urine is collected and utilized for manure preparation. Efficient collection of dung, in Indian context, is somewhat difficult - as most of Indian cattle graze outside the protected house and major part of dung and urine are lost. Semi-nomadic grazing on marginal and uncultivated lands is also a common practice in many villages. Nearly 50 per cent of collected dung is burnt as fuel - which is a criminal waste of valuable resource.

More scientific and practical methods need to be derived to effectively collect the dung and urine from Indian cattle and other livestock - as applicable to our diverse rural situations. For example, collection of urine from shed is the most neglected aspect of shed maintenance. Urine can be scientifically collected and used so that nutrients are not lost. The loss of nitrogen as nitrates from urine has linear relationship with the time for which it is exposed to air. Hence, it is advocated to soak the collected urine as early as possible in a litter or straw bedding or saw dust - so that nutrients are not lost.

5.2.4 Use of Jeevamruta

In recent years, the organic farmers have been found to practice an innovative method of preparation and use of a liquid biological culture called jeevamruta. It is also being called by many local names in different parts of the country. Not withstanding the variation in terminology, this liquid biological culture is found to be of greater use of enhance the microbial activity in soil and there by help in improvement of soil fertility.

Jeevamruta can be prepared very conveniently by every farmer owning any type of cattle in their farm. Jeevamruta is essentially an incubated mixture of cattle dung, cattle urine and small quantity of jaggery and pulse flour. Preparation of jeevanmurtha is scientifically validated and standardised by this author. The details are presented in 4.1 (p. 139 of this book). In a field study conducted by Kasbe (2008), application of jeevamruta resulted in manifold improvement of N-fixing and P-solubilising bacteria in the soil, which is known to improve the fertility. This was due to inherent high population of useful bacteria like *Azotobacter, Azospirillum* (N fixing) and several P solubilising bacteria in jeevamruta. Many farmers have found the beneficial effects of applying the jeevamruta repeatedly to their crops in terms of stabilized yields and improved soil fertility.

Since this practice is highly farmer friendly, less costly and a simple technology, it can be treated as the most versatilie method of enhancing the soil fertility in long term. Although it contains small quantity of nutrients (0.3 to 0.5% N, 0.01 to 0.03 P_2O_5 and 0.05 to 0.08 K_2O) and can be treated as source of supply of nutrients, it need not to be construed as liquid manure. But it is a polent liquid microbial culture to enhance the microbial activity to ultimately improve the soil fertility.

Jeevamruta has been used in many crops by soil application, or supplied through irrigation water, or even sprayed by regular sprayers. It has been found not have any determinantal effect on any crop at any stage, indicating the versatility of liquid microbial culture.

5.3 CROP RESIDUE RECYCLING

Generally, a part of the nutrients taken up by the crops are harvested as economic yield and major portion is available as the crop residues. The nature expects this to be returned to the soil medium to meet the sustainable needs of the eco-system. The removal of the crop residues along with the economic yield is a major human interference and acts to the detriment of natural sustainability of the system.

Hence, the returning of crop residues to the system is an essential aspect of sustainable eco-system. Incorporation of the crop residues for decomposition in the soil not only returns the essential nutrients of the plants but also provides much needed substrates for microbial activity. It is this microbial activity, which is essential for the development and sustainability of the agro-eco-system.

The management of crop residues for returning the plant nutrients is a natural phenomenon in a forest eco-system or any natural vegetation. The residues such as leaves, twigs, fruits will fall on the ground and in the course of time decompose. But in an agro-eco-system, agriculturists have regarded crop residues as materials unwanted at the site. The reasons for such a perception may be many

(a) The residues may come in the way of cultivation
(b) Some residues may not decompose quickly
(c) The crop residues are used as cattle feed
(d) Many of crop residues serve as essential fuel in villages
(e) Residues are necessary to prepare the decomposed manure
(f) Some residues need to be chopped before they are ready for decomposition.

Out of these, the use of residues as cattle feed and fuel are directly influenced by the land : man ratio, while others can be regarded as modifiable human interferences. In most of developing and under-developed countries, geometric increase in population growth has caused acute stress on land : man ratio. As the fuel needs of the human society increase, the crop residues will have to meet their fuel requirements. If the crop residues have to be returned to the soil for sustainability, alternate fuels at rural levels will have to be identified and popularized. In this direction, the popularization of bio-gas technology and other use of non-conventional energy in India deserves the attention, as it has partially satisfied the fuel needs of the rural folk. However, the administrative efforts to popularize this technology needs to be further strengthened.

The necessities to use crop residues as cattle feed is a major constraint in returning them to soil in a country like India, as it is reported that 33 per cent of global cattle population exists in India depending upon meager 3.28 per cent global area under cultivation. In most other countries, the cattle population does not pose pressure of such proportion on the cultivated land. In developed countries such as USA, France, Germany, Canada and farmers need not depend upon such crop residues for the purpose of cattle feed. They can afford to reserve part of their land exclusively to grow fodder and thereby return the crop residues easily in their cropped land. The main reason for increased cattled population in our country is the lack of culling practice of unwanted and unproductive live stock due to sentimental and orthodoxical reasons.

Thus, the practice of returning the crop residues is not a difficult proposition in many developed countries such as USA and European countries as well as many developing countries such as Latin America, Africa and Far Eastern countries, where the cattle population is far less than in India. In India, the return of the crop residues is adversely affected by the reasons of great demand as cattle feed by large cattle population. Ironically, large quantities of crop residues are also often wasted or burnt in some states. Burning of crop residues is a practice representing ancient agriculture found on unscientific principle that ash serves as main source of nutrient after burning. It is even now a part of shifting cultivation or slash and burn agriculture in forest eco-systems of world. But, it is well settled that such practice has no specific advantage in profitable agriculture, as nutrients added by the incorporation of residues are much more than nutrients added by their ash. Whenever they are removed from the site for preparation of manure, they are not properly utilized due to lack of awareness that they are handling valuable resource containing the huge quantity of nutrients.

Despite all these limitations, crop residue incorporation can be practiced on a fairly large area to supply substantial quantity of nutrients. Recently Sarkar *et al.* (1999) have made detailed investigations about the prospects of using the crop residues in India. They have estimated that total crop residue availability for manuring purposes as 170 million tons (out of 355 million tons of residue production) supplying nutrients equivalent to 1.72 million tons of nutrients (Table 5.14).

The benefits arising out of crop residue recycling should not be assessed from their manurial contribution in one season. Many research reports show that crop residues applied to a crop has benefited the succeeding crop. Wagger *et al.* (1987) and Ladd *et al.* (1983) have shown that in wheat-sorghum and Alfalfa wheat systems respectively 12 to 33 per cent and 20 to 30 per cent N mineralized by the crop residue applied to first crop becomes available for succeeding crop. Similar reports are made for P and K also.

Large number of research workers have reported increased crop yields by crop residue incorporation. Some of them are listed in Table 5.15.

Table 5.14. Potentials of crop residues for plant nutrition

Crop	Total crop residue (m. tons)	Crop residue available for manuring (m ton)	Nutrient content (% on oven dry basis)**			Potential nutrient contribution (m tons)		
			N	P_2O_5	K_2O	Total N + P_2O_5 + K_2O	Available for manuring	Fertilizer replacement value (in terms of nutriens)
Paddy	119.25	39.75	0.61	0.18	1.38	2.59	0.86	0.43
Wheat	93.60	31.30	0.48	0.16	1.18	1.71	0.57	0.28
Sorghum	19.80	6.60	0.52	0.23	1.34	0.41	0.14	0.07
Pearlmillet	10.80	3.60	0.45	0.16	1.14	0.19	0.06	0.03
Maize	14.40	4.80	0.52	0.18	1.35	0.30	0.10	0.05
All pulses	17.95	6.00	1.29	0.36	1.64	0.59	0.19	0.10
Groundnut	11.48	11.48	1.60	0.23	1.37	0.37	0.37	0.19
Other oil seeds	29.94	29.94	0.80	0.21	0.93	0.58	0.58	0.29
Sugarcane	28.05	28.05	0.40	0.18	1.28	0.52	0.52	0.07
Potato	2.60	7.60	0.52	0.21	1.06	0.14	0.14	0.07
Barley	2.50	0.83	0.52	0.18	1.30	0.05	0.02	0.008
Total	355.67	169.95				7.44	3.58	1.72

* Sarkar *et al.* (1999) ** Tandon (1995)

Note :

(a) Total crop residue production has been calculated on the basis of economic yield (grain, cane or tuber) by the corresponding economic yield : residue yield ratio using the FAO document No. 15 and No. 24 as cited by Bharadwaj (1995).

(b) Assuming that 2/3 of the residue is used as cattle feed for animal, fuel and other purposes, only 1/3 is considered avai-lable for manuring purposes except oil seeds, sugarcane and potato residue, where entire residue is considered useful for manuring).

(c) Fertilizer replacement value is calculated on the assumption that half of the total N+P+K is mineralized in the same season.

(d) Potential nutrient contribution is calculated for residues available for incorporation.

Although the fertilizer replacement value of the available crop residue in India limits the scope to replace only 1.7 million tons of nutrients out of 20 million tons of nutrients used in the form of fertilizers every year, the importance of crop residue recycling should not be underestimated. Because, the crop residue recycling is basically aimed at building up the soil fertility deteriorated by continuous use of chemical fertilizers for last 30 years. As these crop residues are available every year, their continuous incorporation is bound to improve the soil physical characters in long run. Number of reports are made in global literature about the beneficial effects of crop residue recycling on soil characters. They are summarized in Table 5.16.

Table 5.15. Yield improvement by added crop residues in different crops

Crop	Yield improvement	Reference
Sorghum	Yield improved by 15 to 25 per cent	Patil *et al.*, 1996
Maize	Yield improved by 20 to 28 per cent	Hegde *et al.*, 1982
Sugarcane	Yield improved by 21 to 26 per cent	Rasal *et al.*, 1989
Sugarcanc	Yicld improved by 37.5 per cent	Yadav *et al.*, 1987
Rice	Yield improved by 28 to 41 per cent	Prasad *et al.*, 1999
Wheat	Yield improved by 30 per cent	Anon. 1985
Rye	Yield improved by 1.0 mg/ha	Andraski and Bundy, 1999

Table 5.16. Beneficial effects of residue recycling on soil properties

Crop	Soil character .mproved	Reference	Soil type
Wheat	Organic matter	Knapp *et al.*, 1983	Clay loam
Cowpea/ pearlmillet	Bulk density	Venkateshwaralu 1987	Sandy loam
Jowar	Bulk density	Hadimani *et al.*, 1982	Red soils
Maize	Soil aggregates	Ramussen and Smiley 1989	N.A.
Cowpea/ pearlmillet	Soil aggregates	Venkateswaralu 1987	Sandy loam
Jowar	Infiltration rate	Magar 1982	Vertisols
Cowpea/ pearlmillet	Infiltration rate	Hadimani *et al.*, 1987, Venkateshwaralu, 1987	Vertisols sandy loam
Wheat/rice	Cation exchange capacity	Das *et al.*, 2001	Sandy loam
Wheat/rice	Organic carbon content	Dhiman *et al.*, 1993	Clay loam
Sugarcane	Increased organic carbon, N, P_2O_5; reduced pH	Shinde *et al.*, 1993	Clay
Sugarcane	Earthworm population	Anon 1996	–

–	Straw incorporation increased S.O.M. & increased release of nutrients	Mukherjee and Gaur 1980	–
Maize	Reduced B.D. increased, infiltration rate and WHC	Pandey *et al.*, 1985	Sandy loam
Sorghum	Increased infiltration rate, reduced B : D	Badanur *et al.*, 1990	Vertisols
Wheat/rice	Increased baterial population	Sidhu *et al.*, 1995	Sandy loam
Rice/wheat	Increased organic carbon	Bhat *et al.*, 1991	–
Wheat	Increased N availability	Prasad *et al.*, 1999	–
Silagecorn	Increased organic carbon	Bary *et al.*, 1999	–

In the overall interest of organic farming, the crop residue recycling offers excellent scope to utilize the resource available at the farm. Besides reducing the fertilizer requirement of a crop, it can improve many soil physical characters, which help in better microbial activity and thus help in achieving the goals of organic farming to develop the self supporting sustainable agro-eco-system.

Advantages and limitations of use of crop residues

As compared to other methods of organic farming, the crop residue recycling offers many advantages for easy adoption.

Advantages

(a) They do not involve cost of transportation as in case of fertilizers and FYM

(b) Does not cost a crop/ season as in case of green manure crop

(c) They are not purchased from outside and hence they do not burden the farmers economically.

(d) No contamination by pollutants such as heavy metals as in case of sewage farming or using manure prepared out of industrial wastes.

At the same time, some limitations are also discernible to restrict they use crop residue managment on large scale.

Limitations

1. The poor decomposibility of residues

Due to wider C : N ratio and high lignin content, some residues are not decomposed quickly. For example, the sugarcane crop can generates large residues (about 12-15 t/ha) and it is hardly utilized as cattled feed. But its incorporation in the soil offers limited scope due to its wide C : N ratio (upto 1 : 120). Newer practice of using *Pleurotus* spp., *Phanerochelate chrysosporium* fungal culture to accelerate the rate of decomposition have been recently tried successfully to decompose sugarcane residues, coirpith/areca husks.

2. Soil moisture requirements for decomposition

The incorporation of crop residues can be successfully practiced in irrigated regions. But, in vast arid and semi-arid regions the incorporation of residues is not practical because most of the residue remain undecomposed due to limited moisture availability. Even in humid regions, rainfed areas are not ideal for incorporation because, the wet part of year is used to grow crops and incorporation during dry part will not allow complete decomposition of residues.

3. Physical transformation of residues

Most crop residues require physical transformation (like chopping, cutting) before they are effectively decomposed. It is not a practical proposal to carry the huge quantity of residues away from field to chop them. Many field devices need to be developed to physically transform **the residues on the site itself, so that their decomposition can be** achieved more quickly.

4. Pests and diseases

The residue incorporation poses dangerous threat in perpetuating the insects and pathogen, when they hibernate/sporulate in residues. Hence, incorporation of diseases/damaged residues should be avoided.

Biodigester in farm

In the last decade, a new practice oforganicfarming has emerged to effectively use all farm wastes including left over crop residues and harvesting wastes and even weeds. The concept involves dumping all such wastes in to cement tank with adequate wate rand allow it to decay for a period of 25-40 days. All the liquid wastes from cattle shed including washings and cattle urine could be directed to the tank for effective digestion.Then, the liquid part of it is pumped, filtered and used as source of nutrients along with irrigation. Thedigested liquid looks greeninsh or brownish. The practice of biodigester is quite popular among organic growers due to ease of its operation and effective use of wastes. The biodigester liquid may contain a nitrogen upto 0.8% and phosphorus upto 0.25% and potassium upto 0.4% besides containing useful doe of iron, zinc and manages.

5.4 OIL CAKES

Use of oil cakes as manures for increasing the crop yields is one of the conventional form of organic farming. Prior to introduction of fertilizers, the oil cakes were applied in granulated formulations to agricultural lands, so that the nutrients contained in them were made available to crops. Such method added large organic carbon to the soil, which helped in enhancing microbial activity.

In the present day Indian scenario, use of oil cakes as a source of nutrients is largely limited because :

(a) Most of edible oil cakes are fed to cattle as concentrates and application of oil cakes to agricultural lands should preclude these sources.

(b) Oil cakes, in general, supply limited quantity of nutrient due to better oil extraction methods like solvent extraction as compared to traditional methods.

(c) As extraction process is a part of oil industries located away from agricultural lands, the cost of transport of oil cakes increases the a cost/ kg of nutrient in oil cake.

Nevertheless, the potentiality of oil cakes for being used as manures is clearly perceptible in :

1. Non-edible cakes like castor cake, linseed cake and neem cake
2. Oil bearing trees (most of them remain unexploited)

Non-edible cakes

Table 5.17 indicates the possible non-edible oil cakes along with their nutrient composition and potential nutrient value.

Table 5.17. Potential nutrient contribution of oil cakes

Oil cake	Nutrient composition (%)			Cake	Oil cake production million tons	Potential nutrient value lakh tons		
	N	P_2O_5	K_2O			N	P_2O_5	K_2O
Castor cake	4.3	1.8	1.3	(53%)	2.39	1.30	0.43	0.31
Neem cake	5.2	1.0	1.4	(68%)	2.38*	1.24	0.24	0.33
Linseed cake	4.9	1.4	1.3	(60%)	1.32	0.65	0.18	0.17

*Based on adhoc projection of 35 lakh tons neem seed production

Oil bearing trees

Large number of oil bearing trees are identified in India with a high potentiality to commercially produce oil seeds. Mere collection of oil seeds from existing oil bearing trees may result into large oil seed yield. Alternatively, the oil yielding trees could be used for alley cropping with crop husbandry so that oil production becomes integral part of farm with other crops. A list of oil bearing trees is given in Table 5.18.

In either case, the oil bearing trees are bound to yield substantial recyclable cakes. They can be used either to prepare farmyard manure or incorporated directly.

Table 5.18. List of oil bearing trees

Scientific name	Scientific name
Terminalia chebula (ink nut)	*Mesua ferrea* (Iroa wood tree)
Sapinders mukorassi (soap nut)	*Ximena americana* (Bastard sandal)
Madhuca indica (Indian butter tree)	*Aleurites moluccana* (Wanish tree)
Garcinia combogia (Malbar tamarind)	*Terminalia catappa* (Indian almound tree)
Calophyllum nightianum (Pune tree)	*Mimusops elengi* (Bullel wood tree)
Gurodia odorata	*Salmalia malbarica*
Mallotus phillipensis (canel tree)	*Cochlospermum religiusum* (Silk cotton tree)
Caesalpinia crista (Bomcaknut tree)	*Garcinia indica* (Cocam butter)
Aleurites fordii (Tung oil tree)	*Pongamia pinnata* (Honge)
Terminalia bellitaca (Belliric milloberou)	*Simaruba indica*

Source: Kallapur (1982)

5.5 URBAN AND RURAL WASTES

Human habitation has always created large usable wastes. Such wastes may consists of :

1. **Garbage** - Household wastes, road sweepings, wastes from shops/offices/other establishments and all other dry wastes grouped as garbage.
2. **Sewage** - Consisting of solid and liquid wastes from internal drainage networks in metropolitan and other bigger urban establishments. It may consist of
   (a) Sewage sludge : Solid part, which usually consists of human excreta and other solid wastes.

(b) Sewage water : Liquids generated from water closettes, kitchen, bathrooms and office.

3. **Sullage** - That part of city wastes, wherein separation of liquid and solid part is not possible and consists of flushings of septic tanks and other liquids generated from various human activities.

As most of these wastes are carbonaceous and therefore easily decomposable, they could be easily recycled. Proper methods to handle, collect, transport them to spot where they are processed/ decomposed are crucial in management of urban and rural wastes. Although cities and villages have large potentials to generate the wastes, urban centers are considered as more potential than rural centers due to high density of population. As the population is thinner in rural areas and usually the disposal is not a serious proble in rural areas, rural wastes are generally ignored.

Wastes in cities are handled and collected more for the reasons of disposal from an unwanted locality than for their manurial value. Hence, conventionally most Indian cities and urban agglomerations have programmes to merely handle and collect the wastes to dispose them off in places far off from dwelling places. Despite technological break through in many other fields, the handling and collection of garbage is done with unsafe, inefficient age old technology. In most civic bodies of our country, the garbage is lifted manually and transported openly with no safety measures.

In recent 10 to 15 years, major metropolitan cities have established large scale compost making units to convert huge quantities of city wastes into useful manure. However, they are established to handle garbage only. The use of sludge to produce nutrient rich manure is yet to be practiced by many Indian cities. Wherever city compost units are working, they are recycling 30 to 40 per cent garbage produced in respective city.

All the wastes generated in urban and rural areas contain large quantities of plant nutrients such as nitrogen, phosphorus, potassium, calcium and sulfur besides many micronutrients such as iron, sodium, magnesium, zinc, copper and boron (Goel and Trivedi, 1984). Occasionally they contain heavy metals such as cadmium, chromium, nickel and lead - out of which cadmium and nickel concentrations are upto toxic levels in some Indian cities (Juwarkar *et al.*, 1991). Besides, supplying these plant nutrients, solid sludge and compost from garbage adds about 2.7 per cent organic carbon to the soil (Zothy *et al.*, 1984).

Each group of urban/rural wastes offers an excellent scope of recycling a large quantity of wastes (often with nuisance or health hazard values) into rich manure for useful agricultural production. Recycling provides twin objectives of handling/disposal of wastes as well as use of natural resource efficiently for productive purposes. Both these activities are yet to grow on commercial scale, although many major cities have already established compost units on massive scales.

To underline the manurial importance of rural/urban wastes, it is necessary to know their composition, quantity of wastes generated and their manural contribution. As the reports on composition of rural wastes are scanty, an estimate of manural contribution by urban wastes are presented in Table 5.19.

Table 5.19. Manurial contribution by urban wastes

Type of waste	Quantity generated	Chemical composition (%)				Nutrient contribution (lakh t/yr)				Source
		N	P_2O_5	K_2O	OM	N	P_2O_5	K_2O	OM	
Sewage water	17.4 million cuM/d	60 ppm	25 ppm	40 ppm	–	1.2	0.5	0.8	–	Gaur *et al.*, 1990
Sewage sludge	4 million t/yr	3%	2%	1%	30%	1.2	0.8	0.4	12	Kansal, 1992, Bhar-adwaj & Gaur, 1985
Garbage	12 million t/yr	0.5%	–	0.3%	5.2%	0.6	–	0.36	4	Dahama, 2001
Sullage	No estimation									
	Total					3.0	1.3	1.8	1.6	

Sewage water

Sewage effluent irrigation is practiced for centuries throughout the world (Shuval *et al.,* 1986). Sewage water has been successfully used for long time in India as a source of irrigation for crop production, as sewage water is found to be enriched with plant nutrients. As early as 1895, first sewage farm was established in Ahmedabad. At present, it is estimated that atleast 200 sewage farms covering 50,000 ha of area are using sewage water as the main source of irrigation and nutrients (Juwarkar *et al.,* 1991). These farms are mostly restricted around urban dwellings, cities and metropolitan centers, where large sewage water is generated. The use of sewage water for farming can be successful when civic planning and civic administration manages to collect the sewage by scientific methods. In most cities, the sewage is not collected properly and allowed to be flown in rivers, streams, tanks around urban places. The low utility of rich potential source from Indian cities can be ascribed to poor method of collecting sewage.

In many cities, sewage effluent water has some hazardous chemicals. Maiti *et al.* (1992) have found that Calcutta city sewage contained hazardous level of micronutrients and many metals.

Although it is ideal to use the sewage water after necessary pre-treatment, most Indian sewage farming units, use undiluted and untreated sewage (sometimes even sullage is used). In Mexican and South American countries too, such practice is reported (Strauss and Blumental, 1990). They directly let in the sewage water to a farmed land. Following disadvantages of using untreated sewage can be listed:

1. Bacterial contamination renders the vegetable growth with sewage water unhealthy for human consumption. Ironically, the cultivation of vegetables is most popular than other crops using sewage water - mostly due to the fact that demand for vegetables is enormous and consistent in cities and proximity of sewage farms to cities will have logistic advantage. Israel has imposed stringent treatment of effluent before it is used for vegetable cultivation (Shelef, 1990).
2. Soil is rendered sewage sick due to mechanical clogging by suspended particles in sewage water. This causes poorer root growth due to poor aeration.
3. Large mass of micro-flora applied along with sewage water can create imbalance in nitrogen dynamics of a soil, as the natural equilibrium between nitrification, denitrification and ammonification can be upset.
4. Sewage water is found to be contaminated with many heavy metals like cadmium, selenium, lead which deteriorate soil health due to continuous application of untreated sewage water.

The treatment of sewage water may be done in four steps

I. Primary

Removal of coarse organic/inorganic solids by screen, settling and flotation. It may also involve steps to neutralize the sewage, if it is alkaline (it is seldom acidic in nature).

II. Secondary

Aerobic and anaerobic biological processes to decompose the organic debris through necessary structures to provide more air supply.

III. Tertiary

Chlorination, micro-screening, coagulation, precipitation and activated carbon adsorption.

IV. Quarternary

Upgrading the water to the level of portable water by ultra filtration by processes such as ion exchange, reverse osmosis electrodialysis etc.

Although above steps broadly indicate the processes necessary to treat the sewage water, the actual schedule of various steps depends upon many factors such as pH, percentage of suspended particles, composition of solids, cost of treatment, availability of land for treatment, the need for treated water, problems of transportation and handling the sewage etc. As a result, different schedules of treatments are evolved for different situation, although as much as 80 to 90 per cent of Indian sewage reaches farms without treatment.

Due to prohibitive costs, sewage water is treated with Step I and Step II- wherever it is proposed to be used for farming. When it is used for other industrial processes or for feeding animals, it is advisable to treat it even by using Step III and Step IV.

Generation of sewage water has functional relationship with growth in urbanization and industrialization. As the density of population increases and as the industrialization intensifies, the larger amounts of sewage water are generated. The concerned civic and industrial authorities have an uphill task to collect, handle and dispose off the large quantity of sewage water - as the sewage water generation increases. Although sewage water is generated in villages, Indian villages do not have any system of collecting the sewage. Accoridng to an estimate, the sewage water collection in Indian cities (major) has grown by 125 to 180 per cent in the last 20 years, while its generation has grown by 200 to 450 per cent. Based on this, during the year 2000-01 the sewage water collection was as much as 35 million Cu m/day. It is equivalent to 28,300 cusec and can mean that a sewage water river is available for exploitation.

The composition of city sewage is at great variance from city to city. Similarly, industrial sewage is not similar to city sewage regarding its composition. Moreover, industrial sewage is likely to contain several injurious chemicals and heavy metals (Arsenic, Selenium, Cadmium, Mercury etc.). Hence, when industrial effluent is exclusively used for farming, a pre-treatment with specialized technology is needed.

Several reports of benefits of using sewage water for farming are reported. Excepts from some reports are presented in Table 5.20.

Table 5.20. Response to added sewage water

S.N.	Study	Reference
1.	Calcutta city sewage effluent contained high NH_4-N, NO_3-N and available P	Maiti *et al.*, 1992
2.	U-100 per cent sewage application resulted in increased maize yield	Quasim *et al.*, 2000
3.	Mulberry dry matter yields are reported to have increased by sewage application	Ramos *et al.*, 2000
4.	50 per cent concentration sewage increased the yield of *Vigna mungo*	Kamabiran and Harilal, 1998
5.	Mobility of Mn, Zn, Cu and Fe was increased by heavy metal composition of sewage	Khan Samiullah *et al.*, 1997

Disposal of sewage, in Indian context is neither realized and implemented in its real perspective as ecological necessity nor as an excellent manurial source for increasing production. Complex laws regarding pollution and unending litigations have plagued the industrial effluent treatments and in most cities the sewage finds an easy way of entry into rivers, oceans besides other city wastes. It is a matter of great concern that vast quantity of resources are simply wasted to cause several irreversible ecological upsets. Some industrial giants have come forward to use recycled sewage for their industries. They have a tieup with civic sewage boards to use processed/recycled water. Such a move has motivated the civic sewage boards to establish large scale treatment plants. Similarly, all domestic sewage water is still not being used for irrigation although some sewage farms are using it. The reasons could be lack of pretreatment facilities, lack of awareness, huge costs involved in treating/transportation of sewage and sometimes stigma attached to its use. According to one estimate only 30 per cent of potential is presently being used.

As a result of lack of capital and facilities for treatments, farmers around cities directly use the sewage for cultivation of vegetables. They prefers vegetable cultivation and not grains, because they are much in demand in cities and their transportation to market is easy, cheap and quick. However, growing vegetables with untreated sewage is more dangerous health hazard than that of grain. This is not realized by many sewage farmers.

Sewage water, thus, is no doubt a most potential source of organic farming with large quantity of nutrient contained in it. Allowing it to go waste is indeed a social offence. But their usage should be encouraged only after treatment. Use of untreated sewage may lead to number of health problems (like typhoid, dysentery, meningitis, hepalitis), as found out by Hogg *et al.* (2003).

Sludge

Although large quantity of sludge is produced in India, only an estimated 30 to 40 per cent of sludge is collected in internal drainage networks of major cities. Remaining sludge is not easily collectible and it is either allowed to be dried up and wasted as sullage or dumped in an uncultivated waste area. Out of collected sludge, less than half is converted into manure.

However, sludge is a very rich source of plant nutrients (Table 5.21). Several reports are available from India and other countries to indicate that sludge can supply substantial quantity of nutrients (Goel and Trivedi, 1984; Kansal, 1992 and Zodhy *et al.,* 1984).

Table 5.21. Composition of sludge and its estimated potential nutrient value in India

Component	**Concentration (%)**			**Potential nutrient value (based on medium value)**
	Minimum	**Maximum**	**Medium**	
Organic C	6.5	48.0	30.40	6.65 million tons
Total N	<0.1	17.6	3.3	7.2 million tons
NH_4+-N	<0.1	6.7	1.0	2.19 lakh tons
NO_3-N	<0.1	0.5	< 0.1	0.22 lakh tons
Total P	<0.1	14.5	2.3	5.04 lakh tons
Total S	0.6	1.5	1.1	2.41 lakh tons
Ca	0.10	25.0	3.9	8.54 lakh tons
Fe	<0.1	15.3	1.1	2.41 lakh tons
Na	0.01	3.1	0.2	0.44 lakh tons
K	0.02	2.6	0.3	0.66 lakh tons
Mg	0.03	2.0	0.4	0.88 lakh tons
Zn	101	27,800	1,740	38,000 ton
Mn	18	7,100	260	5,700 ton
Bo	5	39	30	657 ton
B	4	757	33	723 ton

However, the composition of raw sludge varies greatly from city to city.

According to one estimate every 10,000 people in a country can produce 365 tons of dried anaerobically digested sludge per year (Dahama, 2001). In India, 60 crore adult population may be able to generate 21.9 millions of dry digested sludge annually according to this estimate. Based on median concentration, it can supply 6.7 million tons of organic carbon, 0.72 million tons of nitrogen, 0.55 million tons of phosphorus besides laerge quantities of micro nutrients. However, sludge equivalent to only 4 million tons of dry digested sludge is collected at present in India and a part of it indeed utilized.

Although use of sewage sludge is an established practice in many advanced countries like UK (Davies, 1993) its use is on a limited scale in India. Sludge can be used in many ways. It may be used as digested manure after compositing it or raw sludge itself may be used for field application. Vaccum filtration of sludge may yield filter cakes which may be spread on land or incorporated into the soil. The other processes like flocculation and pressure filtration may yield a semi soild (20% solids) substances which may be used as composting material or injected into/spread on land.

In the background of growing concern over environmental pollution, healthy disposal of sludge is an integral part of public health system. Sludge disposal is also to be viewed from resource conservation angle beyond the reasons of public health. In addition, it can be an important source of organic farming. Instead of depending upon costly concentrated fertilizers, resources naturally available around us could be used for increasing crop production simultaneously maintained soil health.

However, raw sludge needs treatment and digestion before use, although in some parts of world unhealthy practice of applying raw sludge to fields still continues. Series of steps involving separation, grit removal skimming, activated sludge tanks treatment, chlorination and aerobic/anaerobic digestion are suggested. In western countries well digested and irradiated sludge is bagged and supplied to farmers. In India, however, the use of sludge as an organic source is in its elementary stage. While, modern techniques of collecting more and more sludge from Indian cities need to be implemented, the collected sludge may be atleast treated with low cost techniques and digested before use. For example, mixing night soil rich sludge with equal volume of ash and 10 per cent powdered charcoal produces odourless solid material with potassium rich dry sludge. Mixing sludge with saw dust (40 to 50% by volume) may yield straight away a dry acidic powdrette containing 2 to 3 per cent N.

Use of sewage sludge involves some problems such as :

(a) Social stigma attached to its use, as sludge contains night soil.

(b) Some pathogens retains in sludge may caused spread of human and animal diseases.

(c) Toxic levels of some heavy metals like cadmium, nickel, arsenic, mercury in sludge are reported to be higher in Indian sludges (Tiwana *et al.,* 1987). They may inhibit the plant growth or enter food chain to cause lethal/sub-lethal harm to human beings/animals.

(d) Accumulated polychlorinated biphenyls (PCBs) and other toxic organic compounds known for their persistence in animal/human bodies may enter food chain to cause extended damage to the health of human beings/ animals.

(e) Exceesive salt accumulation in the soil may cause salinity.

Most of the above problems are linked to the level and type of sludge treatment/digestion. If the sludge has to be more healthily disposed off, the costs are very high. By not using sludge as a source of organic manure, it may not be possible to get away the pollutants contained in them. It is necessary to adopt the cheap and effective techniques available under our conditions and make use of as much sludge as possible.

The beneficial effects of use of sludge are reported by number of research workers. Some of the latest reports are cited in Table 5.22.

Table 5.22. Response to applied sludge

S.N.	Study	Reference
1.	Continuous sludge application as granulated digested manure in Yakohama resulted in high nutrient values in soil in available form as compared to fertilizer	Noguchi and Ito, 1992
2.	Net increase in NO_3 concentration by raw sewage application	Smith *et al.,* 1992
3.	Four years study of sludge application resulted in increased PO_4 - P availability	Brendecke *et al.,* 1993
4.	Increased N uptake by crops due to sludge application	Serna and Pomares, 1992
5.	Increased grain yield of wheat by sludge application @ 20 t /ha	Al-mustafa *et al.,* 1995

6.	Seven year application of digested sewage sludge increased organic carbon from 0.85 to 1.63, N content from 0.7 to 0.15 and CEC from 14.18 to 16.28 meq/100 g soil	Hinesly *et al.*, 1979
7.	Application of sewage sludge released longer NH_4 salts, which where were nitrified quickly in calcareous soils of Spain	Verdu *et al.*, 1992
8.	Application of 6.6 Mg/ha of bio-solid wastes in Wisconsin (USA) for 15 years could consistently improve available N status to 200 kg N/ha and available P status to 200 kg/ha besides increasing the corn yields by 3.3 t/ha	Peterson *et al.*, 1999
9.	Application of sewage sludge resulted increased availability of P, total P in soil depths upto 60 cm	Akhtar and Mc Callister, 1999

Garbage

Collection and disposal of huge quantity of garbage generated in most of urban areas is a matter of great concern for municipal bodies and city corporations. Inspite of good planning, most of our municipal bodies have not been able to collect all the garbage and also not able to dispose it off usefully. Large quantities of garbage from metropolitan and many big Indian cities go uncollected adding to nuisance value and unasethetic value besides causing health problems. In most cities, most of the collected garbage is simply dumped outside the city. Recently many metropolitan cities have realized its manurial impoertance. They have now taken the garbage disposal more seriously to produce large quantity of digested manure. Cities like Bombay, Delhi, Calcutta, Chennai, Hyderabad, Bangalore, Pune, Ahemadabad, Lucknow, Bhopal have already established large scale garbage handling methods and commercial compost preparation on scientific lines.

Application of city garbage waste has resulted into beneficial effect in improving the yield of many crops besides improving soil characters. Large number of reports on beneficial effects of use of city wastes are available. Some latest reports are abstracted in Table 5.23.

Table 5.23. Effect of garbage compost application on soil properties

Sl. No.	Study	Reference
1.	Maize crop yields were improved by 1000 kg/ha, bulk density decreased upto 0.71 g/cc water holding capacity increased from 70 to 150 per cent	Greenwood *et al.*, 1999
2.	Compost provides slow release N for many years after application, as mineralization rate for residual compost N was 2 to 6 per cent per year	Cramer *et al.*, 1999

More than 80 per cent of garbage is bio-digestible, while remaining 20 per cent consisting of metal pieces, stones etc. may have to be removed. Of late, plastic materials in garbage have increased and their bio-digestibility is not easily possible. Unlike sludge and sewage, the disposal of garbage does not need any advanced scientific techniques. After the physical separation of undigestible materials, the garbage is to be composted to yield manure. Some alternate disposal systems such as using the garbage as fuel are also suitable in the background of scarcity of energy.

Table 5.24. Manurial contribution by urban wastes

Type of waste	Quantity generated	Chemical composition				Nutrient contribution (lakh ton/yr)				Source
		N	P_2O_5	K_2O	Organic carbon	N	P_2O_5	K_2O	Organic carbon	
Sewage water	17.1 million Cu M per day	60 ppm	25 ppm	40 ppm	–	1.2	0.5	0.8	–	Gaur *et al.*, 1990
Sewage sludge	4 million tons/year	3%	2%	1%	30%	1.2	0.8	0.4	12	Kansal 1992; Bharadwaj & Gaur, 1985
Garbage	12 million tons/year	0.5%		0.3%	5.2%	0.6	–	0.36	0.4	Dahama, 2001
Sullage	No estimate done									
Total						3.0	1.3	1.8	16	

The basic principle underlying the composting of garbage is bio-digestion by narrowing down C : N ratio. Mostly composting is suggested to be done under anaerobic pits. Compost from city garbage yields manure in the order of 30 to 35 per cent of original weight after digestion and contains about 0.8 per cent N, 0.3 per cent P_2O_5 and 1.5 per cent K_2O. With the garbage collection of 12 million tons from Indian cities every year, it is possible to produce 4.0 to 4.3 million tons of compost containing 32,000 ton N, 12,000 ton P_2O_5 and 60,000 K_2O (Table 5.24).

The compost preparation from garbage may make use of sludge/night soil for decomposition or may use synthetic composting techniques wherein decomposition is encouraged by adding inorganic nitrogenous fertilizers.

5.6 AGRO-INDUSTRIAL WASTES

Large number of agro-based industries were established in our country during the last 40-45 years. As they handle organic plant products, they invariably produce large quantities of organic decomposable wastes. At present, most of agro-industrial wastes are not recongnized to be potential manures and their disposal outside the premises of industries is of greater priority for such industries than utilizing them for manures. Hence, such wastes are dumped somewhere outside the factory premises - where they may decompose over a period of time with a sheer wastage of nutrients contained in them.

Although industries regard them as valueless wastes, they are rich in plant nutrients. Wasting such vast resources can only be halted - if the industries are made to realize their manurial importance and handle them carefully. In most situations, the industries themselves can establish a commercial manure unit to produce rich quality manure for sale. This will solve their problems of disposal besides resulting in additional income. The untapped potential of nutrients in these wastes could be utilized for crop production only with such attitudinal change the attitude managements. The farmers, too, need to change the attitude towards use of such manures in addition to their own source of manures. In general, agro-industrial wastes need some pre-treatment or digestion before they are called as manures. Invariably, they are ideal material for manure preparations.

List of industries with large potentials to produce agro-industrial wastes is presented in Table 5.25.

Table 5.25. List of industries with potentiality to produce large quantities of organic wastes

Name of industry	
Sugar industry	Coconut based industries
Rice industry	Areca based industries
Cotton industry	Tea/tobacco industries
Silk industry	Coffee, pulping industry
Paper industry	Pulping units Food industries Jute industry
Fruit processing industry	

Out of them, sugar industry, rice industry, cotton industry and tea/tobacco industries are known to generate very large quantities of wastes. The forthcoming discussion deals on these industries while all other industries also have potential to contribute the wastes to organic farming in general.

5.6.1 Sugar industries

Sugar industry mainly produces large quantities of begasse (stem tissues left over after extraction of juice) and pressmud (fine soil particles along with carbonaceous materials released during condensation process). Pressmud is also called as filter cake.

India produces around 240 million tonnes of sugarcane annually, out of which nearly 80 million tonnes are crushed for preparing jaggery. Remaining is crushed by 353 sugar mills to produce refined sugar. Sugar mills produce begasse and pressmud, while jaggery units produce begasse alone.

Table 5.26. Generation of wastes from sugar industry

Type of industry	Cane crushed million tons/year	Begasse produced million tons/year	Filter cakes lakh (tons/ year)	Begasse used as fuel million tons/year	Begassc available for manure production million tons/year
Sugar mills	160	65	58	55	10
Jaggery units	80	27	–	23	4
Total	240	92	58		14
		33% cane weight	3% cane weight		

The generation of wastes from sugar industries and their manurial value of wastes generated by sugar industries including jaggery units is presented in Table 5.26 and 5.27.

Table 5.27. Manurial value of sugar industry wastes

Type of industry	Begasse million ton/yr	Average nutrients (lakh tonnes)			Pressmud lakh ton/yr	Average nutrients (lakh tonnes)			
		N	P_2O_5	Organic matter		N	P_2O_5	S	Organic matter
Sugar mills	10#	6	3	25	58*	0.87	1.16	1.25	1.74
Jaggery units	4#	2.4	1.2	10	–	–	–	–	–
		8.4	4.2	35					

* Composition of pressmud 1.25% N, 2% P_2O_5, 2.25% S, 2.5% OC

\# Composition of begasse 0.6% N, 0.3% P_2O_5, 25% OC

5.6.1.1 Use of begasse

In most sugar industries, begassee is used as fuel. Though it is to be regarded as wastage of valuable plant nutrients, it is inevitable as any other form of fuel would be costly and may reduce the profitability of industry itself. Some sugar industries have started generating captive electricity by buring begasse to boil water with steam being used to produce electricity. However, a cursory look at any sugar mill or jaggery unit reveals that a large quantity of begasse is wasted while drying, pressing and transportation. If the nutrient value is the begasse is realised by people who handle begasse, such wastage could be stopped and used for manurial purposes. Further, most sugar mills and jaggery units regard begasse as no cost fuel and therefore do not bother to improve the boiler efficiency. A large amount of begasse could be saved if the boiler efficiency of these units is improved. By more diligent handling method alone, atleast 10 per cent of begasse in sugar mills and 15 per cent of begasse in jaggery units could be saved. Such begasse could be used for manure preparation.

If the manurial values of begasse is fully realised and if the imperatives of using them to prepare maures are fully appreciated, solutions to these difficulties could be found with attitudinl changes of the people who handle begasse. The utility of begasse as a source of manure has some practical difficulties. Some difficulties of using begasse for manure preparation are :

(a) Wide C : N ratio of begasse (1 : 60) could be the major constraint for it easy and fast decomposition. For this purposes, the manure preparation involving begasse may have to use more N-rich residues or urea along with begasse. Alternatively, decomposer cultures could also be for fast decomposition.

(b) The recyclable begasse is produced in different sugar mills and jaggery units spread over vast geographical area. It may be impossible to pool it at a place for mass manure production. However, it is practically feasible to consolidate the begasse at different factory/ jaggery unit level and take up manure preparation at unit level. Depending upon the quantity of begasse generated, the size and number of pits may be decided.

A study at National Sugar Institute, Kanpur indicates that begasse could also be used to produce biogas within 40-45 days by anaerobic process. Besides producing 200 cuM gas with 60 per cent methane, 30 per cent CO_2 and 5 per cent H_2, each ton of begasse can result into large quantity of spent slurry containing 1.5-1.8 per cent N, 1.0-1.3 per cent P_2O_5 and 0.6 - 0.8 K_2O on dry weight basis.

Global literature indicates that begasse is converted into manures in many countries known for sugar production. In India, use of begasse as a source of manure preparation is mostly reported on experimental scale. Mass manure production using begasse is not yet undertaken, although it is practically feasble both in sugar industries and jaggery units. The use of begasse based manure has been reported to have increased in crop yields, improved the soil structure and microbial population.

5.6.1.2 Use of pressmud

At present, sugar mills are finding it very difficult to dispose off large quantity of pressmud generated by them. Looking to its nutrient value, the sugar mills have a system to distribute it to nearby farmers free of cost. Nevertheless, the farmers are not convinced about its utility and offtake of pressmud from sugar factories is tardy.

Pressmud contains large quantities of sulphur and organic carbon besides nitrogen, phosphorus and potassium. Its repeated application can improve the organic carbon status of soil, besides supplying large amount of micronutrients like zinc, manganese, boron. Depending upon the process of sugar manufacturing, it may contain large quantities of lime too. All these ingredients are useful in crop production. However, most of pressmud produced by sugar mills has low pH values (pH <0.5). For this reason, it is advocated to cure it in air for a period of 15-20 days before use. This could substantitally increase pH values. Alternatively, pressmud could be used as material in manure preparation along with other crop residues. At any rate, the farmer's perception about pressmud is decisive in its use and sugar mills can be realized of their problems only when farmers are made to realize the manurial value of pressmud.

In recent years, some sugar industries have started manufacturing bio-compost using pressmud and other ingradients. This is a symbol of attitudinal change of such industries in favour of organic farming.

Apart from manurial values, pressmud is known to be useful in controlling some diseases like root-knot of tobacco.

Many reports on benefiicial use of pressmud and begasse based manure in agriculture are made (Table 5.28).

Table 5.28. Beneficial effects of use of pressmud

Sl. No.	Details of study	References
1.	Application of pressmud increased organic matter content and available P & K	Piedra *et al.* (1992)
2.	Pressmud compst application recorded highest cowpea yield	Srikant, 1997
3.	Pressmud application increased organic carbon content of soil	Yadav, 1995
4.	Application of pressmud produced more grain yield upto 5.29 t ha^{-1} than coir pith (4.4 t ha^{-1})	Pachiaraj *et al.*, 1991

5.6.1.3 Distillery spent wash

The manufacture of sugar release a huge quantity of molasses. A ton of sugar production releases 40 lit of molasses. Such molasses are invariably used to produce ethanol by industry called as distillery. Often such industries are attached to sugar industries themselves. Conversion of molasses in to ethanol (which is the primary objective of distillery) releases a large quantity of liquid waste called spent wash. Processing of each litre of molasses results in 13-14 lit of spent wash. The spent wash in most distillery industry is a source of pollution of both ground and surface flown water. But, it contains sizable quantity of plant nutrients (0.1% N, 0.06% P_2O_5 and 1% K_2O). Many distilleries have found out the productive way of using such a large liquid waste by diverting the spent wash to grow sugarcane, banana, vegetables and many field other crops. The actual potentiality to use spent wash for crop production can be realised, if the size of it production is perceived.

Table 5.29. Potentiality and manurial value of distillery spent wash

Estimated numbers of distilleries in country	Estimated quantity of spent wash (million tr/yr)	Manurial value of spent wash million kg/year		
		N	P_2O_5	K_2O
248	41000	41	24.6	410

*based on average composition of spent wash as 0.1% N, 0.06% P_2O_5 and 1% K_2O

5.6.2 Rice industry

India is one of the largest producer of rice only next to China. On an average, we produce 80 million tons of rice, out of which about 4 million tons is used to manufacture puffed rice/ha (where entire kernel is used). Remaining 76 million tons of rice is invariably milled to produce raw rice and parabolied rice. As 25 per cent of paddy grains make husk, the rice mills produce around 19 million tons husk. In addition, it also results in 5.3 million tonnes of rice bran, which is mainly used by solvent extraction plants to produce rice bran oil or used as cattled feed. But, husk does not find any useful avenue - except that a small quantity of husk is used as polutry littler bed in deep litter system of poultry. In most paddy growing areas, husk is used as a popular fuel.

But, rice husk contains 43 per cent cellulose 20 per cent lignin, 19 per cent pentosans - all being decomposable organic materials. Unlike other crop residues it is a poor source of manure preparation presumably due to high C : N ratio. In laboratory analysis, it is found to have 0.3-0.4 per cent N, 0.2 - 0.3 per cent P_2O_5 and 0.3 - 0.5 per cent K_2O.

Scientific studies confirm that rice husk can be used as good source to prepare manure are scanty. A suitable technology to decompose rice husk at a faster rate needs to be evolved probably by using decomposer cultures or by suitably adding N rich materials during decomposition. No reports are also available to indicate the response of husk based manure on soil properties and crop yields. If the husk is used to prepare the manures, it will have manurial value of 76,000 tons of N, 51,000 tons of P_2O_5 and 95,000 tons of K_2O.

5.6.3 Cotton industry

Due to introduction of intra-hirsutum and interspecific hybrids and due to general improvement of technology to cultivate cotton, the nation is able to produce 325 lakh bales of cotton (each table = 170 kg). This works out to 55.25 million tons. Decomposable organic wastes are produced in substantial quantities on fields as well as during various stages of cotton industry. The organic wastes in cotton cultivation include cotton stalks, while at various stages of industry, linters, hulls, cotton dust and spinning wastes are generated.

5.6.3.1 Cotton stalks

From7.6 million ha cotton area, around 35 million tonnes of cotton stalks are produced annually. At present they are invariably burnt as fuel in most cotton growing villages. These stalk contain as much as 25 per cent organic carbon, 0.4 per cent N 0.3 per cent P_2O_5 and 0.235 per cent K_2O. As the stalks are not easily decomposable, they are never treated as source of manure. CIRCOT, Bombay has found some methods to decompose cotton stalks. Many more methods could be tried to improve the decomposition of stalks (by using decomposer cultures or pulverizing the stalk). But, a vast manurial resource of around 14 lakh tonnes N, 10.5 lakh tonnes P_2O_5 and 8.75 lakh tonnes K_2O besides 8.75 million tonnes of organic carbon should not be allowed to be burnt.

5.6.3.2 Linters and hulls

Linters are small fibres on cotton seed that escape ginning machines, when ginning is undertaken. Hull is an outside coating of cotton seed. Both these are removed from cotton seed before oil is extracted from cotton seed. Approximately 65 per cent of cotton comprises seed and therefore, annual cotton seed production works out to 35.9 million tonnes. On weight basis, 0.5 per cent of it is linter and 3 per cent is hull.

Thus, annual production of 18.0 lakh tonnes of linters and 107.7 lakh tons of hulls are available for manurial purposes. At present, they have never been used as source for preparing organic manure. The composition of linters and hull and there manurial value is presented in Table 5.30.

Table 5.30 Nutrient composition of cotton wastes and their manurial value

	N(%)	P_2O_5 (%)	K_2O (%)
Linters	1.4	0.6	1.2
Hulls	1.1	0.5	1.3

Their manurial value could be worked (tons) ont as

	N	P_2O_5	K_2O
Linters	25200	10,800	21600
Hulls	118470	53800	140010
	143670	64650	161610

Such a vast nutrient resource is never used for manurial purposes. The methods to use them for preparing manure and constrains of their decomposition could be standardized so that a suitable technology is adopted to use them for crop production.

5.6.3.3 Cotton dust

Cotton dust is produced during ginning, pressing processes out of blowing rooms. On an average, around 110500 tonnes of cotton dust is produced from our cotton industry. It contains 70 per cent organic matter (mainly cellulose) and 1.4 per cent N, 0.6 per cent P_2O_5 and K_2O. It is an ideal decomposing matter in any manure pit due to its narrow C : N ratio. A recovery of 70 per cent of manure is expected.

At present, this resource is also not used to prepare manure. It is simply wasted in and around ginning mills. Efforts are needed to collect it and use for preparing the manures out of it.

5.6.3.4 *Spinning wastes and cloth wastes*

India has 766 spinning mills with a total yarn production of 1.337 million tons. It also has 282 weaving mills producing 13452 million mt cloth. Irrespective of type of cloth and yarn, large amount of wastages are expected at spinning and weaving levels due to reasons of rejection of cloth, neps in yarns etc. On a conservative estimation, roughly 20,000 tons of spinning wastes and 40,000 tons of cloth wastes can be expected. All of such wastes contain cellulose and substantial quantity of plant nutrients. All of them are decomposable too. At present they are never recycled in any processes to produce manures. No estimate is available about their nutrient composition. But, such a vast resource should not be allowed to be wasted.

5.6.4 Tea/Tobacco industry

Leaf is the economic part of tea and tobacco crops. During the processing, handling, grading of these leaves in tea/tobacco industries, large quantity of wastes are generated due to rejections, sampling wastes and other reasons. Annually it is estimated that 10,000 tonnes of tea wastes and 60,00 tonnes of tobacco wastes are generated during various stages of handling and processing. Except some part of tea leaves being used to produce caffeine, they are not found used in any way. All of them are decomposable with a narrow C : N ratio. Tea wastes contain 3.5 per cent N, 0.4 per cent P_2O_5 and 1.5 per cent K_2O while tobacco wastes have 1 per cent N, 0.8 per cent P_2O_5 and 0.8 per cent K_2O. If suitable arrangements are available to decompose them, they could yield substantial quantity of nutrient rich manures.

Projections of this nature could be made available to show the large potentialities existing in using the wastes generated in other agro-industries like, silk industry, jute industry, paper industry, fruit processing industry, coconut based/areca based industries, coffee industry and many other food industries. Most of them generate carbonaceous wastes not containing hazardous chemicals.

It only needs shift in mindset of industrialists of know that the wastes generated by their industries are indeed not wastes. They are valuable plant nutrients and they could be used to prepare manures by proper decomposition. If the cost of preparing manure out of these wastes is a deterrent factor to undertake such works, government should come forward to sponsor such programmes to ensure that wastes are converted into useful manures.

Our country is not in shortage of organic sources to make the organic farming successful. To make use of resources available in our society, it only needs change in the attitude of farmers, policy makers, industrialists about possibilities of organic farming. The organic resources available around us could be explored in all possible ways to generate the required quantity of manures. In this connection, efforts made by Kadalli *et al.* (2001) to convert coir dust into manure, coffee pulp into compost deserve attention. Similar efforts in other organic somes could be made in future.

5.7 VERMICOMPOST

Earthworms are regarded on farmer's friends looking to their capacity to enrich the soil by nutrient filled feces left by them during the course of their normal life. In the last 15 to 20 years, this capacity of earthworms is commercially exploited to artificially allow the earthworms to feed on any decomposable organic waste and generate large quantity of irch manure. It is relatively new branch of science-wherein good deal of research work is needed. However, the methodology of manure preparation, synchronization of earthworm's life cycle, nutrient composition of manure are already well studied and documented.

The term `vermicopost' refers to manure prepared out of large scale rearing of earthworms in an artificial or natural pit. Earthworms can be used in specially designed pits or in any ordinary FYM pits. Frequency, a term vermiculture is also used to denote a culture of young earthworms which are let in any undigested compost pit or in an agricultural field directly. They reproduce and grow in number. The fecal wastes from large numbner of earthworm is available as manure. Generally, this method is adopted when soil has large undecomposed organic matter. The establishment of required population of earthworms may take more time to really contribute sizable manure.

The other popular method is to prepare vermicompost in a natural/artificial pit and allowing them to feed on wastes available in the pits. Commerical scale vermicompost production is attempted successfully. It takes around 45 to 60 days to get a well prepared manure. Few turnings are needed in between. Vermicopost has 1.0 to 2.0 per cent N, 0.8 per cent P_2O_5 and 0.6 per cent K_2O and can be successfully used as manure rich in N, P, K_2O as well as micro nutrients in comparison with FYM.

Some latest research works about vermicompost reported in last few years are listed in Table 5.31.

Table 5.31. Recent advances in vermicopost research

S.N.	Nature of study	References
1.	In open field condition, sugarcane trash and sunflower residues were found as suitable substrates for earthworm	Biradar *et al.*, 2000
2.	Application of vermicompost enriched rockphosphate increased the grain yield of cowpea by 490 kg/ha	Shailaja Kumari and Usha Kumari, 2002
3.	Mean earthworm populations were higher in fallow plots of degraded alfisols due to higher litter fall, lower soil temperature and soil moisture	Tian *et al.*, 2000
4.	No till soils had 100 per cent higher mammalliated Vughs (relatively large irregularly shaped pores) due to higher earthworms population.	Vanden Bygaart *et al.*, 2000
5.	Vermicoposting reduced C : N ratio and increased humic acid content, cation exchange capacity and water soluble carbohydrate content of 150 days of composting.	Talashilkar *et al.*, 1999
6.	Grain yields, organic carbon content of available N, P, K micro-nutrients, CEC were higher due to vermicompost application in rice	Vasanthi and Kumaraswamy, 1999

5.8 OTHER LIVE STOCK WASTES

Apart from cattle and buffaloes, large number of other domesticated animals offer great scope for organic farming, when their dung and urine can be collected and usefully converted into manure. Anmals such as sheep, goat, horse, elephants, pigs and camel are in such a large number that their manurial contribution cannot be easily ignored. Although they are conventionally uncommon sources - the manurial importance of these wastes and compelling reasons to recycle all possible organic wastes require the renewed attention on sources.

Out of these animals, sheep and poultry excreta are already utilized as manure. Their excreta is directly applied to cultivated lands. Nevertheless, sheep manure is an underexploited source, as only a small part of it used by pegging a given herd of sheep on the cultivated land prior to cropping season. But, major part of their dung and urine are never collected. So is the case of all other animals.

Table 5.32 and Table 5.33 indicate the manurial value of these livestock wastes in India based on 2001 data.

Table 5.32. Contributions of other livestock

Animal	Population (million)*	Daily excreation		Annual excreation	
		Dung (kg)	Urine**	Dung (lakh tons)	Urine (lakh tons)
Sheep and goat	165.9	0.3	0.2	181.7	121.1
Pigs	12.8	2.0	5.0	93.4	233.6
Horses	0.80				
Donkeys	0.07	3.4	2.9	58.8	38.8
Camels	1.04				
Others	0.41				

* = Anonymous (1997) ** = Gaur *et al.* (1990)

Table 5.33. Manurial value of dung and urine from other livestock.

Animals	Annual excreation (lakh tons)		Chemical composition**						Manurial value (lakh tons)		
			N		P_2O_5		K_2O				
	Dung	Urine	Dung	Urine	Dung	Urine	Dung	Urine	N	P_2O_5	K_2O
Sheep & goat	181.7	121.1	0.65	1.70	0.50	0.02	0.03	0.25	3.24	1.84	0.36
Pigs	93.4	233.6	0.60	0.40	0.50	0.10	0.20	0.50	0.93	0.56	0.65
Other animals	58.8	38.8	0.50	1.20	0.30	–	0.30	1.00	0.54	0.30	0.16
									4.71	2.70	1.17

** Gaur *et al.* (1990)

Each type of animals poses different set problems of collecting the dung and urine.

Sheep and goat : Quantity of dung and urine per animal is very less and it will be a challenging task to collect such small quantity of dung urine from a herd of larger size Hence, practice of pegging is popular to allow the dung and urine from a given herd to fall at the desirable place.

A scientific improvement of the practice of pegging the sheep and goat is needed so that the valuable resource of their dung and urine will not go wasteful.

Pigs : Scienfitic and housed pig rearing is practiced by very few farms in India, while more than 95 per cent of pig population is left open to eat and thrive. They are only caught for meat. Collecting dung and urine from housed farms may not be difficult, if proper cleaning and washing arrangements are made to collect the droppings. But collecting the dung and urine is more challenging when the pigs are left to thrive in nature. They offer a very stiff resistance, when a human being approaches them.

Similarly, other livestock like camel, horses, elephants, rabbits offer problems peculiar to their rearing/ living environment.

Till now, no research Information is reported on methods to collect dung and urine from cattle or other live stock - especially in Indian situation where animals are not housed but are allowed to graze/feed outside for a period ranging from 10-12 hours. Although it is indeed a challenging task to plan and collect this valuable resource and initial efforts may appear to be impractical, it is worth formulating newer methods to collect the dung and urine from this vast biological resource for the benefit of organic farming. After all, it can offer a manurial source of 4.71 lakh tonnes of N,2.70 lakhs tonnes of P_2O_5 and 1.17 lakh tonnes of K_2O annually.

The possibilities of using dung and urine of other livestock depends upon success of collection mechanism, the cost of transportation from generation to utility points, the necessity of dung and urine at a given place for FYM preparation and the availability of biomass at a place for using the dung and urine in order to prepare FYM.

5.9 BIO-CONVERSION OF UN-UTILIZED/UNDER UTILIZED SOURCES

Some unconventional sources of organic farming remain unidentified and unutilized. Although the sectors, which generate these sources, are part of our society- they are never utilized for organic farming on a large scale. Total utility of these sources may not be immediately possible. Whether they could be used as a last source of organic manures or they could be simultaneously exploited with other sources - depends upon the stigma attached to their use.

The are grouped as (a) Human excreta and urine and (b) byproducts of slaughter houses.

5.9.1 Human excreta and urine

Around every village in India, the soil is fairly rich as compared to soils of lands little far off from dwelling places. They have commanded better market rates and have been known for high fertility. It is well known that this phenomenon is due to use of lands in the immediate vicinity of dwelling place as open toilets. So, the experience of manurial contribution by human excreta and urine is not new to Indian society.

Table 5.34. Manurial contribution of human excreta and urine in India (population 1031 million)

Annual excretion (milli. tons)		Chemical composition*						Manurial value (lakh tons)		
		N		P_2O_5		K_2O				
Excreta	Urine@	E	U	E	U	E	U	N	P_2O_5	K_2O
50.05	451.58	1.60	1.00	1.20	0.15	0.55	0.20	53.10	12.77	11.70

E - Excreta U - Urine

@ 0.133 kg excreta and 1.2 litre urine/day/capita

Adapted from Gaur *et al.* (1990)

Nevertheless, the huge resource of human excreta and urine remain unexploited to a great extent. In fact, the extent of human excreta and urine generated in our country and their manurial contribution is so high that - it is hardly possible to ignore this vast source (Table 5.34).

Utilizing the vast nutrient source in excreta and urine of human beings involves many management challenges and necessary technical processing of sources.

5.9.1.1 Management challenges

More than 65 per cent of Indian population being in rural areas and none of the villages provided with internal drainage system, collecting human excreta and urine from the rural lavatories is indeed a stupendous task. Whereever septic tanks are provided, collection of an accumulated and partly decomposed excreta may not be easy. But in vast rural areas, the people use open air toilets and thereby their excretions are spread daily on any open field near their place of dwelling. Collecting them is almost next to impossible.

In cities, where internal drainage is provided, the excreta and urine become part of sludge and sewage. The utility of sludge from cities could be planned, as the collection mechanism is fairly established, in many cities, where septic tanks are provided, the excreta will have to be collected an periodical basis.

The excreta and urine of human beings, thus, offers an excellent scope to manufacture rich organic manure is large quantities. But, the scope is subject to limitations like :

(a) Establishment and strengthening of collection mechanism.

(b) Providing septic tanks and lavatories to a larger rural population.

At any rate, the vast resource of nutrients cannot be allowed to be wasted for the reasons of poor collection mechanism. The present level of use of night soil and urine for manurial purpose is negligible.

5.9.1.2 Processing of excreta and urine

Excreta has a very narrow C : N ratio (due to which its decomposition is fast) and requires very little processing. When handled in pure form or mixed with other organic source in the form of sludge - it requires to be dried in such a way that minimum amount of nutrients are lost. This involves mixing the excreta/sludge with dry soil, saw dust and turning it once or twice. Later, this could be used in preparing FYM or used directly as manure. Once dehydrated and nutrients trapped in dry soil or dust, night soil has no smell and easily to handle.

5.9.1.3 Collection of urine

Human urine is a rich source of nitrogen. Its collection mechanism unfortunately involves a great deal of problems. In most situations, including internal drainage network, urine is mixed with sewage and diluted. Hence, its nutrients values are realized when sewage farming is practiced. As the loss of nitrogen from urine is directly proportional to the time of its exposure to air it is desirable to collect the urine as early possible. An interesting study was carried out by ICAR, wherein collection of human urine was made easy without more loss of nutrient value by redesigning of public urinals. The decomposition of urine was curbed by acidulating materials like superphosphate or hydrochloric acid. The litter was continuously soaked and dried as and when urine flows from urinals by making use of materials like saw dust or charcoal dust. Later the dried litter was converted into human urine cakes on further dehydration. Trials on use of such cakes on wheat crop were also conducted on wheat crop in Delhi - wherein the increase in yield upto 28 q/ha was repaired as compared to use of ammonium sulfate (Gaur *et al.*, 1990).

In summary, it can be concluded that human excreta and urine are rich source of plant nutrients and taking their total manurial value into consideration in a populous country like India such a vast resource can not be allowed to be wasted. Even partial utilization of excreta and urine can replace large quantity of fertilizers.

In recent years, great interest has been shown to human urine as a part of recycling anthropogenic waste for productive purposes. As the collection of urine itself is a challenging task, specially designed toilets under the name ``ECOSAN" have been used to automatically collect the urine from masses. A recent study at Bangalore has indicated that use of human urine with varying dilution resulted in good quality high yields of banana crop on large scale. The market quality and biochemical properties in such study have indicated that use of human urine will not affect the banana crop in any way. The large scale use of human urine for agriculture production not only needs total restricturing of toilets but also revolutionary change in mindsets of people.

5.9.2 Byproducts of slaughter houses

More than 4,000 slaughter hourse are functioning in India with various degrees of slaughtering capacities, levels of health and hygiene maintenance and many methods of disposal of wastes. They handle nearly 40 million sheep/goat and 1.5 million buffaloes annually (estimated by Gaur *et al.,* 1990). Although systematic documentation is not available in most slaughter houses, it is amply clear that large quantities of bones, blood, slaughtering wastes, hoofs and horn meals as well as leather wastes are produced annually from these slaughter houses. Estimates of slaughtering the animals outside slaughter houses are lacking. Wastes of slaughter houses are not reported to have been used for crop production.

Bone meal

An estimated quanity of 4.5 lakh tons of animal bones are available in our country, annually, out of which nearly 1.4 lakh tons are collected and crushed. Remaining quantities of bones are yet to be utilized. They remain unutilized for various reasons such as difficulties in collection, high cost of processing and lack of demand. Well crushed and steamed bones are ideal phosphatic fertilizers as they contain 25 to 30 per cent P_2O_5 and 1 to 2 per cent nitrogen. According to one estimate, annually it is possible to supply 9000 tons of nitrogen and 1.1 lakh tons of P_2O_5, if systematic efforts are made to use all available animal bones in the country to make nutrient grade bone meal.

Use of bone meal as a source of plant nutrition is a traditional method. But, application of scientific knowledge in developing better technology to produce good quality bone meal efficiently is lacking. With strategies to utilize all the sources of animal bones and with newer technologies to produce better quality bone meal, it is possible to recycle valuable natural resource as a part of organic farming.

Blood meal

In most slaughter houses, blood is not properly collected and allowed to be dried and therefore decomposed at a wrong place. As it contains 10 to 12 per cent N, one to two per cent P_2O_5 and one per cent K_2O with a C : N ratio of 1 ; 3-4, it decomposes at a fast rate and most of nitrogen is lost to air if not trapped in a suitable medium. It is estimated that 55,000 tons of animal blood is produced annually in slaughter houses and most of it is unscientifically handled. If the slaughter houses are modified and redesigned to collect the blood, it is possible to utilize manurial value of animal blood more efficiently. It is necessary to provide a good soaking medium like soil, saw dust, chopped straw in the collecting blood and ensure that blood is soaked in the medium before it is decomposed. Later, such medium could be used to prepare the manures. The manurial value of animal blood is estimated to be 5,500 tons of N, 550 tons of P_2O_5 and 550 tons of K_2O.

Meat meal

Man has developed taste to prepare food almost from every part of slaughtered animals. Legs, heads ad other parts are also utilized for food besides muscles and other internal parts of animals.

Nevertheless, slaughter houses produce large quantity of meat not normally used as food. It is estimated that nearly 1.2 lakh tons of decomposable animal wastes are generated from slaughter houses. However, in India, the slaughter houses seldom classify the wastes into decomposable and indecomposable groups thereby allowing large valuable nutrient source to go waste. Meat meals so generated have not found any useful application till now in India. Systematic efforts to collect them and decompose them with other residues will help to prepare nutrient rich manures. As, much of animal wastes from slaughter houses (except bones, hoofs and leather) have a narrow C : N ratio of 1 : 5 or 1 : 7, their decomposability with other crop residues is very high. Hence, they would enhance the decomposition process.

Leather wastes

Although major part of leather reaches leather industry for leather processing, sizable leather wastes are generated at slaughter house level. It is estimated that about 5,000 tons of leather wastes ae generated annually from slaughter houses. They could be used for manure preparation.

5.9.3 Bio-gas slurry

An estimated number of 20 lakh bio-gas plants are working in India. Bio-gas plant being a digester to convert crop residues and animal dung into combustible gas - generates large quantities of bio slurry. It is rich in plant nutrients. On an average it contains 1.4 per cent N, 0.8 per cent P_2O_5 and 0.6 per cent K_2O. Although it is a rich source of plant nutrients, gradual and slow generation of slurry makes its collection difficult. Moreover, if exposed to air for longer time it loses most of its nitrogen. Hence, it is ideal to quickly transfer the slurry to a farm yard manure pit under concealed conveying pipes so that maximum utility of nutrients is harnessed. Otherwise, the slurry will have to be soaked in a soaking medium and later transferred to FYM pit at convenience. No estimate of quantity of biogas slurry generated in India is yet made. However, 7 to 8 kg of slurry can be expected from each bio-gas plant each day. This works out to 52 lakh tons of biogas slurry annually with a manurial value of 72,800 tons of N, 41,600 tons of P_2O_5 and 31,200 tons of K_2O.

5.10 POULTRY MANURE

Poultry manure is one of the most potential organic source for farming activity awaiting fuller exploitation. Poultry industry is especially well developed in states such as Andhra Pradesh, Punjab, Tamil Nadu, Karnataka and Maharashtra. State wise poultry bird population is presented in Table 5.35. As each poultry bird on an average produce 0.025 kg litter (excreta) per day on dry weight basis, 2.8 million tons of poultry manure is produced annually in our country. The average manurial value of this poultry manure works out to 27,553 tons of N, 36,772 tons of P_2O_5 and 21,941 tons of K_2O.

Table 5.35. Potential nutrient value in poultry manure in India

States	Number of poultry birds (million)	Potential litter* production dry basis/year (000 tons)	Potential manurial value (tons/year)		
			N	P_2O_5	K_2O
Andhra Pradesh	49.98	456.0	4560	6384	3648
Assam	16.39	149.5	1495	2093	1196
Bihar	17.65	116.0	1610	2254	1288
Gujarat	5.66	51.6	516	723	413
Haryana	8.58	78.3	783	1096	626
Jammu & Kashmir	4.63	42.3	423	592	338
Karnataka	16.16	147.5	1475	2065	1180
Madhya Pradesh	11.80	107.7	1077	1508	862
Maharashtra	32.19	293.7	2937	4112	2349
Manipur	3.26	29.7	297	416	238
Meghalaya	1.82	16.6	166	232	133
Nagaland	21.70	19.8	198	277	158
Orissa	13.06	119.2	1192	1669	954
Punjab	18.33	167.3	1673	2342	1338
Rajasthan	3.05	27.8	278	389	222
Tamil Nadu	24.55	224.0	2240	3136	1792
Tripura	2.59	23.6	236	331	189
West Bengal	37.41	341.4	3414	4779	2731
Uttar Pradesh	10.79	98.5	985	1379	788
Kerala	21.90	199.8	1998	2797	1598
All India	307.07	2802.0	27553	36774	21941

Composition of poultry manure; 1.0 per cent N, 1.4%, P_2O_5 and 0.8% K_2O.

- Based on generation of 25 g dry litter/bird/day on an average.

There are two systems of poultry farming :

1. Deep litter system

In this system, the birds are allowed to live on a straw bed on the floor and the litter is naturally mixed with straw bed and subjected to partial decomposition. As the manure is mixed with bedding, the nutrients in the litter are diluted. On an average the manure from deep litter system contain 0.6 to 0.8 per cent N, 1.0 to 1.2 per cent P_2O_5, 0.5 to 0.6 per cent K_2O as well as 500 to 600 ppm of zinc, 25 to 300 ppm of iron. The collection of manure in this system requires frequent shifting of birds and therefore coincides with life cycle of a batch of birds.

2. Cage system

In this system, the birds are reared in cages and their droppings are allowed to fall on floor in the form of heap. This heap of poultry litter is regularly collected and used. As this manure is not allowed to mix with other organic materials, the nutrients are more concentrated. As the collection of droppings is independent of birds' life cycle, frequent collection of manure in pure form are possible. On an average, the manure form cage system contain 1.1 to 1.6 per cent N, 1.2 to 1.5 per cent P_2O_5 and 0.7 to 1.0 per cent K_2O along with 800 to 1000 ppm Zn, 400 to 500 ppm Fe.

Generally, the quantity of litter per bird from layers (birds reared for eggs) is less than that from broilers (birds reared for meat) as the litter production is a function of feed consumption and feed consumption is more in broiler farm than layer farm.

5.10.1 Methods of use

The present method of disposal of poultry litter involves more wastage and unscientific handling. Most of manure produced by poultry farmers is not dried scientifically and allowed to undergo decomposition. Sometimes, it is simply dumped in wet condition in a place - where its nutrients are easily lost. In many situations, the nutrients in the poultry litter are lost by volatilization, leaching before it is applied as manure.

Ideally, the NH_4-N contained in poultry manure needs to be trapped by use of suitable drying materials at the time of drying. Chopped straw, saw dust and other materials could be conveniently used. Essentially, the poultry manure has to be dried by collecting it as early as possible from the place of its generation - without losing the nutrients in it. Alternatively, wet litter could also be used to prepare farm yard manure, compost etc. As C : N ratio of poultry manure is as less as 1 : 5 or 1 : 4, it is ideal material for enhancing the decomposition of residues and wastes of higher C : N ratio.

5.10.2 Response to application of poultry manure

Application of poultry manure has been reported to improve the :

(a) Crop productivity and quality

(b) Availability of nutrients in soil

(c) Soil characters

Crop productivity and quality

Grain yields of maize and pod yield of groundnut are reported to have increased in heavy rainfall tract of Meghalaya (Das *et al.*, 1991). Similarly, yield improvement in maize (Mbagwu, 1992), potato (Makarenko *et al.*, 1991) are reported by application of poultry manure. Italian workers have reported that poultry manure was responsible to improve fruit colour of tomato, size of onion and potato (Pimpini *et al.*, 1992).

Availability of nutrients

Madhumita *et al.* (1991) have inferred that application of poultry manure was superior than application of FYM in improving available P_2O_5, K_2O, Ca and Mg in acid soils with low organic matter status. Egyptian scientists have observed that poultry manure was responsibe to increased uptake of Mn and Zn in sandy soils (Elgwla *et al.*, 1990). For the similar reasons, Rainey *et al.* (1992) found in Arkansas (USA) that poultry manure is most suitable amendment in infertile paddy soils exposed by leveling. Although poultry manure contains nitrogen in the range of 0.8 to 1.1 per cent, most of it is likely to be lost by volatilization during drying. Gloria *et al.* (1992) have found that such loss could be prevented effectively by using powdered superphosphate as an additive with poultry manure. Application of poultry manure could increase Zn uptake (Elgwla *et al.*, 1990), improve mineralization of organic carbon (Mbagwu, 1992) and improve uptake P (Das *et al.*, 1999).

Ritcher and Ritcher (1992) found that application of poultry manure increased N mineralization, while Maraikar and Amarasiri (1989) found that poultry manure was as effective as muriate of potash in increasing exchangeable K in red - yellow pod zolic soils of Srilanka. Bijay Singh *et al.* (1992) concluded that application of poultry manure increased DTPA extractable Zn in calcareous soils of Punjab under flooded conditions.

A great deal of work is reported in recent years to show that poultry manure has number of advantages. Few of them are listed in Table 5.36.

Table 5.36. Response to added poultry manure

Sl. No.	Study	Reference
1.	High rates of broiler litter rapidly increase available P in surface resulting in increased yield of cotton, pearl millet, wheat	Gascho (1999)
2.	Peanut yield was increased by poultry litter application as compared to equal quantity of NPK through fertilizers	Adams and Hartzog (1999)
3.	Composting poultry litter results in reduced N mineralization, thereby making N over long period of time	Preusch *et al.* (1999)
4.	Long term application of poultry manure tends to increase bio-available soil phosphorus	Tapadar *et al.* (1999)
5.	Dry matter yields increased upto 2.4 t ha^{-1} in tall fascus (fodder crop) by applying 18 mg ha^{-1} of poultry manure	Ruiz-diego *et al.* (1999)

Improved soil characters

A Nigerian study has indicated that the application of poultry manure at 50 Mg ha^{-1} can improve the physical properties of soil like volumetric water retention, infiltration rate and total poultry manure to correct soil acidity. He inferred that 5 to 10 g of poultry manure per kg soil is equivalent to 6 to 7 cmol/kg.

5.10.3 Prospects of use of poultry manure on large scale in India

Poultry enterprise in India is not as developed as in Europe and USA as indicated by per capita availability of eggs. In India per capita availability of egg is as less as 39 as compared to 270 in USA and 242 in UK. Hence, the production and availability of poultry manure is on a limited scale looking the vast cultivated land that India has.

Nevertheless, poultry manure can be considered as an important organi source of plant nutrients in the vicinity of its production. Most poultry farms are situated in and around metropolitan and other urban countries. As concluded by Weik *et al.* (1979), logistic handling of poultry manure dictate that it is better utilized economically near to its source due to loss of comparative advantage over other sources - when transported to far off distances.

Hence, the large scale use of poultry manure may be restricted to agricultural area near cities - where poultry units are located. If this is the state of organized poultry enterprise, large number of birds are spread over the country in unorganized sector with size of herd as small as 8 to 10 / farmer. The birds are reared by many small and marginal farmers in and around their place of residence. Collecting the poultry manure from these birds is indeed onerous task although on a yearly basis collection of manure from even small size of heard also leads to substantial poultry manure.

5.10.4 Limitation on use of poultry manure in India

Although manurial value of poultry litter is well established and large quantity of poultry manure produced in our country deserves to be exploited as an important organic source of plant nutrients, many practical limitations restrict their large scale use in our country. Suitable strategies will have to be developed to overcome them gradually before poultry manure could be really used on large scale.

1. As cage system is more capital intensive, most Indian poultry forms adopt deep litter system. Hence, production of good quality poultry manure is restricted to large scale poultry farms run by rich farmers and corporate farmers.
2. Awareness of people who handle the poultry manure with regard to its nutrient contents is at its lowest level. Generally, they lack proper training in handling poultry manure scientifically.
3. Tranportation of poultry manure from small and big poultry farms posses practical problems and cost.
4. No systematic infrastructure is developed to produce, preserve and market the poultry manure.
5. The awareness of farmers to use poultry manure as a rich source of plant nutrients need to be strengthened by regular training programmes.
6. The pouiltry manure may not be applied directly on any field and it needs a air curing for 10 to 15 days before it is used on field for direct application. However, poultry manure may be directly used in preparation of manure.

5.11 BIO-FERTILIZERS

In recent years, a new class of fertilizers called `biological fertilizers' or `bio-fertilizers' has come to forefront as a means of supplying nutrients to the crops. They neither refer to synthetic source of nutrients (inorganic chemical fertilizers) nor to organic source of nutrients (manures, residues etc.).

Bio-fertilizers refer to either :

(a) Culture of micro-organism applied to soil/seed in order to encourage fixation of N or release of P_2O_5

(b) Biomass of certain plants - which contain large quantities of N by way of their association with N fixing organism.

Mainly, following micro-organisms are regarded as bio-fertilizers.

(a) Symbiotic N-fixing bacteria - Rhizobium

(b) Asymbiotic N -fixing bacteria - Azotobacter, Azospirillium

(c) P-solubilizing fungi-mycorrhiza - Vascicular Arbiscular Mycorrhizae (VAM)

(d) P-solubalizing bacteria

(e) Anabena - Azolla association

Although a great deal of research work is reported on utilizing above sources as a replacement to fertilizers, their practice in practical agriculture is on a limited scale. Use of bio-fertilizers in India on commercial scale is limited to use of Rhizbobium culture on pulses and on a very limited scale use of Anabena-Azolla manure in paddy cultivation.

As an important component of organic farming, bio-fertilizers can attain a very prominent position because it is most efficient and least damaging way of supplying nutrients to the crops. As N available from such sources is slowly and gradually available after mineralization, it suits the crop requirement and do not cause pollution and do not result into losses. Similarly VAM can released fixed P from soil and may substantially reduce any P additon from outside sources.

5.11.1 N-fixing rhizobia

Association of legume roots and their capacity to enrich soil is well reflected in age old practices of using pulse crop in various cropping systems. However, the association of rhizobium (inside the root nodules of leguminous crops) and their capacity to fix the atmospheric nitrogen has drawn great attention in the last 30 years. Such symbiotic association between legume roots and rhizobia is exploited profitably by applying a bacterial culture to the soil/seed so that bacterial population proliferates in the soil. Various steps involved in symbiotic N fixation are :

- Application of rhizobium culture to soils
- Proliferation of rhizobial population
- Rhizobia infects the crop roots

- Nodule formation
- N-fixation
- N-utilization by host
- Decomposed roots enrich soils

As rhizobium bacteria exhibits crop specificity, different strains of rhizobia will have to be used for different crops. Invariably, the bacterial culture (supplied in polyethylene packets) is mixed up with granulated FYM and spread on the field and stirred by suitable cultivation. Bacterial culture mixed with FYM can also be applied in bands or around individual plants. Seed coating along with slurry of culture is also used. Some reports indicate that culture is applied in rows as slurry.

The application of rhizobial culture to enrich soil N and to improve the performance of crop - is restricted to leguminous crops alone. Hence, the manurial value of bio-fertilizer to other crops can be realized only as residual content in soil. Various cropping systems are in practice to exploit the N-fixation mechanism by pulses.

1. Intercropping (groundnut + sunflower, maize + redgram, jowar + bengalgram, cotton + cowpea)
2. Sequential cropping (greengram-wheat, bengalgram-Jowar, groundnut-rice, cowpea-sunflower)
3. Relay cropping (rice-cowpea, rice-greengram).

They are the common methods through which N-fixation by Rhizboia is utilized by other crops. A crop subsequent to a legumnious crop is always reported to have performed better due to accumulated nitrogen. The benefits of using rhizobial culture are well documented since many decades. The increased yields, enriched soil, improved soil physical conditions are reported on large number of crops. Few latest works are listed in Table 5.37.

Table 5.37. Response of rhizobium inoculation on crop yields

Sl. No.	Nature of study	Reference
1.	Rhizobium inoculation increased total bacteria, actinomycetes besides increasing rhizobium population. It increased the yield of Mungbean by 10 per cent	Singh and Tarafdar, 2002
2.	Innoculation of brady rhizobium Japonicum to Soybean roots increased nodule weight and seed yield by 140 mg/plant and 3-4 q/ha respectively	Dubey, 1999
3.	Dual inoculation of VAM and rhizobia resulted in increased root infection increased N-uptake by greengram	Das *et al.*, 1999

4.	Rhizobium inoculation of frenchbean increased the grain yield by 3.3 q/ha	Pant *et al.*, 1999
5.	Strain A150 of Rhizobium leguminosarum was found to be suitable for inoculating Vicia faba, as it could increase total N in root and shoot	Schulze *et al.*, 2000
6.	Rhizobium inoculation increased nodulation, grain yield and N-uptake in chickpea	Kumar *et al.*, 1998
7.	Rhizobium incoulation of pigeonpea increased the grain-yields by 300 kg/ha	Patel and Patel, 1994

In practical agriculture, the benefits of use of rhizobium culture are dependent on number of factors such as viability of culture at the time of use, initial status of rhizobium in soil, organic carbon status of soil, molybdenum content of soils as well as pH of soil. While, each one can act as decisive factor in ultimate success of practicing the biological N fixation, the viability of bacterial culture itself is most important. Large number of government, private and industrial lboratories in India have facilities to provide such bacterial culture on large scale. But, storing them in the temperature of 5° to 15°C till it is finally used, transporting them to various farmers fields have posed many practical problems in their use. No systematic distribution mechanisms through retain outlet are yet established as available in case of fertilizers. Over and above all these constraints, the actual benefits to the host crop is generally not more than 10 to 15 per cent yield increase. The residual benefit by succeeding crop is subject to many variations and is unpredictable by many management problems.

Nevertheless, N-fixation by rhizobia could be practiced successfully as one of the organic method along with other methods. In USA, it is projected that annual benefit of 1067 million $ could be obtained by biological N fixation through reducing N fertilizers by 15 to 47 lakh tons (Taues, 1989).

5.11.2 Non-symbiotic N fixation

Nitrogen is found to be fixed by non-symbiotic process in number of other crops like jowar, wheat, cotton, pearlmillet, finger millet and paddy. The application of bacterial culture remains same as in case of rhizobium culture. But, the fixation of N is not through nodulation but the bacteria will proliferate directly in soil and fix the atmospheric N. Mainly, Azotobacter and Azospirillum are used to fix N by non-symbiotic process.

Large number of reports are available to show the benefits of these organisms. Some of latest reports on beneficial effects of using Azotobacter and Azospirillum are reported in Table 5.38.

Table 5.38. Response of non-symbiotic bacteria inoculation on crop yields

S.N.	Nature of study	Reference
1.	Azotobacter along with FYM and green manure increased paddy yields upto 31 q/ha (ab. Control - 10 t/ha) and availability of N, P_2O_5 and K_2O to 106 kg/ha, 14 kg/ha, 140 kg/ha, respectively.	Surinder Singh *et al.*, 2002
2.	Uptake of N was increased by inoculation with Azospirillum in sesame	Singarvel *et al.*, 1998
3.	Azotobacter inoculation at 30 days and 60 days of composting increased NH_4+ and NO_3 and total N contents and decreased C : N ratio	Thakur and Sharma, 1998
4.	Innoculation of Azospirillum increased microbial biomass in rice	Kaloyanova and Kostov, 1991
5.	Innoculation of Azotobacter chroocoum increased the grain yield of pearlmillet by 0.3 t/ha	Marula *et al.*, 1991
6.	Nitrogen fixing activity was promoted in soil with plant residues where inoculated with Azospirillum	Krovchenko and Makarova, 1990
7.	Azotobacter inoculation of tomato by root dipping in slurry increased tomato yield by 6.6 t/ha	Kumaraswamy and Madalgeri, 1990
8.	Innoculation of Azospirillum increased the grain yield of wheat under stress conditions	Panwar, 1992

5.11.3 P-solubilising organisms

A part of applied P is available to a crop and it is necessary to repeatedly apply P every time to keep labile P level so that crop could utilize it. However, P fixation from applied P fertilizers is a recurring feature and in the course of fertilizer application for a long period fixed P levels in a soil are substantially high.

Table 5.39. Beneficial effects of P-solubilizing organism

S.N.	Nature of study	Reference
1.	Bacterial isolate of phosphate solubilzing property showed P solubilizing efficiency equivalent to 30 kg P_2O_5/ha	Dubey *et al.*, 1999
2.	Dual inoculation of VAM and rhizobium increased available P status from 0-6 kg/ha to 5-4 kg/ha	Das *et al.*, 1999
3.	Inoculation of VAM increased P uptake and sulphur uptake in S deficient soils	Banarjee *et al.*, 1999
4.	VAM (*Glomus fasciculatum*) inoculation increased root nodulating grain yield and N uptake in chickpea	Kumar *et al.*, 1998

In order to release the fixed P from fertilizer sources, the use of mycorrhiza caled Vasicular Arbiscular Mycorrhizae (VAM) is commonly employed. They are available in the form of microbial culture and can be applied to soil. Application of VAM culture is known to be highly beneficial to release fixed P and such application can replace application of additional P fertilizers. Bacterial P solubilizing agents like *Aspergillus awamorvi* are used. Many reports are made about the benefits of using VAM and bacterial culture. Some of them are summarized in Table 5.39.

5.11.4 Azolla-Anabena association

Anabena, a blue green algae known for atmospheric nitrogen fixation is associated with aquatic weed Azolla. Growing Azolla in paddy fields upto 40 to 45 days and incorporating the biomass of Azolla in paddy fields has been recommended to make use of N rich biomass of weed as manure. Anabena, a N fixing cyanobacteria is associated with Azolla. Therefore, biomass of Azolla itself serves as bio-fertilizer. It is indeed an excellent eco-friendly method to supply nitrogen to paddy crop. The utility of Azolla as a method of supplying N to paddy depends upon :

1. Availability of time to grow Azolla and transfer the biomass to paddy field.
2. Availability of separate space for cultivation of Azolla.

Large number of workers have reported the benefits arising out of using Azolla in paddy fields. Few latest reports are summarized in Table 5.40.

Table 5.40. Benefits of Azolla incorporation

S.N.	Nature of study	Reference
1.	Inoculation of Azolla increased grain yield, N content in plant and N uptake in sodic soils	Gopalaswamy and Kannaiyan, 1998
2.	N use efficiency was increased by integrated nutrient management system using N fixing Azolla	Kumar *et al.*, 1995
3.	Azolla-anabena association can fix N upto 150-200 kg/ha	Kannaiyan, 1998
4.	Azolla increased microbial activity and activity of soil enzymes	Kumar, 1994
5.	Azolla can fix N upto 224 kg N/ha theoretically. But, in practical term it can fix N upto 20-150 kg/ha	Roger and Ladha, 1990

5.12 CULTURAL METHODS

Organic farming in conjunction withc other methods can be practiced by cultural methods too. Many cultural methods are effective in augmenting organic sources of nutrients as well as in releasing nutrients and are therefore regarded as organic methods.

They do not involve any additional materials but mere shift in method of cultivation/sequence or pattern of cropping/ systems of soil and water management.

As a matter of fact, many of these methods are part of our agriculture since ancient times. Some are of recent origin. Their importance in terms of nutrient contribution is being realized in recent times.

Important cultural methods include :

1. Effective and useful rotation of crops.
2. Ideal intercropping with complimentary interactions.
3. Alley cropping with N-fixing trees
4. Establishment of vegetative bunds
5. Practice of homestead farming
6. Soil and water conservation practices
7. Fallowing

5.12.1 Crop rotation

Rotation of choice crop may lead to increased nutrient content in soil or increased availability of nutrients. The sequence of cropping may yield substantial quantity of nutrients reducing the addition of fertilizers to second/third crop.

The beneficial effect of rotation is due to :

1. Impact of nutrients fixed/released by first crop on subsequent crop.
2. Impact of underground residues left by first crop on subsequent crop
3. Impact of organic practices of first crop on subsequent crop due to the slow nature of availability of nutrients.

In harnessing the benefits of crop rotation for organic farming, the choice of crop should be such that preceding crop helps in making the nutrients available to subsequent crops.

The inclusion of many N-fixing legumes enriching the soil in terms of soil organic matter, available nitrogen as well as number of other nutrients, is an important organic cultural practice to improve the availability of nutrients. Including the pulse crop in sequence cropping is a regular method followed in most countries. The reports about the benefits of such practice are available since many decades. Few of latest reports are summarized in Table 5.41.

Table 5.41. Impact of crop rotation on nutrient availability

Crop sequence	Effects on soil fertility and yield	Reference
Rye-potato	More S.O.M. decomposed (20 yr study)	Schmidt *et al.*, 2000
Rape-peanut	Legume crop cultivation was an important source of NO_2	Nanzyo *et al.*, 2002
Soybean-maize	Phosphorus uptake and maize growth were positively correlated with Arbuscular mycorrhizal colonization - which was enhanced by soybean	Karasawa *et al.*, 2000
Rice-wheat	General decline in N, P, K status by continuous rice-wheat system	Hedge, 2000
–	Citrate and oxalate (root exudates) have greater ability to mobilize P	Gerke *et al.*, 2000
Maize	Higher share of carboxylic acids found in soils after maize cultivation	Gransee and Wittenmayer, 2000
Brassica Fagopurm	Brassica and *Fagopyrum esculentum* have preferential root proliferation for additional P absorption	Nanzyo *et al.*, 2002
Tomato	Osmotic potential increased by roots of tomato allowing it to increase water uptake due to increase of betaine allehyde dehydrogenase	Moghaiels *et al.*, 2000
Wheat	Phosphoto phenolpyruvate carboxylase activity increased helping in assimilation of NH^+_4 ions	Koga and Ikeda, 2000
Rice-chickpea	Highest energy output and higher water use efficiency	Parihar *et al.*, 1999
Pearlmillet-wheat	General decline in fertility due to continuous cropping	Singh *et al.*, 1999b
Legume-rice	Legume-rice substantially reduced N requirement as compared to cereal-cereal	Singh and Verma, 1999

The decision about sequence of crops should also consider combined nutrient demand of two crops as well as capacity of soil to supply them. Some crops are exhaustive and such crops will have to be followed by nutrient enriching crops. The proper choice of crop will retain the organic matter status and help in making the nutrients available. Maize and sunflower are exhaustive crops and should be invariably followed by such crops which help in building up the nutrients rather than further exhaustion. Sometimes, crops may exhaust a particular nutrient resulting into dearth of such nutrient alone to subsequent crops Ex. Groundnut exhausting phosphorus.

Recently, the beneficial impact of underground decomposable residues left by first crop on second and sub-sequent crop is investigated sufficiently. As such roots decompose slowly and they increase the soil organic carbon status, they can make a definite contribution in terms of plant nutrients generally to subsequent crops.

5.12.2 Intercropping

Nutritional complimentarity in intercropping may lead to supply of nutrients by organic methods, as such crop combinations are responsible for fixation / release of some nutrients to companion crop. Such contribution may come from (a) nitrogen fixed by an intercrop, (b) release of fixed nutrients (like P) by intercrops. Both the concepts help in partially supplying the nutrient needs of main crop. Invariably, such benefits of intercropping are available when a legume is chosen as an intercrop. Even with legume, only few combinations have been reported to have nutritional complementarity. Reports of nutrients benefits due to intercropping are scanty.

Rana nad Saran (1999) have reported such complementarity with Brassica and potato. Sugarcane + french bean is also another example to show the complementality.

5.12.3 Alley cropping

Use of tree crops in agricultural lands is practiced for variety of reasons such as protection from wind, providing shade, conservation of soil etc. But, it can also be used as a method of nutrient supply to crops. Nutrient supply by N-fixing trees in alley cropping is possible due to perennial N-fixation and its slow and gradual utility by crops. When alley cropping is used for supplementary N addition to cultivated crops, the location of rows of trees will be different than the arrangement, found in alley cropping aimed to achieve soil conservation. As the purpose is not conservation of soil, the trees need not be planted on sloppy situations but could be planted in any convenient rows. Similarly, population of such trees could be decided based on nature of agricultural crop, and its planting pattern.

Large number of N-fixing trees are identified. Their adaptability in varying agro-ecological situations and their potentiality to serve as

N-fixing agent need to be tested. The use of N-fixing trees for nutritional purposes of agricultural crops can be a viable proposal only on long term basis, as substantial N contribution can be expected to be fixed only after 5 to 8 years. The list of N fixing trees is presented in Table 5.42.

Table 5.42. List of N-fixing trees for alley cropping

Acacia auriculiformis	*Moring oleifera*
Acacid decurrens	*Madhuca indica*
Vachellia farnesiana	*Larger stroemia speciosa*
Acacia nilotica	*Eucalyptus fereticoruis*
Acacia arabica	*Encalyptis grandis*
Terminalia bellirica	*Diospyros melanoxylas*
Termanalia arjuna	*Pongamia pinnata*
Tamarindus indica	*Dulbergia sisoo*
Samanea saman	*Cassia florida*
Pithecellobium dulve	*Careya arboroe*
Prospis cineraria	*Barringtonia acutangub*
Parkinsowa aculeate	*Angeissus latifolia*

Singh *et al.* (1991)

The practice of growing leguminous trees in coffee garden of Mexico is recorded by Grossman (1999).

5.12.4 Establishment of vegetative bunds

Nutrient conservation is observed by growing vegetative bunds on contour. The type of grass/perennial plant, the distance between two bunds, gradient and population of bund vegetation may decide the efficiency of this practice.

The mechanism of conservation of nutrients could be in the form of stoppage of physical carriage of nutrients by soil practicles by networks of roots of vegetation on budns.

5.12.5 Practice of homestead farming

The principle behind homestead farming is to utilize all available natural recyclable sources available at a place in a integrated manner. This practice is in consonance with basic theme of organic farming. As the house of farmer is located in side the farm itself and cattle are on the farm permanently, the best possible utility of all available recyclable wastes could be planned and implemented more efficiently in homestead farming. Cattle dung, cattle shed waste, washings and urine from cattle shed as well as household/kitchen waste could be profitably recycled along with harvesting wastes, crop residues to practice the organic farming. Due to proximity of site of generation and site of utility, the method of organic farming is more practically feasible in homestead farming.

5.12.6 Soil and water conservation practices

Various soil and water conservation practices such as contour bunding, leveling, trenching scooping, water harvesting, cultivation across the slope, mulching etc. are extremely useful in conserving the nutrients at a site. Not following such practices is to the detriment of organic farming in the sense that naturally available nutrients are allowed to be lost and crop production in eroded soil is bound to suffer. These measures help in water conservation too. Conserving the moisture to retain the soil near field capacity for longer period will help in higher microbial activity which in turn helps in building up of organic carbon.

Hence, the practice of organic farming is desirably combined with various soil and water conservation practices.

5.12.7 Fallowing

Substantial quantities of soil nutrients are removed by every crop harvest. Hence, continuous cropping is bound to deplete the nutrients in the soil. On the contrary, if the land is kept as uncultivated fallow, the benefits of both nutrient saving and nutrient buildup could be helpful to improve the fertility. Especially, when moisture conservation practices are associated with fallowing the benefits could be drastic. Fallowing can improve organic carbon status of soil by improved microbial activity.

5.13 ORGANIC METHODS OF PEST CONTROL

Synthetic pesticides used to control the pests viz., insects, pathogens and weeds have caused a great deal of ecological imbalance in an agro-eco-system. The imbalance initiated by them has, infact, made the control of pests more difficult because :

(a) Pests have developed resistance and do not respond to pesticides (requiring higher dosages/newer pesticides)

(b) Pesticides have killed natural enemies/predators and the damage by pests has been more virulent due to such effects

(c) Many pests can adopt the technique of avoidance by altered life styles or proliferated progenies.

(d) Pests can shift to alternate hosts and attack the crop.

Hence, the control of pests by synthetic pesticides has not offered *a de facto* solution to the damage by pests. As number of pesticides and their dosage have increased, the damage by pests has enlarged. Besides ineffectiveness of synthetic pesticides to achieve the pest control, their use has caused number of ecological problems and pollution to environment. Such problems are discussed in detail in chapter II.

In the last 15-20 years, great deal of work is done to identify various non-chemical organic methods of pest control has been reported in many crops. Such works have opened up newer natural methods of pest control out side the ambit of conventional chemical control. In fact, these crop specific and pest specific measures were identified and tried along with chemical control in the popular INTEGRATED PEST CONTROL (IPM) technology. IPM technologies were tried and popularized for crops like cotton, tomato, cabbage and many other crops, whose pest control offered more challenges. In IPM technologies, these non-chemical measures were sought to be combined with limited chemical measures for effective control. IPM technologies were popularized as eco-friendly effective measures to obviate the total reliance on pesticides. In the process of identifying these measures, many painstaking efforts to study the life cycles of various pests were made.

But in the new found meaning of organic farming, use of synthetic chemicals is completely avoided. Even the organic certification necessitates the disuse of synthetic chemicals. Considering the dangerous impacts they had on soil, water, food, microbes and the entire ecosystem, discouraging the use of synthetic chemicals is undoubtedly a positive signal. Most antagonists of organic farming easily argue that pest control sans synthetic chemicals is next to impossibility. Almost to defeat such arguments, many organic methods of pest control have been identified drawing some of them from IPM technologies.

The field of organic methods of pest control is relatively new and more challenging than chemical control. Very little scientific work is reported in this field, although success stories are plenty.

From the view point of sustainable agriculture, chemical and organic methods pest control offer contrasting features as depicted in Table 5.43.

Table 5.43. Contrasting features of organic and chemical pest control methods

Organic pest control method	Chemical pest control
The process of control is slow but steady	The process of control is instantaneous but not lasting for long period
The control does not lead to ecological complications such as resurgence or imbalance in food chains	Pest resurgence can be more devastating than original pest damage it self. This leads to vicious cycle of use of more chemicals
The organic materials used to control the pest do not cause any hazards including various types of pollution	Have potential to upset the biological equilibrium, reduce the microbial activity an also can cause different types of pollution
Deal with pest control on eco-system on holistic approach invariably dealing with the whole farm or whole crop.	Adopt one pest - a pesticide approach or non-selective approach, killing a range of organisms including useful organisms
Do not harm any non-pest organisms	May harm non-pest organisms, including annelids, predators
No possibility of any resistance being developed in pests	Pests are likely to develop resistance leading to use of more virulent chemical
No uniformity of pest control practice can be evolved, as the set of practices should be decided based on local ecological features, crop and bio-resources like predators etc	Uniform recommendation of a chemical and its dosage can be made
It is usually a combination of many practices synthesized in an effective module to achieve control of a pest	It consists of a single practice to apply a chemical
The success of organic pest control depends on overall practices of farming like green manuring, fertility management, mulching etc	The success depends upon the type of chemical, its concentration and its effective application

Organic methods of pest control involves

(1) Biological control

(2) Cultural methods for pest control.

5.13.1 BIOLOGICAL CONTROL OF PESTS AND DISEASES

The concept of biological control came into existence mainly to harness more natural methods of pest control, in vicw of innumerous problem associated with use of synthetic pesticides. In the last 30-35 years, this new field of agriculture has gained a large popularity in view of its ecofriendly principles. Biological control is now accepted as a principal segment of organic farming. It includes two main approaches viz.,

1. Use of natural enemies of the pests
2. Use of plant extracts or plant based chemicals.

5.13.1.1 Use of natural enemies(called as biological control)

The beneficial use of natural enemies to control the pests is based on the common principle of ``Enemy's enemy is a friend". A pest lives as a part of agricultural ecosystem maintaining dynamic population balance with other living components. Many living members of the ecosystems are natural enemies of pests. When a pest population overtakes the population of others, it damages the crop. But if the population of natural enemies is built up to overtake the pest populations, the pest damage can be easily and ecofriendly minimized. The field of biological control has used this concept to control the pests in a natural way. A batch of laboratory bred natural enemies released in the crop infested with pests will upset the dominance of pest population and reduce the pest damage.

The natural enemies tried so far with varying levels of success belong to following categories :

1. ***Predators*** *:* They are free living animals which attack and eat the entire body of pest quickly at any stage of prey.The predators reduce the population of pests by surviving on the egg or larva of the pests. Therefore they are further classified into

(a) *Egg predators* :

Ex : *Chrysoperla carnea, Coccinellide beetles,* ants

(b) *Larval predators :*

Ex : *Chylocarvus nigrita Cheilomenes sexamaculata Coccinella septumpunctata*

2. ***Parasitoids*** *:* They parasitize during their larval stages and complete their life as free living during their adulthood. These can reduce the damage by pests, by killing/ irritating/weaken them. Usually they arc ovipositioned near or inside the prey's body.

They are further classified into

a. *Egg parasitoids* :

Examples *Trichograma chilornis*

Trichograma japonicum

Trichograma brassicaceae

Trichograma pretiosum

Trichograma embryophagum

Telenomus remus

b. *Larval parasitoids* :

Example *Iotima javensis*

Cotesis flavipes

Hyposoter didymatar

Bracon habtor

c. *Pupal parasitoids :*

Example *Platugaster oryzae*

Brachymeria nosatoi

Tetrasticus pupivora

Tetrasticus israelie

*3- **Pathogens*** *:* They are microorganisms infecting the pests to suppress their activity or kill them. They may be bacteria, fungi, virus or nematode. Some examples of pathogens of pests are :

- Viral pathogens like Nuclear Polyhydrate Virus against Helicoverpa and Spodoptera in various crops
- Bacterial pathogens like *Bacillus thuringensis* (Bt), *Bacillus sphaericus* and *Clostridium brevifaciens* (against Spodoptera, *Helicoverpa armigera* and *Plutella xylostella*)
- Fungal pathogens like *Beauveria bassiana* (against potato beetle, corn borer pod borer or mealy bugs), *Metarhizium anisopliae* (against BPH, Pyrilla and Rhinocerus) and *Verticillium lecanii* (against Helicoverpa and Spodoptera)

The effective control of pests by use of natural enemies is favored only under following circumstances

a. ***Host specificity*** : Each natural enemy is specific to a particular species of the pests. Hence the effectiveness of the control depends on whether a properly selected enemy is used.

b. ***The adequacy of population of natural enemy*** *:* Unless the population of natural enemies is sufficiently built up (either by release of by in situ proliferation) the pest population cannot be suppressed.

c. ***Favorable climatic conditions*** *:* The damage to the pest population by predator, parasitoids or pathogens can only be expected under favorable climatic conditions. The rainfall, temperature, relative humdity, sunlight have a well known impact on the populations of both pests and their enemies.

Under these circumstances, the successful biological control on widely grown crops through for use of natural enemies necessitates the mass scale rearing and breeding of these natural enemies under laboratory conditions.

There are number of bio control laboratories established in our country, both under private and government initiative. Among all the natural enemies, the most popular natural enemies produced in large scale are various species of Trichograma, chrysoperla and Nuclear Polyhedrate Virus. Numbers of other natural enemies are in the process of popularization.

The success of use of natural enemies is reported in number of crops both at experimental level and commercial level. Some such reports are presented in Table 5.44.

Table 5.44. Control of pests by natural enemies

S.N.	Details of study of natural enemy	Reference
1.	Parasitic nematodes were controlled by *Mylonchulus minor*	Choudhary and Sivakumar (2000)
2.	Isolates of *Paecilomyces lilacinus* inhibited the emergence of juvenile eggs of *Meloidogyne incognita* (nematode)	Khan and Goswami (2000)
3.	*Drosicha dalbergia* (green) of almonds was controlled by natural enemy *Coccinell septumpunctata, Chrysopera carnea*	Kaul *et al.* (2000)
4.	Control of gundhi bugs in upland rice achieved by fermented snail	Hisra (1999)
5.	*Staphylinid* predator *Paederus fuscipes* was observed to feed viraciously on leaf folder larvae of rice	Rai *et al.* (2000)
6.	*Bacillus thunngensis* formulations were very effective against *Helicoverpa armigera*	Gujar *et al.* (2000)
7.	*Bacillus thuringensis* formulations (Del in) was effective in reducing population of leaf webber [*Crocidolomia binotalis*] of cabbage	Malathi *et al.* (1999)

Some successful practices of biological control through natural enemies

- Release of *Trichograma chilonis* @ 50,000 adults / ha/ release. Totally eight releases with the gap of ten days starting from 45th day after planting can reduce sugarcane shoot borer.

- Release of *Trichograma japonicum* in the similar schedule starting from 60 days after planting can reduce the instances of top borer of sugarcane.
- Soil application of *Bacillus papillae* @ 0.5 Kg/ha along with manure can reduce incidences of white grubs.
- Release of *Trichogramma chilonis* or *Trichogramma japonicum* @ 50,000 adults /ha/ release. Totally six releases with the gap of seven days starting from 20th day after planting can reduce paddy yellow stem borer.
- Release of *Chrysoperla cornea* @ 14000 larvea synchronizing with occurrence of sucking pests can protect the cotton crop against early damages due to sucking pests.
- Application *Bacillus thurengensis* (Delfin) @ 1 Kg/ha at the appearance of *Erias* spp. can protect the cotton crop from boll worms.
- Need based application of Ha NPV @ 3 X 10^{12} (500 LE/ha) can successfully reduce the damages by *Helicoverpa armigera* in cotton.
- Application of *Verticillum lecanii* @ 5X10^{11} spores/ha is successful against green scales of citrus.
- Release of *Telenomus renus* @ 40,000 / ha three weeks after planting in tobacco can reduce the damages by *Spodoptera litura*.
- Release of *Trichogramma pretiosum* or *T. brasilensis* @ 50,000/ ha or application of Ha NPV @ 1.5 X 1012/ha at 40 DAT is effective in controlling *Heliothis armigera* in tomato.

5.13.1.2 Plant extracts (botanical pesticides)

Extracts of leaves, stems, seeds and fruits of many plants either repel, deter feeding, disrupt behavior or even directly toxic to physiological activity of pest's body. They reduce the damage by pests substaintially by reducing thepopulation or activity of pests. Sometimes, plant products like oil cakes (neem kernel cake) are also known to have beneficial control of pests. They include plants like sweet flag, custard apple (agaist aphids), Garlic (against Spodoptera), neem (against Heliothis), sweet citrus (against *S. frugiperda*), clerodendron and parthenium (against *Spodoptera* spp)

Table 5.45. Efficacy of plant extract and bio-agents

S.N.	Name of efficacy of plant extract/bioagents	Reference
1.	Neem products reduced fruit borer attack (*Erias* spp.) of okra	Ambedkar *et al.* (2000)
2.	Methnol extract of *Azadirachta indica* and hexane extract of *Thevatia nerifolia* leaves showed antifeedal and insecticidal effects against *Achaea janata* (on castor)	Babu *et al.* (1997)
3	Spider population in rice was worst affected with monocrotophos and least affected by neem seed kernel extract	Baitha *et al.* (2000a)
4.	Neem oil (2-4%) controlled the larvae of peaf roller of rice to the tune of 56-60 per cent	Baitha *et al.* (2000b)
5.	Powdered neem leaves were found effective in controlling the population of beetle *Callosobruchus chinensis* on greengram	Dayakar and Ray (1999)
6.	Chloroform extract of *Pongamia pinnata* @ 5 per cent controlled the damage by the mosquito bug (*Helopellis theivara)*. Antifeedant property of *lantana camera* extract on tea mosquito bug was also demonstrated.	Deka *et al.* (1998)
7.	Growth disrupting effects of hexane extract of neem seed kernel resulted in delayed development and mortality of *Spodoptera litura*	Kaur *et al.* (2001)
8.	Treating pigeonpea seeds with neem seed kernel powder (4% w/w) prevented egg laying of *Collosobruches chinensis* (Pulse beetle)	Singhal and Chauhan (1997)

5.13.2 Cultural methods of pest control

Pests include all organisms, which inhibit the growth of crop and reduce its yield directly or indirectly. They may be bacteria, fungi, insect, a non-insect, a mammalian, a plant (weed) or any other organism. Organic agriculture acknowledges the fact that crop should grow in nature's cradle and be a part of en eco-system, in which it is grown. It also recognises that a crop can attain dominance in the eco-system and result into economic yield only when the competition by other members of the eco-system is minimised. But, all such members cannot be singly targeted and killed (as is being done say in use of pesticides against an insect). Alternatively, they have to be managed together in such a way their attack on the crop is minimised. For these reasons organic farming focuses more on cultural ways to control the pests than chemical control. The different between cultural and chemical control of pest attack are discernable as follows :

Chemical pest control	Cultural pest control
Taken up after damage is seen or as prophylactic measure, in anticipation	Taken up to avoid the pest. But taken up as routine practice, even when pest is not expected
Taken up against specific pest	Taken up against a range of pests
Pesticide can cause damage to the ecosystem or cause pollution to food and water	No synthetic pesticide is involved. When bio pesticide/plant extracts are used, they do not cause pollution or damage the eco-system
The attack may recur, once the effect of pesticide is reduced / removed (also called as resurgence)	The attack does not recur, as cultural control is a part of readjustment of component of eco-system
Generally, involves higher cost due to chemicals and labour	Involves minimum cost-mostly as labour cost
Are immediately effective	Are effective slowly over a period of time
Do not favour development of resistance / tolerance to pests in plant body	Generally result into built up of resistance /tolerance to pest attack in plant body
Slight variations in predisposing factors for pest attack can cause severe damages	Generally the predisposing factors of pest attack are ineffective or least effective, when ideal cultural methods of pest control are followed.

The cultural methods of pest control are not the same in all the crops or in all situations. They must be recognised by experience for each situation and crop separately and implemented as a blend with other cultivation practices. They should become part of routine practices followed by the farmers, rather than deliberate attempt to control the pests. In recent years, cultural methods of pest control have been integrated with chemical, biological and physical methods of pest control to achieve a high level of success. This is popularly termed as Integrated Pest Control. The success of use of chemical in control of pests is often restricted by the unfavourable situation-which may be the result of wrong cultural practice or climate favourable to proliferation of pest. Such effect is sought to be mitigated - when chemical/biological methods of pest control are integrated with cultural methods.

The current difficulties in control of pests of many crops are mainly due to least importance attached by the farmers to follow the cultural methods of pest control and to blindly follow the chemical method. Farming community has almost assumed that pest control can be achieved only use of chemicals. But, many pests could be avoided or their damage could be mitigated, if appropriate cultural method is adopted. By ignoring the cultural options to control the pests, the farmers have allowed the pests to flourish, as they have advantage of several pre-disposing factors to proliferate and attack the crops - when these cultural methods are ignored.

The cultural methods to control the pests differ greatly from crop to crop and pest to pest and sometimes situation to situation. This implies that such methods lack of universality and uniform recommendations to adopt these practices can hardly be made. The adoption of any cultural practice has to be based on the local conditions, severity of pest, pre-disposing factors and the practical success of each practice.

Cultural methods of pest control comprise following categories of practices followed almost as routine practices of growing crops. They may become a part and parcel of other cultivation practices. The following groups of practices may be considered as cultural practices for pest control.

1. Choice of date of sowing / planting
2. Adoption of suitable spacing / maintaining appropriate plant population
3. Suitable water management practices
4. Suitable nutrient management practices
5. Choice of varieties and hybrids for a given area
6. Intercropping /mixed / trap cropping practices
7. Physical and other cultural methods of pest control
8. Application of green manures and other manures
9. Removal of weeds and alternate hosts
10. Methods of sowing and planting

5.13.2.1 Choice of date of sowing/planting

For a given agro-ecological situation, crop or its variety/hybrid have to be ideally sown/planted in a particular season or in a particular fortnight. For an example, in south India, hybrid cotton should not be sown beyond July or annual sugarcane should be planted in January to March period. Similarly in north India, wheat should be sown in October-November. Some crops like maize, ragi are highly flexible and could be grown in all three seasons. But, some crops like Bengal gram, safflower and wheat are rabi specific while red gram and sesamum are kharif specific under south Indian conditions.

Beyond the seasonal impact on their yield potentiality, the seasonal features greatly influence on incidence of pest can attack or its severity. The strategy to adopt the right date of sowing is a best no cost technology available to control the pests or disease. Within seasonal specificity, sowing them in particular fortnight/week is critical to decide their performance. For example, late planted paddy is prone to blast/ stem borer; late sown jowar/maize is prone to shoot fly; late sown cotton / red gram is prone to attack by Heliothis.

In fact, the delay/early sowing has adverse impact on growth and development of the crop due to climatic affect. But, the adverse effect of untimely sowing/planting on pest damage can escalate the climatic effect compounding the damage in terms of reduced growth and yield. Hence, it is advisable to follow the schedules of sowing/planting fixed for a given an agro-ccological region to successfully avoid the possible pest damage than allowing the pest to attack in all its virulence and later attempt to control the pests.

Many evidences in the literature suggest the benefit of sowing/ planting the crop at right time in terms of reduced pest damage. Some such evidences are presented in Table 5.46.

Table 5.46. Effect of date of sowing on control of pests and diseases

S.N.	Crop	Pest/Disease	Nature of control	Reference
1	Soybean	Pod borer	2nd week of June sowing is helpful to reduce pod borers	Raman gouda H Patil, (2002)
2	Sorghum	Shoot fly	Early sowing reduced the incidence	Gandhale *et al.*, 1983
3	Paddy	Blast	Early sowing from 1st July to 15th July reduced the incidence	Ram singh *et al.*, 1999
4	Tur	Pod borer	Mid June sowing helped the control	Sharma & Chaudhary, 1984
			Incidence in kharif is less than other season.	Satpute *et al.*, 1993
5	Wheat	Smut	Early and late sowing are ideal than normal sowing.	Kumar *et al.*, 2004
		Rust	Early sowing is useful to control the disease.	Mishra *et al.*, 1992
6	Sugar-cane	Red rot	Early planting in 2nd week of March helped to reduce incidence	Anon., 2004

7	Potato	Late blight	21st March planting reduced late blight.	Sikka and Chauhan, 1973
			20th and 30th September reduced the incidence.	Sekhon and Sokhi, 1999
			Ridge method is ideal.	Sood, 1988
8	Ground-nut	Tikka disease	Early sowing upto the end of July helped to control Tikka disease.	Raveendra, 1993; Lewin *et al.*, 1972

5.13.2.2 Adoption of suitable spacing/maintaining appropriate plant population

The spacing between rows and between plants - ultimately deciding the plant population per unit area - not only decides the yield but can influence the extent of attack by pests. The crowding of plants leading to poor air movement or increased humidity or poor interception of sunlight may provide predisposing environment for many pests like jassids, thrips or aphids. They may be ideal environment for escalated attack by moulds. Similarly, extraordinarily wide spacing provided between rows/plants may not be good to curb the insect attack, many pests have an extended cycles of their lives or may have chance to complete extra life cycles due to availability of their food for long period. In such situations, the crops may get damaged by diseases like leaf spots, blasts, borers, flies, etc.

As such, for every crop, the spacing between rows/plants is standardised in every agro-ecological region, based on study of its population dynamics. If the farmers attempt to grow more crowded crop in the false belief that a crowded crop can yield more, they encounter two fold damages. A crowded crop induces unhealthy competition for nutrients and water and the crop is bound to suffer in terms of growth and yield. But, an over-crowded crop invites more pest attack than a normally planted crop and therefore such crops will suffer double damage. In ultimate analysis, the farmer ends up in reduced growth and yield coupled with increased cost of cultivation. In addition to this, the excessively crowded crop always leads to weaker leaves, flowers and fruits - which are more easily attacked by the pests. When the pest is a weed, wider spacing always leads to more weed growth and increases the possibility of attack by other pests, as such crop will always be lanky and weak.

For these reasons, it is always advisable to adopt the recommended spacing to not only obtain the higher yield but to minimise the damage by pests and thereby avoid making the crop prone to pest attack, besides reducing the cost of cultivation. Several instances are cited in the literature to prove that adopting higher/ lower spacing (or increasing / decreasing the plant population per unit area) have had adverse effect on performance of the crops. Some such instances are cited below in Table 5.47.

Table 5.47. Impact of spacing/population levels on control of pests and diseases

S.N.	Crop	Pest/Disease	Nature of control	Reference
1	Ground nut	Tikka disease	Wider spacing is helpful to control the disease.	Ghewande 1982
2	Wheat	Rust	Lower seed rate can help reduce the disease.	Hassabnis *et al.*, 1997
3	Tur	Pod borer	Row spacing of 37.5 cm is ideal.	Singh and Singh, 1978

5.13.2.3 Suitable water management practices

The timely availability of water through efficient method of irrigation or through different management practices in rainfed situations is crucial to the success of crop respectively under irrigated and rainfed conditions. The successful performance of the crop and expression of the genetic yield potential of specific variety/hybrid in question are wholly dependent upon timely availability of either rain water or irrigation in adequate quantity.

But, it is seldom recognised that water management practices in both irrigated and dry land conditions have important role to play even in aggravating or mitigating the damage due to pests. The pests, including weeds, show definite response to availability or shortage of moisture in the soil. Most weeds flourish well under irrigated conditions and cause major problems to the crop production. Similarly, improvement of soil moisture status can be the reason for control of some pests like mites, while improved humidity due to higher moisture status can be the reason to aggravate some pests like aphids. The moisture deficit situations can aggravates powdery mildew. Besides soil moisture status, the proliferation, survival and rate of reproduction of pests often depends upon the predisposing factors such as humidity and moisture related temperature regimes. The changing soil moisture status influences these pre-disposing factors to ultimately influence the control of pests.

The crop specific pests and their reactions to either deficit or excess moisture need to be recognised, before any recommendations on control of pests through better water management practices could be evolved. In this direction very little work is reported on water management practice - as a strategy to control the pests. But, many studies and observations are documented to show that moisture relations can significantly alter the pest's life cycles indicating that improved water management practices could be used to avoid or mitigate the pest damage. Some of such documents found in the literature are presented in Table 5.48.

Table 5.48. Effect of varying water relations on control of pests and diseases

S.N.	Crop	Pest/Disease	Nature of control	References
1	Sorghum	Shoot fly	Drier soil is ideal to reduce the pest	Men *et al.*, 1986
2	Coconut	Black headed caterpiller	Drip irrigation is ideal than other methods to reduce damage	Anon., 2004
3	Wheat	Smut	Reduced no. of irrigations helped control of smut.	Mongia *et al.*, 1993
4	Ground-nut	Tikka disease	Lower soil and leaf moisture reduced tikka disease	Porter *et al.*, 1987
5	Coconut	Coconut mite	High soil moisture status helped to control of mites	Kannaiyan *et al.*, 2002
6	Sugar-cane	Red rot	Drip irrigation method is ideal to reduce red rot disease.	Rajeev *et al.*, 1999

5.13.2.4 Suitable nutrient management practices

The practice of supply of concentrated sources of nutrients in the form of fertilizers is a part of modern agriculture to mainly to increase the crop yields. But supply of nitrogen in adequate quantity (to presumably meet the heavy nutrient demand) or not in proportion to other nutrients can make the crop more susceptible to the pests. Similarly, supply of adequate quantity of potassium to the crop roots can induce sufficient resistance to many pests. It is ironical that factor responsible to increase the crop yield itself is also the reason to aggravate the pest damage. The supply of excess nitrogen has been main strategy to increase the crop yields - without adequately recognising that more nitrogen supply itself can be the cause of aggravated pest attack to ultimately reduce the crop yield. Nitro-positive crops like cereals are more promoted to receive excess nitrogen, unmindful of its negative effect. But, suitable potassium doses to match with nitrogen may not really aggravate the pests, due to positive impact of potassium on pest control. But, the right proportion of nitrogen and potassium is hardly practiced by the farmers.

The role of nitrogen and potassium in pest control leads to many corollaries such as

- Varying doses and proportions of N and K may either aggravate or control the pests
- The factors governing the availability of nutrients (like soil texture, structure, microbial load and organic carbon status etc) may favourably or unfavourably act on pest survival.
- The fertility status of soil will decide the pest survival

Numbers of published evidences suggest that nutrient status or nutrient management practices can indeed decide the fate of some pests. Extracts of some of such published literature are presented in Table 5.49.

Table 5.49. Impact of nutrient management practices on control of pests and diseases

S.N.	Crop	Pest/ Disease	Nature of control	Reference
1	Sorghum	Shoot fly	Application of low N dosage helps control of shooti fly.	Kundu *et al.*, 1978
			Application N with K reduces the incidence.	Balasubramanian *et al.*, 1986
2	Paddy	Blast	Increasing N dose increase the neck blast.	Maharudrappa *et al.*, 1993
			Two later splits of N are useful to reduce blast.	Bonman *et al.*, 1998
			Silicon reduced the spread of disease.	Seebold *et al.*, 2000
3	Coconut	Black headed caterpiller	Higher K in upper leaves reduce the incidence. High K application is needed.	Nadarjan, 1997
4	Wheat	Smut	Increase in level of N and P decrease the incidence	Mongia *et al.*, 1993
5	Ground nut	Tikka disease	Lower dosage of N and P can reduce Tikka.	Chandrashekar *et al.*, 1983
			Application of K reduces the spread	Balasundaram *et al.*, 1976
			Increased level of P and K help reduce Tikka.	Ghewande (1982)
			Lower dosage of RDF is useful to control Tikka	Raveendra (1993)
6	Rust	Wheat	Lower dosage of N or higher dosage of K is suitable to reduce rust.	Hassabnis *et al.*, 1997
7	Coconut	Coconut mite	Higher dosage of K and Neem cake reduces the spread.	Anon., 2004
				Muthiah and Bhaskaran (1999)

5.13.2.5. Choice of varieties and hybrids for a given area

Every crop and for that matter every variety/hybrid of a crop has certain inherent character as far as its reaction to pest or disease. While breeders look to achieve the resistance/tolerance to major pests and diseases of a crop at the time of breeding, when they are tried under varying agro-climatic conditions, each of them show different reactions to pest/disease. The reaction to pests and disease is dynamic and changes according to time, features of agro-climatic situations, predisposing factors, severity of pests and disease as well as the stabilization/break down of original genetic character.

It is absolutely necessary to understand the reaction of a particular variety/hybrid of a crop to the pests and diseases of such crop specific to agro-climatic features of the locality, where it is grown. The uniqueness of such reaction may vary vastly from one pest to other pest and therefore acquiring this knowledge may become complicated looking to multple numbers of pests/diseases and multiple numbers of crops at varying locations. Nevertheless, it is absolutely essential to acquire this knowledge not only to effectively control the pests culturally but even to know the limitations of chemical control of such pests. Large numbers of incidences of control of pests and diseases by use of resistant varieties/hybrids are reported in many crops at various locations of our country. Few such reports are presented in Table 5.50.

Table 5.50. Incidence of pest/disease control by use of resistant varieties/hybrids

S.N.	Crop	Pest/ Disease	Resistant variety / hybrid	Reference
1	Sorghum	Shoot fly	SPV-1015 is resistant	Prem Kishore, 1990
2	Paddy	Blast	Jaya, KRH-2, Mangala are resistant. Intan is resistant variety.	Nagaraju *et al.*, 1991. Chandrashekar, 1983
3	Tur	Pod borer	Prabhat is resistant	Shrma and Chaudhary (1983)
4	Wheat	Smut	Bansi pali 808 and Bansi CP are tolerant. WH283, HD2329 are disease free.	Mathur *et al.*, 1960 Kumar *et al.*, 2004
5	Potato	Late blight	Kufri muthu and kufri kanchan are resistant.	Azariah, 1973 Phadatare *et al.*, 2000
6	Coconut	mite	Cochin, China, Andaman Are tolerant	Muthiah and Bhaskaran (1999)

5.13.2.6 Intercropping / mixed / trap cropping practices

Selection of a crop and cropping practices like intercropping / mixed cropping or trap cropping is not to be decided only with the sole objective of increasing the production or profitability but also in the interest of efficient and natural control of pests and diseases. When crops are grown together - a number of micro level changes can be expected, with reference to life cycles of pests and diseases. Hence, the strategy of proper choice of cropping practice could be a no cost and effective measure to control the pest or disease. The choice of right combination of crops and their proportionate populations are of great importance in deciding the success of this strategy. The mechanisms of actual control of pests by such measures may be like transfer of pests to crops due to special affinity of pests (trap crops), suppression of pest due to differing growth rates of crops (intercrops) or due to change in microclimatic effect, created by growing the crops together. Large numbers of reports are made in scientific literature to justify such an option to control the pests. Few of such reports are presented in Table 5.51.

Table 5.51. Reports of control of pests/disease by cropping practices

S.N.	Crop	Pest/Disease	Nature of control	Reference
1	Soybean	Pod borer	Intercropping with cowpea and chilli can reduce pod borers	Ramana Gouda H Patil, 2002
2	Sorghum	Shoot fly	Garlic as intercrop can reduce shoot fly	Halepyati *et al.*, 1987
3	Tur	Pod borer	Intercropping with pearl millet is ideal	Agarwal *et al.*, 2003
4	Sugarcane	Red rot	Companion cropping with garlic reduces red rot incidence.	Rao, 2002
5	Groundnut	Tikka disease	Crop rotation is ideal.	Smith and Littrell, 1980
			Mixed cropping, intercropping and rotation are ideal.	Ghewande, 1982; Raveendra, 1993
6	Wheat	Rust	Sowing after Rice results in no disease.	Singh, 2004
7	Cotton	Heliothis	Growing Bhendi	Anonymous, 2008

5.13.2.7 Physical and other cultural methods of pest control

Sometimes, physically catching/trapping of insects or physical treatment to seed materials helps in control of pests/diseases. Such efforts will help in ensuring the physical contact of insect/pathogen with the plant in question. Such attempts mainly concentrate on physically separating the pathogen/insect from being in contact with the crop. Most of these treatments are harmless to both crops and environment. Some such methods are anticipatory, but offer enough insurance against the possible attack. Many such reports are available in different crops. Some of them are presented in Table 5.52.

Table 5.52. Physical and other cultural methods to control the pests and diseases

S.N.	Crop	Pest/Disease	Nature of control	Reference
1	Coconut	Black headed caterpiller	Light traps are efficient in reducing the damage by BHC	Padmanabhan and Sukumaran, 1998
2	Tur	Pod borer	Predatory birds can reduce the pod borer population	Jananeshwar, 1998
			Putting bird perch is ideal to control pod borer.	Yelashetty Suhas, 1999
3	Sugarcane	Red Rot	Hot water treatment	Anonymous, 2008
4	Cotton	Heliothis, Pectyno-phora	Use of pheromone traps to physically trap the male moths	Shivayogeshwara, 2002
			Burning the stems of affected crop, having cocoons	

5.13.2.8 Application of green manures and other manures

Conventionally, it is held that application of manures not only enhances the availability of nutrients but improves the physical characters of the soil. But some manures may have inhibitory effects on the activity and survival of pests. Such effects may be direct due to poisonous ingredient of such manure (ex : neem cake) or may be due to indirect through aberration of one of the predisposing factors of the pests. For example, application of any manure will improve the water retention capacities - which may make the hibernation of pests more difficult or easy, depending on the nature of pest. It is also common experience of the farmers that regular application of manures will impart tolerance capacities to the crop against many pests, as well grown healthy crop is naturally tolerant to the pests than weak and pale crop. Hence, regular application of manures should be a part of routine practce to control the pests more efficiently. Numbers of evidences in the literature suggest the successful control of pests by application of manures. Some such reports are presented in Table 5.53.

Table 5.53. Green manuring and other manure incorporation influencing the pests and diseases

S.N.	Crop	Pest/ Disease	Nature of control	Reference
1	Coconut	Coconut mite	Sun hemp intercrop reduces the spread.	Muthiah and Bhaskaran (1999)
2	Sugar-cane	Red rot	In situ green manuring with dhaincha reduced redroot incidence.	Sezhian, 1999
3	Sunflower	Cut worm	FYM and Vermicompost are good to control cutworm.	Rajasekhara Rao, 1996
4	Wheat	Smut	Poultry manure reduces the spread.	Kumar *et al.*, 2004
5	Sugarcane	Red rot	Press mud reduces the incidence.	Sharma *et al.*, 1994

5.13.2.9 Removal of weeds and alternate hosts

It is established beyond all doubts that many insect pests and disease causing pathogens thrive on weeds during the non-cropping seasons to continue their life cycles. Some pests hibernate in the specific weeds/other alternate hosts. The polyphagous pests like Heliothis may have the capability to attack many weeds/plant species to keep their life cycle and proliferate. T ransfer of pathogen to alternate hosts may actually aggravate the diseases in many crops. Due to these reasons, removal of weeds and alternate hosts attains importance in the interest of non-chemical control of pests. A clean cultivation without weeds, thus, has twin benefits - one from avoiding the competition to the crops in respect of nutrients and water and second, to avoid the chance of pest attack. The advocates of ``nature farming'' are against such practices, in principle. Because, no man made efforts to weed (by hand weeding or intercultivation) are allowed in nature farming. Nevertheless, removal of weeds or alternate hosts is one of the best, eco-friendly cost effective options to avoid/mitigate the pest attack.

5.13.2.10 Methods of sowing and planting

The survival / death/ proliferation of pest population or pathogen depends on number of micro-envrionmental parameters of crop environment. A slight change in such parameters may either help the proliferation of pest or may be totally fatal to the pest. Such a phenomenon is useful in choosing appropriate method of sowing/ planting to reduce the possiblity of specific pest incidence. A change in method of planting/sowing a crop may help in changing the moisture relations or aeration - which may ultimately be useful to control the pests. Such beneficial effects are reported by number of researchers in some crops. Important works are presented in Table 5.54.

Table 5.54. Effect of sowing/ planting methods on incidence of pest and diseases

S.N.	Crop	Pest/Disease	Nature of control	Reference
1	Tur	Pod borer	Ridge sowing can reduce the incidence.	Agarwal, 2003 Satpute *et al.*, 1993
2	Wheat	Smut	Ridge sowing helps to reduce the smut diseases	Hassabnis *et al.*, 1997
		Leaf rust	Flat bed sowing is ideal method than ridge sowing to control leaf rust.	Hassabnis *et al.*, 1997
3	Sugar-cane	Red rot	IISR and STP planting techniques can reduce the red rot disease	Vishwanathan and Saniappan, 1999
4	Potato	Late blight	Ridge sowing helps to control late blight.	Sood, 1988
5	Sun-flower	Cut worm	Ridge sowing is reduces the attack of cut worm.	Bakhatia and Arora, 1995

5.14 LINKAGE OF ORGANIC FARMING WITH SUSTAINABILITY

Organic farming, in its new wider perspective with the use of all possible avenues of naturally available materials, is not aimed to replace chemical farming as a choice. It is not aimed to produce more than what is produced by chemical farming. But, organic farming is an imperative than a choice; it is much nearer to nature and natural processes than chemical farming; it can achieve greater sustainability than chemical farming.

Sustainability of present level of crop productivity is at stake in many crops with continuous use of fertilizers. Sustainable increase in productivity has not been achieved in a few other crops. Both these yield sustainabilities are essentially dependent upon sustainable fertility and eco-system. The role of organic farming is not restricted to merely supply the nutrients (as in case of fertilizers). Its primary role extends to improve soil quality - measured in terms of physical and biological characters. The sources of organic farming are able to naturally fit themselves into an agro-eco-system than chemicals and therefore, are able to mould the soil system to long term sustainability. The biological bases of sustainable crop yield lie in sustainable fertility and sustainable eco-system. Increase in productivity or maintenance of productivity of crops can be achieved if fertility and eco-systems are able to be maintained at their best natural forms. Such possibility exists only with the practice of organic farming rather than chemical farming.

Achieving sustainability of eco-system and soil fertility is a slow process- as the lost equilibria are to be reestablished; organic carbon status should be gradually improved; soil macro-aggregates are to be increased and retained; the food chains of eco-systems are to be reformed. The maintainability or improvement of crop yields depends upon such changes in eco-system and soil fertility. Sudden improvement in crop yields cannot be expected (like what was recorded by use of fertilizers) as the organic farming methods can improve the eco-system and soil fertility by slow process of rebuilding.

The practice of organic farming to achieve the sustainability is a holistic approach aimed to improve the crop yields through the sustainability of eco-system and fertility. Undisputedly, such advocacy is in far variance from any short term recommendations to achieve quick solutions such as :

- Recommending the use of synthetic pesticide against a pest.
- Recommending a fertilizer to correct the nutrient deficiency.

Such ad hoc recommendations were made in chemical farming only to achieve short term interest ignoring the natural equilbria and natural components. Repeated use of short term measures of this nature is the cause of unsustainability of fertility and eco-system. To achieve sustainability, organic methods of nutrient supply and pest control are necessary to be followed not withstanding that such methods may take more time to achieve the results in terms of increased crop yields. Advocates of chemical farming frequently show an importance of crop yield even with unsustained fertility and eco-systems. But, such improvements are short run - as they are not formed on natural basis of production.

In understanding the linkage between sustainability and organic farming, the essentiality of supporting and augmenting natural processes in an agro-eco-system needs to be realised. Organic farming practices strongly make use of natural processes such as natural enemies, predators, food chain etc. for pest control as well as decomposition. N-fixation, structural improvement of soil etc. for supply of plant nutrients. All these processes are in tune with holistic improvement of soil fertility and eco-system which are essential components of sustainability. This is diagrammatically represented in Fig. 5.3.

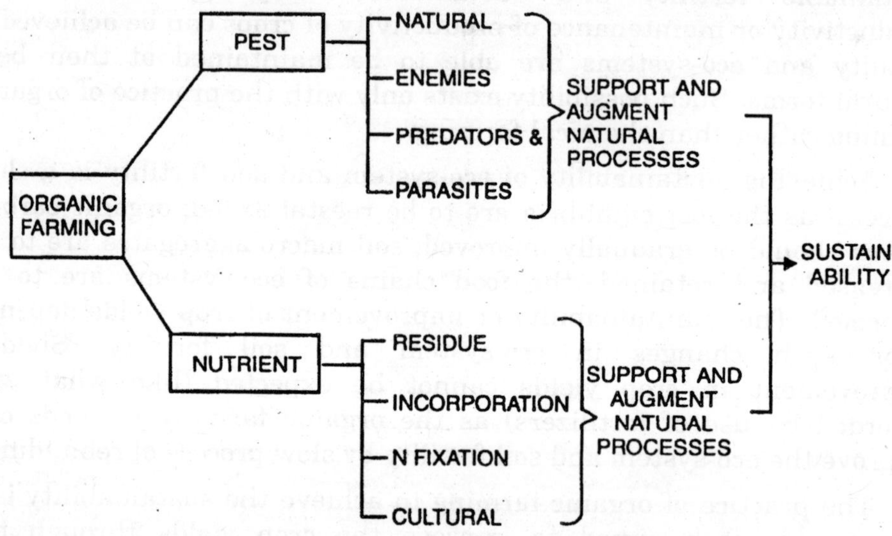

Fig. 5.3. Diagrammatic representation of linkage of organic farming with sustainability

The protagonists of chemical farming lose the sight of such holistic approach of sustainability and end up restricting the meaning of sustainability to crop yields alone. Under such curtailed considerations, the organic farming may or may not lead to sustainable increase in crop yield depending upon the level of soil fertility and eco-balance already achieved. Under extreme conditions of depleted fertility and impaired eco-balance, organic farming may reduce the crop yields in initial years. Under well managed conditions, crop yields may be maintained or improved in the course of time.

But, sustainability of crop yields is only a small part of total sustainability - although in short term interest crop yields can decide economic sustainability. The basis of improved crop yield on long term basis lies in sustained soil fertility and well balanced eco-system. Hence, sustainable soil fertility and eco-system deserve more concern than short term improvement of crop yields. Organic farming alone has potentiality to achieve such sustainability, as all the organic farming practices are in tune with natural processes of an agro-ecosystem.

Natural farming systems like forest system or grass land systems have remained more sustainable without any sort of interferences by man only because, natural processes of eco-systems are well established and uninterfered. Although in agricultural systems, it is not possible to follow complete non-interference, it is at least necessary to support and strengthen the natural processes to emulate the natural farming systems. In this direction, practices of organic farming and practices of chemical farming are in stark contrast (Fig. 5.4) in causing sustainability.

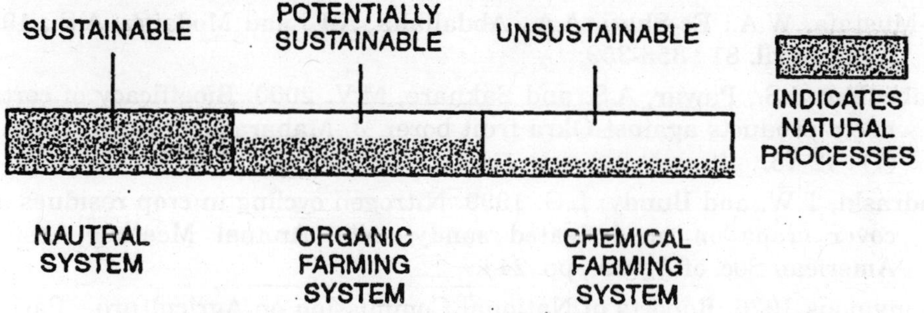

Fig. 5.4. Diagrammatic representation of natural processes

Thus, the linkage between organic farming and sustainability is not only strong and complete, but it is natural and obvious. Without organic farming, sustainability cannot be achieved at all. The practice of organic farming is not a choice but imperative to achieve sustainability. Even achieving sustainability (of not only net returns and crop yields) of the whole eco-system and soil fertility is not an option but a natural necessity and a social requirement. Without sustainability, perhaps the whole production system will collapse gradually.

The time frame to achieve sustainability through organic farming may be varying according to the level of soil fertility and eco-balance available in question. Sustainability of yield may taken 4 to 7 years, while sustainability of fertility (as indicated by soil aggregates) may take 5 to 20 years and eco-balance may taken 8 to 25 years. Short term assessment of sustainability in 1 to 2 years may lead to erroneous results to underestimate the concept of sustainability.

5.16 REFERENCES

Abrol, I.P. and Palanippan, S.P. 1988. Green manure crops in irrigated and rainfed low land rice based cropping systems in South Asia. Proc. Symposium on Sustainable Agriculture : The green manure crop in rice farming system, IRRI, Philippines, pp. 71-82.

Adams, J.F. and Hartzog, D.L. 1999. Poultry litter effects on yield and grade of runner peanut. Annual Meeting Abstr., American Soc. of Agron., p. 258.

Agarwal, S.C. 2003. *Indian J. Agri. Sci.,* 73 : 291-293.

Akhtar, M. and McCallister, D.L. 1999. Leaching potential and fractionation of phosphorus in sewage sludge treated plots. Annual Meeting Abst. of ASA, pp. 342.

Allen, B.C. and Jolley, V.D. 1999. Mycorrhizal colonization unrelated to yield and uptake of field grown dry bean in manured top soil and sub-soil. Annual meet Ab. American Soc. of Agron., pp. 243.

Al-Mustafa, W.A., Ec-Shati, A.A., Abdallaha, A.E. and Modaish, A.S., 1995. Expt. Agril. 81 : 355-359.

Ambedkar, J.S., Pawar, A.S. and Sakhare, M.V. 2000. Bioefficacy of certain neem products against Okra fruit borer. J. Maharashtra Agril. Univ. 25 (1) : 42-43.

Andraski, T.W. and Bundy, L.G. 1999. Nitrogen cycling in crop residues and cover crops on an irrigated sandy soils. Annual Meeting Abst. of American Soc. of Agron., pp. 244.

Anonymous 1976. Reports of National Commission on Agriculture - Part X : Inputs, pp. 90-94, Pub. Govt. of India, Ministry of Agriculture and Irrigation.

Anonymous 1985. Annual Report of CRRI, Cuttack.

Anonymous, 1969-70 to 1976-77. Annual Reports, All India Co-ordinated Agronomic Research Projects, ICAR.

Anonymous, 1996. Annual Report of Dryland Agriculture Project, UAS, Bangalore, pp. 28-57.

Anonymous, 1997. Directory of Indian Agriculture, CMIE, Pub., p. 233-36.

Anonymous, 2004A. http/www.agri coop. inc.in.

Anonymous, 2008. Package of practices for improved cultivation of crops, *University of Agri, Sciences Bangalore.*

Aoyama, M. and Tenimai, 1992. Organic matter and its mineralization in particle size and aggregate size fraction. *Japanese J. soil Sci. & Pl. Nutrition,* 63 (5) : 571-580.

Azariah, M.D. 1973. Potato varieties to combat late blight. *Indian farm.* 5-26. *Dis.*, 17 : 136-140.

Babu, P.B.S., Nair, M.S. and Sunitha, B. 1997. Antifeedant and insecticidal effects of some plant extracts against castor semi-looper Achaea janata. J. Insect Sci. 10 (2) : 179-80.

Badanur, V.P., Poleshi, C.M. and Balachandra Naik, K. 1990. Effect of organic matter on crop yield and physical and chemical properties of a vertisol. *J. Indian Soc. Soil Sci.,* 38 : 426-429.

Baitha, A., Hameed, S.F., Singh, R. 2000a. Effectiveness and economics of various treatments and their impact on spider population in rice ecosysetem. Annals of Plant Protection Sci., 8 (1) : 13-17.

Baitha, A., Hameed, S.F., Singh, R. 2000b. Relative toxicity of neem products against larvae of rice leaf folder. Indian J. Entomol. 62 (1) : 66-68.

Bakhetia, D.R.C., Sukhwinder, K. and Sandhu, I.S. 1995. Cultural control of cutworm in sunflower. *J. Res. PAU.,* 32 (2) : 137-141.

Balasubramanian, M., Shankaran, K. and Manivkam, T.S. 1986. *Madras. Agric. J.* 73 (8) : 471-473.

Balasubramaniyan, P., Palaniappan, S.P. and Francis, H.J. 1991. Effect of green manuring and inorganic N-K fertilization or nutrient uptake and yield of lowland rice. Indian J. Agron., 36 (2) : 293-95.

Balasundarum, C.S., Shanmugam, M., Krishnamurthy, K.K. and Purushothaman, D. 1976. Influence of potassium nutrition on the incidence of tikka disease of peanut. *Potash. Rev.* 23 (1) : 1-3.

Banerjee, M.R., Chapman, S.J. and Killham, K. 1999. Uptake of fertilizer sulfur by maize from soils of low sulfur status as affected by Vesicular-arbuscular mycorrhizae. Canadian J. Soil Sci., 79 (4) : 557-59.

Bary, A.J., Cogger, C.G., Fransen, S.C. and Myhre, E.A. 1999. Yard trimmings as a nutrient source and soil amendment in agriculture. Annual meet Ab. American Soc. of Agron., pp. 251.

Bhagwat, V.R., Sison, M.L.J., Shanower, T.G. *Int. Pigeon pea News 1.,* 17: 38-39.

Bharadwaj, K.K. and Gaur, A.C. 1985. Recycling of organic wastes. Pub. ICAR, New Delhi, pp. 104.

Bharadwaj, K.K.R. 1995. Recycling of crop residues, oil cakes and other plant production in agriculture. In Recycling of crop, animal, human and industrial wastes in agriculture. Ed. H.L.S. Tandon, Pub. Fertilizer Development & Consultation Organization, New Delhi, pp. 9-27.

Bhat, A.K., Beri, V. and Sidhu, B.S. 1991. Effect of long term recycling of crop residues on soil productivity. J. Indian Soc. Soil Sci., 39 : 380-382.

Bhattacharya, P., Dey, S.K., Math, S. and Banik, S. 1990. Organic manures in relation to rhizosphere effect - V. Effect of organic manures on qualitative and quantitative distribution of bacteria, actinomycetes and fungi in rhizosphere soils of rice. Indian Agriculturist, 34(3) : 157-162.

Bijay Singh, Yadvinder Singh, Sadana, W.S. and Meelu, O.P. 1992. Effect of green manure, wheat straw, organic manures on DTPA extractable Fe, Mn, Zn, Cu in calcareous sand loam soils at field capacity and under water logged conditions. J. Indian Soc. Soil Sci., 40 (1) : 114-118.

Birader, A.P., Sunitha, M.D. and Teggihalli, R.G. 2000. Evaluation of crop residues for vermiculturing. J. Maharashtra Agril. Univ., 25 (1) : 39-41.

Bonman, J.A. and Kurschner, B.L. 1998. Effect of nitrogen timing and split application on rice blast incidence in upland rice. *Plant dis.,* 76 (1) : 950-953.

Boparai, B.S., Yadavinder Singh, Sharma, B.D. 1992. Effect of green manuring with *Sesbania aculeata* on physical properties of soil and on growth of wheat in rice-wheat and maize-wheat cropping system in semi-arid regions of India. Arid Soil Res. & Rehabilitation, 6 (2) : 135-143.

Boyd, A.E.W. 1980. Development of potato blight after planting infected tubers. *Ann. Appl. Biol.,* 95 : 301.

Brendecke, J.W., Axelson, R.D. and Pepper, I.L. 1993. Soil microbial activity as an indicator of soil fertility : Long term effects of Municipal Sewage sudge on an arid soil. Soil Biology & Biochemistry, 25 (6) : 751-58.

Cancio, T., Pena, F. and Chaviano, R. 1991. Effect of green manures on California Wonder sweet pepper plants growing in greyish brow soils with average erosion. Soil & Fert., 56 (10) : 1268, Ab. 10331.

Chandrashekar, V., Narayanaswamy, R. and Ramabadran, P. 1983. Effect of foliar spray of potash and neem seed extract on the tikka leaf spot of groundnut. *Indian phytopath.* 46 : 189-190.

Choudhary, B.N. and Sivakumar, C.V. 2000. Biocontrol potential of Mylonchulus minor against some plant parasitic nematodes. Annals of Plant Protection Sci., 8 (1) : 53-57.

Clarholm, M. 1985. Interaction of bacteria protozoa and plants leading to mineralization of soil N. Soil Biol. Biochem., 17 : 181-187.

Clarholm, M. 1994. The microbial loop in soil. In : Beyond the biomass, Ed. K. Litz. J. Dighton, K.E. Giller, pp. 221-238, Pub. Wiley Eastern Pvt. Ltd.

Cobo, J.G., Barrios, E., Kass, D.C.L. and Thomas, R. 1999. N mineralization andcrop uptake from green manure applications to a tropical hillside soil. Annual Meeting Abstr., American Soc. of Agron., p. 260.

Cramer, T.N., Bary, A.J., Sullivan, D.M., Cogger, C.G., Myhre, E.A. and Fransen, S.L. 1999. Residual nitrogen availability following a high rate compost application. Annual Meeting Abstr., American Soc. of Agron., p. 254.

Dahama, A.K. 2001. Use of traditional additivies for organic farming. In : Organic farming for sustainable agriculture. Pub. Agrobios (India), pp. 121-122.

Das, K., Medhi, D.N. and Guha, B. 2001. Indian J. Agron., 46 (4) : 648-53.

Das, M., Singh, B.P., Ram, M., Dwivedi, B.S. and Prasad, R.N. 1991. Effect of P fertilizer amended organic manures on P nutrition of crops under mid-altitude of Meghalaya. Annals of Agril. Res., 12 (2) : 134-141.

Das, P.K., Sethi, A.K., Jena, M.K. and Patra, R.K. 1999. Effect of P sources and dual inoculation of VAM and Rhizobium on dry matter yield and nutrient uptake by greengram. J. Indian Soc. Soil Sci., 47 (3) : 466-470.

Davies, R.D. 1993. Agricultural Utilization of sewage sludge. In : Agriculture & Environment. Ed. Jones, G.J. Chichester. Pub. Ellis Harwood, pp. 17-25.

Dayakar, S. and Ray, S.M. 1999. Comparative effectivenss of some plant products and oils against pulse beetle, cellosobruchus chinensis on greengram. Andhra Agril. J., 46 (3 & 4) : 143-145.

Deka, M.K., Singh, K., Handique, R. 1998. Antifeedant and repellant effects of Pongamia and wildsage on tea mosquito bug (*Helopeltis theivara*). Indian J. Agril. Sci., 68 (5) : 274-76.

Dhiman, S.D., Mandal, D.P. and Hariom 2000. Productivity of rice wheat cropping system as affected by residue management and fertility lévels. Indian J. Agron. 45 (1) : 1-5.

Dubey, A.V., Vaishya, U.K., Bapat, P.M. and Tomar, V.S., 1999. Phosphate solubilizing efficiency of some micro-organism in vertisols. J. Indian Soc. Soil Sci., 47 (1) : 161-64.

Dubey, S.K. 1999. Nodule count, dry matter accumulation and seed yield of soybean as influenced by *Brodyrhizobium japonicum* inoculation amended with Mo in Lithic Ustorthents under rainfed condition. J. Indian Soc. Soil Sci., 47 (2) : 362-365.

Elgwla, A.M., Aly, O.M. and Elsiklhry, E.M. 1990. Effect of certain soil amendments on availability of Fe, Mn, Zn, Ca to sorghum plants grown in sandy soils. Egyptian J. Soil Sci., 30 (1-3) : 301-302.

Ellito, E.T., Coleman, D.C. and Core, C.V., 1979. The influence of amoeba on the uptake of N by plants in gnobiotic soil. In : The soil root interface. Ed. J.L. Harley and S. Russel, pp. 221-229, Academic Press, London.

Flynn, R.P., Lindermann, W.C. and Remmengo, M. 1999. Total N content of dairy manures in New Mexico. Annual Meeting Abstracts of CSSA, pp. 253.

Ganal, B.A. and Singh, C.M. 1988. Effect of FYM applied in rice-wheat rotation on physico-chemical properties of soil. Indian J. Agron., 33 (3) : 327-329.

Gascho, G.J., 1999. Irrigated multicropping using broiler litter and supplemental fertilizers in conservation tillage. Annual meet Ab. American Soc. of Agron., pp. 242.

Gaur, A.C. and Mathur, R.S. 1979. Effect of straw, neem cake and farm yard manure on yield and nitrogen uptake by maize crop. Indian J. Agron., 24: 509.

Gaur, A.C., Neelakanthan, S. and Dargan, K.S. 1990. Livestock and human wastes - Characteristics and manurial value. In : Organic manures, Pub. ICAR, pp. 13-95.

Geethakumari, V.L. and Shivashankar, K. 1991. Studies on organic amendments and CO_2 enrichment in ragi/soybean intercropping system. Indian J. Agron., 36 (2) : 202-206.

Gerke, J., Beibner, L. and Roma, W. 2000. The quantitative effect of chemical phosphate mobilization by carboxylate anions on P uptake by a single root - I. The basic concept and determination of soil parameter. J. Plant Nutrition & Soil Sci. 163 : 207-212.

Ghandhale, D.N., Salunkhe, G.N. and Naik, L.M. 1983. *J. Maharashtra. Agric. Univ.*, 8 (3) : 294-295.

Ghewande, M.P. 1982. Diseases of ground nut and their management. Annual report. NRCG, *Junagad* Pp : 34-42.

Gill, H.S., Rao, D.L.M. and Meelu, O.P. 1991. Effect of anerobic incubation on N-mineralization from urea and FYM in asodic soil. J. Indian Soc. Soil Sci., 39 (1) : 64-67.

Gill, K.S., 1987. Linseed. Pub. ICAR, pp. 362.

Gloria, M.A., DA, Barretto, M.C.V., Moraes, C.J. and Mattiazzo-Prezotto, 1992. Evaluation of gypsum and some phosphates on inhibitors of ammonia in manures. Abstrated in Field Crop Abstracts, 56(1): 75, Abstr. 644.

Goel, PK and Trivedi, R.K. 1984. Some consideration on sewage disposal to fresh waters and resultant effects. Pollu. Res. 3 : 7-12.

Golakiya, B.A., Patil, M.S. and Sutaria, G.S. 1990. Interaction effect of active lime, FYM and moisture region on yield and on P composition of wheat. Guj. A.U. Res. J., 16 (1) : 33-35.

Gopalaswamy and Kannaiyan, S. 1998. Utilization of Azolla hybrids for bio-reclamation of sodic soils and rice growth. *Indian J. Agron.* 43 (3) : 511-517.

Goyal, S., Mishra, M.M., Dinkar, S.S., Kapoor, K.K. and Batra, R., 1993. Microbial biomass turnover and enzyme activity following application of FYM to field crops with or without long term application. *Biology & Fertility of Soils*, 15 (1) : 60-64.

Gransee, A. and Wittenmayer, L. 2000. Qualitative and quantitative analysis of water soluble root exudates in relation to plant species and development. *J. Plant Nutrition & Soil Sci.* 163 : 381-385.

Greenwood, Gallaher, R.N. and McSorley, R., 1999. Recycling of urban yard waste to improve soil quality and sweet corn yield. Annual meet Ab. American Soc. of Agron., pp. 252.

Grossman, J.M. 1999. Use of leguminous trees in coffee agro-eco-system in Chipas, Mexico, Annual Meeting, Ab. ASA 1999, pp. 41.

Gujar, G.T., Kalia, V. and Kumari, A. 2000. Bioactivity of *Bacillus thuringensis* against American bollworm *Helicoverpa armigera*. Annals of Plant Protection Sci., 8 (2) : 125-131.

Gutheridge, R.C. 1992. Evaluation of leaf of a range of free legumes as a source of N for crop growth. Experimental Agri., 28 (2) : 195-202.

Hadimani, A.S., Hegde, B.R. and Satyanarayana, T. 1982. Management of red soils. Review of Soil Research in India - Part II. 12th Int. Congress of Soil Science, Feb. 8-16, 1982, New Delhi, pp. 686-702.

Halepyati, A.S., Hosamani, S.A. and Hunshal, C.S. 1987. *J. Maharashtra, Agric. Univ.*, 12 (2) : 395-396.

Hanes, J. and Mucha, V. 1991. Effect of green manure on the respiration activity and some chemical properties of orthic Lirisol. Zentralblatt fur Mikrobiologie, 146 (7/8) : 525-28. Soils & Ferti., 55 (8) : 1057, Ab. 8843.

Hao, X., Chang, C. and Travis, G.R. 1999. Effects of long term manure application of soil properties and triticale quality. Annual Meeting Abstr., American Soc. of Agron., p. 256.

Hassabnis, S.N., Kulkarni, S. and Patil, B.N. 1997. Role of nitrogen fertilization and host plant density in the management of leaf rust incidence in wheat. *J. Maharashtra. Agric. Univ.*, 26 (3) : 254-256.

Hegde, B.R., Havanagi, G.V., Munikrishnareddy, N., Venugopal, N., Vishwanath, A.P. and Satyanarayan, T. 1982. Studies on the incorporation of maize residue on soil properties and yield of maize residue on soil properties and yield of maize under rainfed conditions. Indian J. Agron., 27 : 254-258.

Hegde, D.M. 2000. Long term effect of fertilizer use on productivity and soil fertility in an irrigated rice-wheat system on different type of soils. J. Maharahstra Agril. Univ., 25 (2) : 117-123.

Helen, E., Estewart., Mccalmount, D.C. and Woslie, R.L. 1983. The effect of harvest date and the interval between harvest and inoculation on the assessment of the resistance of the potato tubers to late blight. *Potato. Res.* 26 : 101-107.

Hickisch, B. and Mulla, G. Jr. 1990. Effect of fertilization on micro-organisms in long term field experiment. Agrokemia & Talajtan, 39 (3-4) : 415-418.

Hinesly, T.D., Ziegler, E.L. and Barsett, G.L. 1979. Residual effects of irrigating corn with digested sewage sludge. J. Environ. Qual., 8 (1) : 35-38.

Hisra, H.P. 1999. Control of gundhi bug in upland rice. Indian J. Entomol., 61 (2) : 169-172.

Hogg, T.J., Weiterman, G. and Tollefsa, L.L. 2003. Effluent irrigation: Saskatchewan Perspective, Internet Site : Cside@agr.gc.co.

Hue, M.V. 1991. Correcting soil acidity of a highly weathered alfisols with chicken manure and sewage sludge. Communications in Soil Sci. and Pl. Analysis 23 (3-4) : 241-264.

Hundal, H.S., Dev, G. and Biswas, C.R., 1991. Effect of green manure on kinetics of dissolved and labile P release in anoxic soil. J. Indian Soc. of Soil Sci., 39 (2) : 260-65.

Jnaneshwar, 1998. M.Sc. thesis, UAS, Dharwad. measures. *Indian coconut. J.* 24 (3) : 1-5.

Joffe, J.S. 1955. Green manuring viewed by pedologist. Adv. Agron. 7 : 141.

Joshi, M. 1999. Nutrient demands and supplies in hybrid cotton (Unpub. data).

Juwarkar, A.S., Deshbharatar, P.B. and Bal, A.S. 1991. Exploitation of nutritional potential of sewage and sludge through land application. In :

Asia experiences in Integrated Plant Nutrition, RAPA-FAO, Bangkok, pp. 178-201.

Kadalli, G.G., Susheela Devi, L., Sidramappa, R. and Patil, C.R. 2001. Quality and efficiency of value added coirdust based compost. J. Indian Soc. Soil Sci., 49 (1) : 141-143.

Kalidurai, M. and Kannaiyan, S. 1991. Sesbania as a biofertilizer for rice. Bioresource Technology, 36 (2) : 141-145.

Kallapur, S.K. 1982. Oil bearing new plants. Pub. UAS, Bangalore, pp. 18-121.

Kaloyanova, N. and Kostov, O., 1991. Dynamics of oxidation and reduction potential, total biological activity of microbial biomass in the course of rice ontogenesis. Soil & Fertilizer, Abst. No. 1270, 56 (2) : 145.

Kamabiran, B. and Harilal, C. 1998. Studies on effect of domestic sewage on the growth and yield of *Vigna mungo*. Pollu. Res., 17 (1) : 33-37.

Kannaiyan, S., 1998. Biofertilizers for sustainable rice production. Bull. of IISS, No. 2, pp. 14-55.

Kanniyan, S. 2002. Integrated packages for the management of coconut eriphyid mite. Interactive workshop on eriphyid mite, Tanu, Combatore.

Kansal, B.D. 1992. Effect of waste disposal on soils and plants. In : Scenario of our environment. Ed. G.S. Dhaliwal, B.S. Hansra and N. Jerath, Pub. PAU, Ludhiana, pp. 197-206.

Karasawa, T., Arihara, J. and Kasahara, Y. 2000. Effects of previous crops on Arbuscular Mycorrhizal formation and growth of maize under various soil moisture condition. Soil Sci. & Pl. Nutrition, 46 (1) : 53-60.

Kaul, V., Mosood, M.A., Bhagat, K.C. and Bhat, O.K. 2000. Biology behaviour and biocontrol agents of Drosicha dalbergia (Green) infesting almonds in Kashmir. Annals of Plant Protection Sci., 8 (1) : 24-26.

Kaur, J.J., Rao, D.K., Sehgal, S.S. and Seth, R.K. 2001. Effect of hexane extract of neem seed kernel on development and reproductive behaviour of Spodoptera litura. Annals of Plant Protection Sci., 9 (2) : 171-178.

Khan Samiullah, Quereshi, M.A. and Singh Jasbir, 1997. Influence of heavy metals composition on the mobility of some micronutrients through soil. Indian Environ. Health, 39 (3) : 217-223.

Khan, M.R. and Goswamy, B.K. 2000. Effect of culture filterates of Paecilomyces lilacinus isolates on hatching of Meloidogyne incognita eggs. Annals of Plant Protection Sci., 8 (1) : 62-65.

Knapp, F.D., Elillett, L.F. and Campbell, G.S. 1983. Carbon, N and microbial biomass inter-relationship during decomposition of wheat straw - A mechanistic simulation model. Soil Biol. & Biochem., 15 : 455-461.

Koga, N. and Ikeda, M. 2000. Methionine sulfoxine suppressed the stimulation of dark reaction by ammonium nutrition in wheat roots. Soil Sci. & Pl. Nutrition, 46 (2) : 393-400.

Krishnamoorthy, K.K. and Ravikumar, 1973. Permanent manurial experiment conducted at Coimbatore. Res. Rev., TNAU, Coimbatore, pp. 1-54.

Krovchenko, L.V. and Makarova, M.M. 1990. Nitrogen fixing activity of *Azospirillum brasilense* applied to soil with plant remain. Soils and Fertilizer, Abst. 7061, 56 (7) : 873.

Kuikman, Jansen, A.G., VanVeen, J.A. and Zehnder, A.J.B. 1990. Protozoan predation and the turnover of soil organic carbon and N in presence of plants. Biol. Fert. Soils, 10 : 22-28.

Kumar, K. 1994. Studies on development of Azolla mutants and their N fixing potentiality, microbial decomposition and biofertilizer value for rice. Ph.D. Thesis, TNAU, Coimbatore, pp. 125-159.

Kumar, K., Thangaraji, M. and Kannaiyan, S. 1995. Use of nitrification inhibitors and N use efficiency in low land rice soils. In : Rice Management Biotechnology, Ed. Kannaiyan, Pub. Assoc. Pub. Co., New Delhi, pp. 68-145.

Kumar, M., Yadav, K., Thakur, S.K. and Mandal, K. 1998. Effect of Vesicular Arbuscular Mycorrhizal fungi and Rhizobium inoculation on nodulation, root colonization, nitrogen fixation and yield of chickpea. *J. Indian Soc. Soil Sci.,* 46 (3) : 375-378.

Kumar, S. and Mandakini Mayak, 1997. Cell wall-bound phenolics in resistance to rice blast. *J. Mycol. Pl. Pathol.,* 27(1): 1-5.

Kumar, V.R. and Singh, Bhim. 2004. Management of flag smut in wheat. *Ann. Pl. Protec. Sci.* 12 (1) : 92-95.

Kumaraswamy, D. and Madalageri, B.B. 1990. Effect of Azotobacter inoculation on tomato. South Indian Hort., 38 (6) : 345-346.

Kundu, G.G., Jotwani, M.G. and Premkishore. 1978. *Indian J. Ent.,* 32 (3) : 215-217.

Ladd, J.M., Amalo, M., Kackson, R.B. and Butter, J.H.A. 1983. Utilization by wheat crop of N from legume residue decomposing in soil in the field. Soil Biology & Biochem., 15 : 231-238.

Lewin, H.D., Natarajan, S. and Govinda Rajan, K.S. 1972. Correlation between sowing time and weather factors on the intensity of ground nut diseases. *Indian J. Myol. and Pl. Patrhol.* 3 : 26-32.

Madhumita Das, Singh, B.P., Munna Ra, Dwivedi, B.S. and Prasad, R.N., 1991. Influence of organic manures on native plant nutrients availability in acid soils.

Magalhaes, J.C., De, Vieiro, R.F., Pereira, J. and Peres, J.R.R. 1992. Efficiency of green manure on availability of P from phosphorus in a crop sequence on a soil under `Cerrado' vegetation. Soils & Fert., 56 (1) : 15, Abst., 119.

Magar, S.S. 1982. Soil water management aspects related to crop production in black cotton soils (Vertisols). In : Proc. Workshop on Water Resources - Planning & Management, Jan. 22-24, 1982 at Walmi, Aurangabad, Maharastra, pp. 5-18.

Maharudrappa, K., Sharanappa and Chandrashekar, S.C. 1993. Performance of rice genotypes under varying levels of nitrogen in blast endemic area of hill zone of Karnataka. Current. Res. 21 (1) : 1-2.

Maiti, P.S., Sah, K.D., Gupta, S.K. and Banerjee, S.K. 1992. Evaluation of sewage sludge as a source of irrigation. J. Indian Soc. of Soil Science, 40 (1) : 168-172.

Makarenko, S.V., Chebotarev, M.T., Korneev, Yu, I. and Khoklovo, L.A. 1991. Effect of OSV (sewage sludge treatments) on yield and quality.

Malathi, S., Sriramulu, M., Babu and Ramesh, T. 1999. Evaluation of certain eco-friendly insecticides against Lepidopterous pests of cabbage. Indian J. Entomol., 61 (2) : 127-133.

Maraikar, S. and Amarasiri, S.L. 1989. Effect of cattle and poultry dung addition on available P and exchangeable K of red yellow podzolic soils. Tropical Agriculturist, 145 : 51-59.

Marula, N., Nijhawan, D.C., Laxminarayana, K., Kapoor, R.L. and Verma, O.P.S. 1991. Response of pearlmillet to soil isolates and analogue resistant mutants of Azotobacter chroococcum. Indian J. Agril. Sci. 61 (7) : 484-487.

Mathur, R.S., Althega, S.C. and Jain, J.S. 1960. Varietal resistance of wheat to loose smut in UP. *Curr. Sci.* 29 : 353-354. 113-118.

Mbagwu, J.S.C. 1992. Improving productivity of degraded ultisols of Nigeria using organic and inorganic amendments- Part I - Chemical properties and maize yield. Bioresource Technology, 42 (2) : 149-54.

Men, U.B., Kandalkar, H.G. and Mundiwale, S.K. 1986. *Sorghum Newsl.* 29 : 58-59.

Mishra, A.N., Varma, P.K. and Ruwali, K.N. 1992. Rust incidence in late sown wheat. *Indian J. Agric. Sci.* 64 (4) : 266-267.

Mitta, T. 1992. The stimulation of root activity and reduction in root diseases of crops through improvement in rhizosphere microflora. Res. J. of Foods & Agri. 13 (6) : 15-20.

Moghaiels, R.E.A., Tanaka, N., Saneoka, H., Hussain, H.A., Yosef, S.S., Ewada, M.A.F., Aly, M.A.M. and Fujita, K. 2000. Expression of betaine dehydrogense gene in transgenic tomato hairy roots feeds to accumulation of Glycine betaine and contributes to maintenance of osmotic potential under salt stress. Soil Sci & Pl. Nutrition, 46 (4) : 873-883.

Mongia, K. and Beniwal, M.S. 1993. Effect of different N and P doses and irrigation levels on loose smut expression in wheat. *Crop. Res.* 6 (1) : 138-141.

Mukherjee, D. and Gaur, A.C. 1980. A study on the influence of straw incorporating on soil organic matter maintenance, nutrient release and asymbiotic nitrogen fixation. Zbl. Bakt. Abt. II, 135 : 663.

Muthaiah, C. and Bhaskaranan, R. 1999. Survey bioefficiency and management of eriphyid mite on coconut. Paper presented in group meeting on coconut eriphyid mite. TNAU, Coimbatore.

Nagaraju, H.L. and Vasantha Kumar, B.M. 1991. Management of paddy blast through cultural practices, host resistance. *Haryana Agri. Univ. J. Res.* 29 : 112-114.

Nandarajan, L. 1997. Studies on integrated control of coconut black headed caterpillar. *Ph.D. thesis.*

Nanzyo, M., Shibata, Y. and Wada, N., 2002. Complete contact of Brassica roots with P fertilizer in a phosphorus deficient soils. *Soil Sci. & Plant Nutrition,* 48 (6) : 847-53.

Nimje, P.M. and Seth, J. 1988. Residual effect of P and FYM on growth, yield and quality of rabi maize. *Farming Systems,* 4 (3-4) : 81-88.

Noguchi, H. and Ito, H. 1992. Long term experiment of applying sludge fertilizer to agriculture land. *J. Institution of Water & Environ. Management,* 6 (5) : 576-82.

Omar, M.S. 1990. Effect of farm yard manuring on soil aggregation under potato production in alluvial loamy soils. *Egyptian J. Soil Sci.,* 30 (9) : 607-613.

Pachiaraj, S.P. and Venkataraman, N.S., 1991. Influence of irrigation regimes, organic amendment and source of P on low land rice. *J. Indian Agric.* 36 (Supplement) : 14-17.

Padmanabhan, N. and Sukumaran, A.S. 1998. Coconut pests and their control measures. *Indian Coconut J.,* 24 (3) : 1-5.

Pandey, S.P., Harishankar and Sharma, V.K. 1985. Efficacy of some organic and inorganic residues in relation to crop yield and soil characteristics. *J. Indian Soc. Soil Sci.* 33 : 179-181.

Pant, L.M., Dwivedi, G.K. and Katiyar, A.K. 1999. Survival and response to inoculation of *Rhizobium leguminosarum* bv. *phaseoli* on frenchbean grown under temperate condition. *J. Indian Soc. Soil Sci.,* 47 (3) : 558-560.

Panwar, J.D.S. 1992. Effect of VAM and *Azospirillum inoculation* on metabolic changes and grain yield of wheat under stress condition. *Indian J. Plant Physiol.* 35 (2) : 157-161.

Parihar, S.S., Pandey, D., Shukla, R.K., Verma, V.K., Chaure, N.K., Choudhary, K.K. and Pandya, K.S. 1999. Energetics yield water use and economics of rice based cropping system. Indian J. Agron. 44 (2) : 205-209.

Patel, J.R. and Patel, Z.G. 1994. Effect of irrigation Rhizobium inoculation and nitrogen on yield and quality of pigeonpea. *Indian J. Agron.,* 39 (4) : 659-661.

Patil, H. Ramangouda, 2002. Evaluation of insect pest managment components in soybean ecosystem. Ph.D. thesis, 84-88.

Patil, J.D., Patil, A.J., Gaikwad, B.G. and Sonawane, S.S. 1996. Recycling of crop residues in soil and its effect on performance of rabi sorghum under dryland condition. J. Maharashtra Agric. Univ., 21 (2) : 186-89.

Peterson, A.E., Speth, P.E., Wright, T.H. and Schlecht, P.L. 1999. Residual effect of 15 years of biosolid application on crop yield soil analysis and ground water. Annual Meeting Abst. of ASA, pp. 30.

Phadtare, S.C., Gopal, J. and Singh, P.A. 2000. Kufri Kanchan - A wart immune and late blight resistant red tuber potato variety. *J. Indian Potato Assoc.* 27(3-4): 113-118.

Piedra, C., Valdes, M., Herrera, C.M. and Nua, A. 1992. Influence of different doses of filter pressmud on first pear of sugarcane cropping in an improved saline soil. Soils and Fertilizers, Abst. No. 5276, 56 (5) : 656.

Pimpini, F., Giardini, L., Barin, M. and Gianquinto, G. 1992. Effect of poultry manure and mineral fertilizers on quality of crops. *J. Agril. Sci.*, 118 (2) : 215-221.

Porter, D.M., Wright, F.S. and Powell, N.L. 1987. Effect of sprinkler irrigation on peanut disease in Virgina. *Plant Dis.* 71: 512-515.

Prasad, R., Gangish, B. and Aipe, K.C. 1999. Effect of crop residue management in a rice-wheat cropping system on growth and yield of crops and on soil fertility. *Expt. Agric.*, 35 : 427-435.

Prem Kishore, 1992. *J. Ent. Res.*, 22(2): 101-105.

Preusch, P.L., Ferier, D., Adler, P.R., Sikora, L.J., Takeda, F. and Twarkoski, T.J. 1999. Nitrogen mineralization and phosphorous availability of composted and uncomposted poultry litter. Annual Meeting Abstr., American Soc. of Agron., p. 314.

Punia, S.S. and Malik, R.K. 2003. Zero till sowing of wheat for sustainability of rice-wheat cropping system. *Indian Far.* pp. 19-23.

Quasim, M., Faridullah, Shah, S., Himayatullah and Rahim, D. 2000. *Pak. J. Biol. Sci.* 3 (11) : 1979-1980.

Rai, M.K., Ramamurthy, V.V., Gupta, S.L. and Dey, D. 2000. Preliminary investigation on Paederus fuscipes, a predator in rice agro-eco-system. *Annals of Plant Protection Sci.*, 8 (1) : 8-12.

Rainey, A.S., Cochran, M.J., Miller, D.M. 1992. Derived demand for poultry litter as soil amendment in rice. *Arkansas Farm Res.*, 41 (6) : 10-11.

Rajeev, K., Upadhyay, K.G., Mukherji and Dubey, O.P. 2003. *IPM system in agriculture,* Aditya Books Pvt. Ltd., New Delhi, India Vol. 6 cash crops, pp : 83-100.

Rajesekhara Rao, K. 2003. Influence of host plant nutrition on the incidence of spodoptera litura on sunflower. *Indian J. Ent.*, 37: 188-191.

Rajput, R.P. and Shastry, P.S.N. 1988. Effect of soil amendments on physico-chemical properties of sandy loam soils - II. Structural and hydrophysical properties. *Indian J. Agri. Res.*, 22 (4) : 209-216.

Ramos, T.O., Lana, L.P.E., Rivera, L.JA and Sanginer, G.J.R. 2000. Internet Abstract WWW. fao.org/DOCREP/005/X9895E/X9395e. Os.htm. 39k.

Ramsingh, S., Sundar, D.P. and Nandal., 1999. Management of paddy blast through cultural practices, host resistance. *Haryana, Agric. Univ. J. Res.*, 29 : 109-114.

Ramusseen, P.E. and Smiley, R.W. 1989. Long term management effect on soil productivity and crop yields in semi-arid regions of Oregeon. Station Bull. 675, Columbia Basin Agril. Res. Stn., Pendleton, Oregeon, pp. 57.

Rana, D.C. and Saran, G. 1999. Comparative performance of *Oliciferous brassica* species in intercropping with potato staggered sowing and fertility levels. *Indian J. Agron.* 44 (2) : 181-188.

Rasal, P.H., Shingte, V.V. and Patil, P.L. 1989. Effects of sugarcane trash on crop yields and soil properties. *J. Maharahstra Agric. Univ.*, 14 : 79-82.

Raveenera, K.R. 1993. Studies on management of tikka disease of ground nut. M.Sc. (Agri) thesis, UAS, Bangalore.

Ritcher, G.M. and Ritcher, J. 1992. N-mineralisation and NH_4-dynamics with manure application. *Abstracted in field Crop Abstracts*, 55(2): 203, Abst. 1688.

Roger, P.A. and Ladha, J.K. 1990. Estimation of biological N_2 fixation and its contribution to nitrogen balance in wetland rice fields. In: Transmission of 14th International Congress of Soil Science held at Kyoti, Aug. 1990, Vol. III : 128-133.

Rose, D.A., 1990. Effect of long continued organic manures on some physical properties of soil. In : Advances in soil organic matter research. Proc. Symposium Colchester, U.K. held on 3-4th September, Royal Society.

Ruiz-Diego, C., Collins, M., Ditsch, D., Coyne, M.S. and Grove, J. 1999. Ground water and crop response to poultry litter application of surface mind land meeting. Annal Ab., CSSA, pp. 330.

Sagare, B.M. and Guhe, Y.S., 1991. Response of hirsutum cotton on topic chromosters to enriched FYM products. *Annals of Pl. Physiol.*, 5(1): 123-125.

Sarkar, A., Yadav, R.L., Gangwar, B. and Bhatia, P.C. 1999. Crop residues in India. Pub. PDCSR Modipuram, pp. 1-31.

Satpute, R.G., Vega, O.P. and Raut, N.O. 1993. *Indian J. Agric. Res.* 27: 40-44.

Schlegel, A.J. and Alam, M., 1999. Soil chemical properties following animal waste application. Annal Meeting Abstracts. *American Soc. of Agronomy*, pp. 330.

Schmidt, L., Warnstorff, K., Dorfel, H., Leinweber, P., Lango, H. and Merbach, W. 2000. Influence of fertilization and rotation on soil organic matter and plant yields of long term eternal rye trial in Halle, Germany, *J. Plant Nutrition & Soil Sci.,* 163 : 639-648.

Schulze, J., BeSchon, H., Adgo, E. and Merbach, W. 2000. Efficiency of N_2 fixation in Vicia faba in combination with different Rhizobium leguminosarum strain. *Plant Nutrition & Soil Sci.*, 163 : 367-73.

Seed Bold, K.W., Kucharek, T.A. and Datnoff, L.E. 2001. The influence of blast in susceptible, partially resistant and resistant cultivar. *The American Phytopathological Society*. 91(1): 63-69.

Sekhon, P.S. and Sokhi, S.S. 1999. Effect of planting on development of late blight of potation Punjab. *Indian Phytopath.* 52(3): 267-269.

Serna, M.D. and Pomares, F., 1992. Indexes of assessing availability in sewage sludges. *Plant & Soil.* 139 (1) : 15-21.

Setty, R.A. and Channabasavanna, A.S., 1991. Optimum combination of nitrogen and Sesbania rice production. *Curr. Res.* 20 (10) : 207-209.

Sezhian, N. 1999. Soya fights red rot in sugarcane. *Kisan world.,* 4: 44-45.

Shailajakumari, M.S. and Ushakumari, K. 2002. Effect of vermicompost enriched with rockphosphate on growth and yield of cowpea. *J. Indian Soc. Soil Sci.*, 50 (2) : 223-224.

Sharma, S.K. and Chaudhary, J.P. 1994. *Haryana. Agric. Univ. J. Res.* 3: 361-363.

Shelef, G. 1990. The role of waste water reuse in water resources management in israel. *Water Sci. Tech.* 23 : 2081-2089.

Shinde, D.B., Navale, A.M. and Jadhav, S.B. 1993. A study on incorporation of sugarcane trash in ratoon sugarcane. *J. Maharashtra Agric. Univ.* 18 (2) : 264-65.

Shivayogeshwara, R. and Sarkar, R.K. 2001. A Histopathological study of viral leaf syndrome in chilli. *J. Intercadenuia.* 5: 71-73.

Shuval, H.I., Adin, A., Fattal, B., Rawitch, E. and Yekutiel, P. 1986. Waste water irrigation in developing countries : Health effects and technical solution. World Bank Tech. Report. pp. 511.

Sidhu, B.S., Beri, V. and Gosal, S.R. 1995. Soil microbial health as effected by crop residue management. In : National Seminar on development in Soil Science, Nov., 2-5, PAU, Ludhiana.

Sikka and Chauhan, 1973. Effectof date of planting on the performance of potato varieties. *J. Indian Potato. Assoc.* 2: 43-45.

Singarvel, S., Govindasamy, R. and Balasubramanian, T.N. 1998. Influence of humic acid, nitrogen and Azospirillum on the yield and nutrient uptake of sesame. *J. Indian Soc. Soil Sci.* 46 (1) : 145-46.

Singh, A.K., Amgan, L.P. and Sharma, S.K. 2000a. Root chickpea, soil physical properties and yield of rice as influenced by integrated nutrient management in rice-wheat system. *Indian J. Agron.* 45 (2) : 217-222.

Singh, A.P. Mitra, B.N. and Tripathi, R.S. 1999a. Effect of soil environment in conjunction with organics and check fertilizers and rice-chickpea cropping system. *Indian J. Agron.*, 44 (2) : 223-27.

Singh, G.P. 2004. *Indian Fmg*. April, 2004. pp. 12-14.

Singh, J.P. and Tarafdar, J.C. 2002. Rhizospheric microflora as influenced by sulphur application, herbicide and rhizobium inoculation in summer mungbean (*Vigna radiata*). *J. Indian Soc. Soil Sci.,* 50 (1) : 127-129.

Singh, M., Sharma, R.C., Lal, S.S., Singh, A.K. singh, P., Govind Krishna, S.S. 2000b. Long tern effect of fertilization and manuring on productivity of some potato based cropping system. *J. Indian Potato Association,* 27 (3-4) : 69-76.

Singh, M.P., Jain, B.P., Srivastava, J.L. and Trivedi, R.M. 1991. Nitrogen fixing and multipurpose tree species for afforestation. Pub. Today and Tommorrow Printer & Publishers, New Delhi, pp. 137-279.

Singh, N.B. and Verma, K. 1999. Impact of selected rabi crops on productivity and N economy in rice based cropping system. *Oryza,* 36 (1) : 89-91.

Singh, R.N. and Singh, K.M. 1978. *Indian J. Ent.* 40: 361-375.

Singh, R.P., Mundra, M.C., Gupta, S.C. and Agarwal, S.K. 1999b. Effect of integrated nutrient management on productivity of pearlmillet wheat cropping system. *Indian J. Agron.* 44 (2) : 250-53.

Singh, V.P. 1999. Effect of organic and inorganic sources of nutrients on rainfed wheat. *Indian J. Agron.*, 44 (2) : 347-352.

Singhal, S.K. and Chauhan, R. 1997. Effects of some plant products and other materials on development of pulse beetles, Callosobruchus chinensis on stored pigeonpea. *J. Insect. Sci.* 10 (2) : 196-197.

Smith, S.R., Tibbeti, M. and Evans, T.D. 1992. Nitrate accumulation potential of sewage sludge applied to soil. *Aspects of Applied Biology*, No. 30, pp. 157-61.

Sood, M.C. 1988. Water optimization and its requirement of potato crop. *annual Scientific Report,* CPRI, Shimla, pp. 87-99.

Srikanth, 1997. Enrichment of compost and its effect on soil-properties and crop growth. Ph.D. Thesis submitted to University of Agricultural Sciences, Bangalore.

Strauss, M. and Blumental, U.J. 1990. Human waste in agriculture and aquaculture : Utilization practices and health propspectives, IRCWD Report No. 9/90. Pub. International Research Centre for Waste Disposal Deubendorf.

Surinder Singh, Singh, R.N., Janardhan Prasad and Binod Kumar, 2002. Effect of green manuring, FYM and biofertilizers in relation to fertilizer nitrogen on yield and major nutrient uptake by upland rice. *J. Indian Soc. Soil Sci.*, 50 (3) : 313-314.

Talashilkar, S.C., Bhangarath, P.P. and Mehta, V.B. 1999. Changes in chemical properties during composting of organic residues by earthworm activity. *J. Indian Soc. Soil Sci.* 47 (1) : 268-71.

Tandon, H.L.S. 1995. Fertilizer organic manures, recyclable wastes and biofertilizers (Ed.) Fert. Dev. & Consultation Organization, New Delhi, pp. 36-51.

Tapadar, S., Kovar, J.C. and Robbins, M.L. 1999. Changes in bio-available phosphorus in a waste amended soil during a three year period. Annal Meeting Abstracts. *American Soc. of Agronomy*, pp. 316.

Taues, L.W. 1989. Economic impact of future biological N fixation technology on US agriculture. *Plant & Soil*, 119 (2) : 261-270.

Tennakoon, N.A. 1990. Goat manure as soil amendment and yield stimulant in coconut. *Cocos*, 3 : 26-32.

Thakur, S.K. and Sharma, C.R. 1998. Effect of rock phosphate enrichment and Azotobacter inoculation on the transformation of nitrogen and phosphorus during composting. *J. Indian Soc. Soil Sci.*, 46 (2) : 228-231.

Tian, G., Olimah, J.A., Adeoye, G.O. and Kang, B.T. 2000. Regeneration of earthworm population in degraded soil by natural and planted fallows under humid tropic condition. *Soil Sci. Soc. America J.*, 64 : 222-28.

Tippannavar, C.M., Rajashekhara, E. and Sreenivasa, M.N. 1990. Effects of FYM extract on *Rhizobium* sp. and *Azotobacter chrococcum* studies. *J. Maha. Agric. & Univ.* 15 (2) : 237-38.

Tripathi, D. and Minhas, R.S. 1991. Influence of fertilizer P and FYM on transformation of inorganic phosphate. *J. Indian Soc. Soil Sci.* 39(3): 472-76.

Tripathi, S.B., Hazra, C.R. and Srivas, M.C. 1991. Effect of nitrogen sources with and without P on oats. *Indian J. Agril. Res.* 25(2) : 79-84.

Vageesh, T.S., Jagannath, V., Veeranna, V.S., Naik, H.H. and Panchaksharaiah, S. 1990. Integrated nutrient supply in rice - rice sequence and its influence on soil properties. *Current Res.*, 19(6) : 96-98.

Vaishnav, P.R. and Dixit, S.K. 1998. Trends of groundnut productivity in long term experiment. *Gujarat Agril. Univ. J.*, 24 (1) : 73-82.

Vanden, Bygaart, A.J., Fox, C.A., Fallow, D.J. and Protz, R. 2000. Estimating earthworm influence soil structure by morphometric image analysis. *Soil Sci Soc. America J.* 64 : 982-988.

Vasanthi, D. and Kumaraswamy, K. 1999. Efficacy of vermicompost to improve soil fertility and rice yield. *J. Indian Soc. Soil Sci.*, 47 (1) : 268-71.

Venkateswaralu, J., 1987. Efficient resource management systems for dryland of India. *Adv. Soil Sci.* 7 : 166-221.

Venugopal, M.S., Mani, M. and Balasubramanian, M. 1997. *Madras Agric. J.* 64(5): 342-343.

Verdu, I., Gomez, I., Burlo, I., Mataix, J. 1992. Influence of P on N mineralization of sewage sludge in two calcareous soils. *Seulo. Y. Planta*, 2(1): 151-161.

Verma, H.S., Singh, A. and Agarwal, V.K. 1986. Environmental factors in relation to loose smut in wheat seeds. *Indian Phytopath.* 39: 423-426.

Viswanathan, R. and Samiyappan, R. 1999. New planting methods in sugarcane red rot disease management. *Kisan world.* 6: 55-56.

Wagger, M.G., Kissel, D.E. and Smith, S.J. 1987. Mineralization of N from N^{15} labelled crop residue under field conditions. *Soil Sci. Soc. Am. J.* 49 : 1220-26.

Weik, R.R., Kroontje, W. and Jones, G.D. 1979. Inorganic N and soluble soils in Davidson clay loam used for poultry manure disposal. *J. Environ. Quality*, 8 (1) ; 86-94.

Westermann, C., Falsen, J.C. and Tindall, T.A. 1999. Dairy compost effects on crop production and soil quality. Annual Meeting Abstr., *American Soc of Agron.,* p. 260.

Whalen, J.K. Chang, C. and Clayton, G.W. 1999. Manure and lime amendments to increase soil pH and improve crop production on acid soils. Annual Meeting Abstr., American Soc. of Agron., p. 253.

Xiong, S.G. and Cheng, C.Y. 1991. Effects of ploughing in green manure on humus forms and physical properties of sandy meadow soils. *Soils & Fert.,* 56 (10) : 1197, Abst. 9743.

Xiong, Z., Xing, G., Tsuruta, H., Shen, G., Shi, S. and Du, L. 2002. Field study of nitrous oxide emission from upland cropping system in China. *Soil Sci. & Pl. Nutrition,* 48(4) : 539-546.

Yadav, D.V. 1995. In : Recycling sugar factory pressmud in agriculture, pp. 91-108.

Yadav, D.V., Todisingh and Srivastava, A.K. 1987. Recycling of nutrients in trash with N for higher can yield. *Biological wastes*, 20 : 133-141.

Yadav, R.L. and Prasad, S.R. 1992. Conserving organic content of soil to sustain sugarcane yield. *Experimental Agriculture*, 28 (1) : 51-62.

Yashoda Hedge, J.K., Kalappanawar and Mohan Kumar 1998. Sowing dates and varietal resistance to blast of rice condition in Karnataka. *Karnataka. J. Agric. Sci.* 11(3): 687-689.

Yelashetty, Sushas 1999. Ph.D. thesis, UAS, Dharwad.

Zodhy, U., SMS Badr El-Din, MA Khalah and HK Abd EL Makund, 1984. Evaluation of digested slurry as a fertilizer. *Egypt J. Microbiol.* 19 : 131-138.

Chapter 6

ORGANIC CERTIFICATION

6.1 BACKGROUND

Certification in organic agriculture ensures that a product is produced, processed and packed under organic conditions. This is an assurance to consumer that the product being bought is 'organic'. Such certification can help in distinguishing between organic and non-organic products. This distinction is made by visible by 'organic tag' or 'logo' displayed on each organic product. Besides identification, certification helps the organic grower to claim premium prices as well as facilitates exclusion of unscrupulous non-organic grower from selling non-organic products as organic. A consumer is also facilitated by certification in the form of genuineness of the product. As organic products developed greater demand, such certification became essential and now integral part of organic agriculture.

The certificate that a product is 'organic' will have validity only when it is issued by a third party or outside organization. Hence, independent certifying organizations were established. Further, certification needs to take into consideration many aspects of production, processing and packing. Different consumer groups seek different qualities and hence each country involved in import of organic product needed its own standard for certification. Therefore, different organic standards were developed by different countries, looking to the standards sought by consumer groups in respective countries.

The awareness to use pollution free food is directly linked to literacy percentage and standard of living. European countries were early to use pollution free organic food products due to these factors. In fact, early attempts by health aware consumer groups in Germany and France to use pesticide free food in 1980s stimulated the wide scale promotion of `alternate agriculture' – without the use of pesticides. This development naturally resulted in establishment of certifying bodies- primarily meant for quality assurance of products grown by organic farmers. However, it was soon realized that few organic farmers cannot meet the demand for organic products, which witnessed alarming growth in short span of time. This necessitated establishment of offices of certifying agencies outside European countries (mostly in developing countries, where pesticide load was less). The production of organic food was simultaneously promoted in these countries during 1990s through government/diplomatic sources in the style of '*opportunity to export*' or '*opportunity to earn foreign exchange*'. This led to worldwide promotion of organic farming. India is not an exception to such global phenomenon. By this time, USA had joined European union in consuming organic food in a large way and very soon became major importer of organic food products from other countries. As a result of these developments, certification has been instrumental in promoting, popularizing and commercializing organic agriculture- helping both producers and consumers.

6.2 ESTABLISHMENT OF CERTIFYING AGENCIES

In early 1980s, governmental patronage to promote organic farming was negligible. Private companies were established for third party certification- obviously supported by independent inspection of organic farms. In absence of any government approved standards, they were following their own set of standards- evolved on the basis of interests of consumer groups. As organic farming became popular due to consistently growing demand, the relative inadequacy of certifying agencies was realized and large numbers of certifying agencies were established both in private sector and government sector- different parts of world.

Presently, 576 certifying bodies (as of 2012) are functioning in world. Most certification bodies are located in the European Union, South Korea, Japan, the United States, China, India and Canada. Other countries have depended on certifying agencies functioning in these countries. For the first time, Asia now has more organic certification bodies than Europe. There has been a slight decrease in the number of certification bodies in most regions of the world, although the number has increased rapidly in South Korea. Many of these certifying bodies are working at international level (following the standards of importing countries), while many others are working for domestic consumption. Governments in many countries have established their own certifying agencies as a part of promoting organic agriculture. The initiatives of private organizations in establishment of certifying agencies are now taken over by many government agencies.

The certifying agencies mainly dealt with not only production but also processing, packing and handling in the process of certification to prohibit the use of undesirable chemical being used or accidentally exposed. Similarly, the certification included not only the crops (both dry and perishables) but also various animal products used as human food – which included milk, milk products, meat and meat products, fish and fish products, poultry products etc. The certification also extended to various food products, which directly or indirectly use agricultural or animal products. Many such items included vegetable oils, bakery products, confectionaries, ice creams, beverages, ready to eat products etc. Thus, certification now includes the process of primary production by farmers, various processes followed by different industries using agricultural products and their handling and packing processes. Even, non food product like cotton products is also in the ambit of organic certification to include primary production by farmers as well as various stages of producing cotton products. Large array of industries is covered under the certification process.

Important certifying agencies working in India are:

Natural Organic Certification Association
Address:Row House Banglow No-2, E-10 Bldg. Sun Empire, Survey No. 7, 9 (Part), Vadgaon-Budruk, Sinhgad Road, Pune-411051 Maharashtra, India

Website: www.nocaindia.com

Bureau Veritas Certification India Pvt. Ltd., Mumbai
BVQI (India) Pvt. Ltd.
Address: Marwah Centre, 6th Floor, Opp. Ansa Industrial Estate, Krishanlal Marwah Marg,Off Saki-Vihar Road, Andheri (East), Mumbai-400 072 Maharashtra, India
Website: http://www.bureauveritas.co.in/

Ecocert SA
Address: Sector-3, S-6/3 & 4, Gut No. 102, Hindustan Awas Ltd. Walmi-Waluj Road, Nakshatrawadi, Aurangabad – 431 002 Maharashtra, India
Website: www.ecocert.in

Control Union Certifications, Mumbai
Address: "Summer Ville", 8th Floor 33rd – 14th Road Junction Off Linking Road, Khar (West) Mumbai – 400052 Maharasthra, India
Website: www.controlunion.in

International Resources for Fairer Trade
Address: Sona Udyog (Industrial Estate), Unit No. 7, Parsi Pandhayat Road, Andheri (East), Mumbai 400 072 Maharashtra, India
Website: http://www.irft.org/

IMO Control Private Limited
Address: No. 1314, Double Road, Indiranagar 2nd Stage Bangalore-560 038 Karnataka, India
http://www.imo.ch/index.php?seite=imo_index_en

Skal International (India)
A Division of CU Inspections
India Pvt. Ltd.
Address: No. 191, 1st Main Road
Mahalaxmi Layout
Bangalore – 560 086
Karnataka, India
Website: www.skal.org

Aditi Organics Certification Pvt. Ltd.
Address: #531/A, Priya Chambers,
Dr. Rajkumar Road, Rajaji Nagar
Bangalore – 560 010
Karnataka, India
Website: http://aditicert.net

APOF Organic Certification Agency, (AOCA)
Address: 126, 1st Floor,
Govindappa Road,
Off D.V.G. Road, Gandhi Bazar,
Bangalore-560 004
Karnataka, India
Website: www.aoca.in

Indian Organic Certification Agency (INDOCERT)
Address: Thottumugham
P.O. Aluva-683 105,
Ernakulam, Kerala, India
Website: http://www.indocert.org

Lacon Quality Certification Pvt. Ltd
Address: Chenathra, Theepany,
Thiruvalla – 689 101
Kerala, India
Website: www.laconindia.com

Rajasthan Organic Certification Agency (ROCA)
Address: 3rd Floor, Pant Krishi Bhawan,
Janpath, Jaipur 302 005
Rajasthan, India
Email: rocajpr.cb@gmail.com

OneCert Asia Agri Certification Private Limited
Address: Agrasen Farm, Vatika Road,
Off Tonk
Jaipur-303 905,Rajasthan, India

SGS India Pvt. Ltd.
Address: 250, Udyog Vihar
Phase – IV
Gurgaon – 122 015 Haryana, India
Website: www.sgs.com

Uttaranchal State Organic Certification Agency (USOCA)
Address: 12/II Vasant Vihar
Dehradun-248 006
Uttaranchal, India
Email: uss_opca@rediffmail.com,
info@usoca.org

FoodCert India Pvt. Ltd.
Address: Plot No.452, Panchavati Colony,
Road No.10, Banjara Hills,
Hyderabad 500 034
Website: www.foodcert.in

Vedic Organic Certification Agency
Address: Plot No. 54,
Ushodaya Enclave,
Mythrinagar, Miyanagar,
Hyderabad – 500 050
Andhra Pradesh, India
http://www.vediccertification.com/

Chhattisgarh Certification Society, India (CGCERT),
A-25, VIP Estate, Khamhardih,
Shankar Nagar, Raipur - 492 007
Chhattisgarh.

Tamil Nadu Organic Certification Department (TNOCD),
1424 A, Thadagam Road,
Coimbatore - 641 013 Tamil Nadu.

Intertek India Pvt. Ltd.,
E-20, Block B - 1,
Mohan Cooperative Industrial Estate,
Mathura Road,New Delhi - 110 044

TUV India Pvt. Ltd.,
801, Raheja Plaaza – 1
L.B.S.Marg, Ghatkopar (West),
Mumbai - 400 086 Maharashtra.

MP State Organic Certification Agency,
Vasundhara, B-II Office Complex,
Gautam Nagar,
Bhopal - 642 023 Madhya Pradesh.

Biocert India Pvt. Ltd.,
74, Sarthak Residency,
Sector - D, Shivdham,
Khandwa Road,
Indore - 452 007 Madhya Pradesh.

Uttarkhand State Organic Certification Agency,
12/II, Vasant Vihar,
Dehradun - 248 006 Uttarkhand.

Odisha State Organic Certification Agency (OSOCA),
Plot No - 326, Baramunda,
Bhuvaneswar, Odisha - 751003.

Rajasthan Organic Certification Agency (ROCA),
3rd Floor, Pant Krishi Bhawan,
Janpath, Jaipur - 302 005 Rajasthan.

ISCOP (Indian Society for Certification of Organic Products),
Rasi Building,
162/163, Ponnaiyarajapuram,
Coimbatore - 641 001 Tamil Nadu

IFOAM (International Forum for Organic Agriculture Movement) has been regarded as international agency to award the accreditation of different certification agencies in world. The standards adopted by agency, the method of certification, period of conversion, logo to be used, validity of certificate etc are covered under accreditation programme.

The certification is awarded separately to the concerned for processes of production, processing, handling and packing and even transportation. At each level, due consideration is given to the relevant laws governing in respective country as well as the standards relied upon.

Hence, the certification is not only necessary to be awarded to the farmers (who produce the organic goods) but also processors (various manufacturers) and transport companies. The processing, handling and even packing should conform to minimum standards to ascertain that adverse chemical does not enter process and pollute the product.

The range of products for organic certification includes entire range of annual and perennial crops (including fruits, vegetables, pulses, cereals, nuts, spices, coarse grains and tuners). It also includes various live stock products like eggs, milk and milk products, different varieties of meat, various meat products, chicken, animal fats as well as diversified range of marine products. The processing industries involved in processing, modifying, and value-adding the agricultural/ live stock products are part of certification process.

6.3 STANDARDS FOR CERTIFICATION

Each certification agency has mandate to conform to certain standard, based on the country of its origin and its legislative and consumers' requirements. In the initial stage, few certification agencies developed standards for certification mainly to cater to the quality considerations of organic products in their respective countries. As the certifying agencies proliferated, the issue of standards became more complicated.

Organic standards define organic practices. Organic standards are sets of requirements applicable to organic operators (including farmers, processors, and traders). Operators who want to sell their products as organic must, in many cases, comply with at least one organic standard, and be verified against this standard.

However, important standards adopted for organic certification globally are:

1. Codex alimentairus
2. European standards (EU standards)
3. IFOAM standards
4. Demeter standards
5. JAS standards (Japanese standards)
6. NOP standards (American standards)

Codex standards

Well before organic agriculture was popularized and necessity of organic standards for certification was recognized, Food and Agriculture Organisation (FAO) and World Health Organization (WHO) jointly developed standards in 1961 to process, pack and label the food products in thc form of *Codcx Alimcntarius.* Heightened consumer interest in food quality and safety as well as increased concern about the potential for food standards to be applied as trade barriers led to the establishment of the *Codex Alimentarius*

Commission (Codex). Its primary objectives are to protect consumer health and to ensure fair practices in food trade through the elaboration, harmonization and publication of food standards and other related texts.

The Guidelines for the Production, Processing, Labelling and Marketing of Organically Produced Foods (Codex Guidelines) were later developed in light of the growing production of and international trade in organically produced food, with a view to facilitating trade and preventing misleading claims. The Guidelines are intended to facilitate the harmonization of requirements for organic products at the international level, and may also provide assistance to governments wishing to develop national legislation in this area. They were adopted in 1999.

EU standards

As the organic movement was started in Germany, France, many other European countries realized the importance of organic products and developed different standards for certification. With the formation of European Union, adopting common laws, common European organic standards were developed during 1980s. All the European countries adopt this standard to certify either to produce or import organic products. Many certification agencies follow these standards.

IFOAM

Initially IFOAM adopted what is popularly called as 'IFOAM Basic Standards' (IBS) during 1984. It included standards to be adopted to produce, process, pack, and label and market the organic products. This basic standard was adopted by many certifying agencies in different parts of world. IBS also became a model to many countries to formulate legislations to establish domestic certifying agencies. IFOAM basic standards were adopted as a part of ' Organic Guarantee System' (OGS), which also included IFOAM accreditation program.

The IFOAM Basic Standards (IBS) has been replaced by the IFOAM Standard, which has been approved through electronic membership vote ending on the 5th of August 2012. The IFOAM Accreditation Criteria for Bodies Certifying Organic Production and Processing has been renamed the "IFOAM Accreditation Requirements for Bodies Certifying Organic Production and Processing".

The IFOAM Standard is an internationally applicable organic standard that can be used directly for certification. It offers a global platform for organic standard setters to discuss standards and create synergies for standard development and harmonization at the global level, thereby taking over the limited role that the IFOAM Basic Standard (IBS) had in the past in this regard.

Demeter standards

Demeter International is the largest certification organization for biodynamic agriculture, a type of organic agriculture. Its name is a reference to Demeter, the Greek goddess of grain and fertility. Demeter Biodynamic Certification is used in over 50 countries to verify that biodynamic products meet international standards in production and processing. The Demeter certification program was established in 1928, and as such was the first ecological label for organically produced foods.

The certification is the oldest traditional organic certification in Europe and is regarded as the highest grade of organic farming in thc world. Demeter's "biodynamic" certification requires biodiversity and ecosystem preservation, soil husbandry, livestock integration, prohibition of genetically engineered organisms and viewing the farm as a living "holistic organism". The certification verifies the fulfillment of the standards on behalf of the farmers, which in turn guarantees high quality food products to the consumers. This is rewarded by receiving a higher price for food certified with the "Demeter" label, ranging from 10-30% on average.

The origin of Demeter is a Cooperative for the processing of products of the biodynamic agriculture created in Berlin, Germany, in 1927. The trademark Demeter was registered in 1928. Demeter was administered by the German agronomist Erhard Bartsch who also directed the Experimental Circle of anthroposophical (biodynamic) farmers, and who had chosen the name Demeter, jointly with the German chemist Franz Dreidax. Dreidax was responsible for the development of the Demeter criteria and the quality control. Demeter ceased temporarily to exist in 1941 when the Nazi Government dissolved the Union for biodynamic agriculture. It was reestablished in Germany after the Second World War. In 1997, 19 independent Demeter organizations came together to establish Demeter International. All continents were represented.

JAS standards

The **Japanese Agricultural Standards** (JAS) are standards for the agriculture industry maintained by the Japanese Government.

They are comparable to Japanese Industrial Standards but for food and agricultural products. Once a product has passed tests conducted by government-backed rating bureaus the JAS seal (right) is appended to the product.

The JAS originally covered only horticultural products such as fruit and vegetables, livestock feed and processed products of horticultural origin. This has been extended to include livestock products such as animals, eggs and processed animal products such as cheese. This Standard does not currently cover apiculture (honey & bee products) or aquaculture. It also does not cover non-agricultural products such as textiles or health & beauty products.

NOP standards

USDA's National Organic Program regulates the standards for any farm, wild crop harvesting, or handling operation that wants to sell an agricultural product as organically produced. **National Organic Program** (NOP) is the federal regulatory framework governing organic food. It is responsible for administering and enforcing the regulatory framework. The Organic Food Production Act of 1990 (7 U.S.C.A. § 6501-22) required that the USDA develop national standards for organic products. The NOP Final Rule was first published in the Federal Register in 2000.

Most of these standards prescribe following aspects:

1. Introduction, aims of standards, definitions used in the standards
2. Inspection and certification systems, regulations governing imports and exports
3. Principles of organic production separately for plant/ plant products [including pest control] and livestock/livestock products [including cleaning/disinfection]
4. Conversion process and duration of conversion to organic
5. Specifications regarding feed stuffs, nutritional standards and health care aspects required for livestock production
6. Specific requirements (for ex: beekeeping and bee products etc.)
7. Handling, storage, transportation, processing and packing
8. Lebelling and logos
9. Detailed methods, steps to be followed for certification and revalidation

10. Substances allowed to be used, not allowed to be used in organic production, processing and packing

Harmonization of standards

A large number of organic standards exist throughout the world. It is estimated that there are hundreds of organic standards. Some are government regulations, others are private standards. Some have a much localized scope of application (e.g. one country or even one region in a given country), whereas, others are applied internationally. Trade between systems that are based on different standards requires these systems recognize and accept each other.

To facilitate such acceptance, IFOAM has developed the IFOAM Family of Standards, which is a list of all standards officially endorsed as organic by the international organic movement. In a sense, the IFOAM Family of Standards draws the line between what is organic and what is not, from a standards point of view. Standards that are approved in the IFOAM Family of Standards are organic standards and can be trusted as such. Approval is based on a detailed technical assessment of each standard, and follows a defined procedure with the IFOAM Organic Guarantee System

The Common Objectives and Requirements of Organic Standards (COROS) were developed as a joint venture of the IFOAM Organic Guarantee System (OGS) and the GOMA (Global Organic Market Access) project undertaken by FAO, IFOAM and UNCTAD. The concept of COROS was first developed by the International Task Force on Harmonization and Equivalence (ITF) through the Annex of the Guide for assessing Equivalence of Organic Standards and Technical regulations (EquiTool) in 2008 (www.goma-organic.org). The document was compiled on the basis of the IFOAM Basic Standards and Codex Alimentarius as the two pre-existing international reference organic standards, and through the review of a significant number of existing standards and regulations across the world.

COROS is intended for use in international equivalence assessments of organic standards and technical regulations. As an annex to the Equitool it is proposed as a template to guide governments and other stakeholders in conducting objectives-based equivalence assessments of two or more organic standards for production and processing. EquiTool encourages equivalence assessments to be based on reference objectives, an approach that is consistent with WTO guidance for judging equivalence. By basing equivalence discussions and assessments on objectives and related basic requirements, parties may avoid tedious line-by-line assessments of one detailed standard against another. Instead parties can assess whether the details of one standard, although they may vary from those of another standard, meet common objectives and baseline requirements.

COROS is provided in the form of a standard document, which facilitates easy reading and comprehension of the objectives and requirements related to them. However, a high degree of functionality is provided in the form of an electronic spreadsheet version containing three worksheets: the first is for data entry; the second re-organizes the data for objectives-based assessment; the third facilitates a summary analysis

Despite many efforts to streamline the certification process, different countries required different certification programme of organic products even today. In absence of such a global uniformity in organic standards and certification, the export of organic goods has been often affected for the reasons of certification. In the context of Indian organic agriculture, certification has remained mainly as a facility to export organic goods than to supply of pollution free food to domestic markets. For example, an Indian organic fruit producer needs to seek certification acceptable through European Union if he wants to export fruits to any country in Europe. But the same farmer may not be able to export to Japan, if he does not seek certification under Japanese certification standards. In India, the same fruits can be sold for domestic consumption with or without certification.

The process of certification and the standards adopted for certification require legal support at domestic and international level. Every country involved in organic production, processing or marketing organic products require enactment of new laws or modification of existing laws to support the cause of organic farming. The acts pertaining to pesticides, fertilizers, use of chemicals in processing, packing, marketing of agricultural goods, import and export of agricultural goods need to be modified to support the process of certification and marketing. Similarly, at international level, support for international trade in organic goods needs to be strengthened by respective provisions of WTO as well as international acceptance of certification.

6.4 ACCREDITATION PROGRAMME

IFOAM introduced the concept of accreditation of certification bodies to ensure that certifying bodies follow minimum set of procedures and follow prescribed standards. Every certifying agency adopting IFOAM standards has to seek accreditation from IFOAM. Later, many government bodies have taken the role of accreditation in most countries. In India, for example, APEDA is accreditation authority to ensure that certifying agencies are accountable in respect of standards and methodology.

6.5 PARTICIPATORY GUARANTEE SCHEME

In 2004, IFOAM, and MAELA, the Latin American Agroecology Movement, jointly organized the first International Workshop on Alternative Certification that took place in Torres, Brazil. It is at that workshop that the concept of "Participatory Guarantee Systems" was adopted. At this event, an international working group on PGS was established, which later became an official Task Force under the umbrella of IFOAM. The Task Force worked on further defining PGS, and established the key elements and key features of PGS in a document entitled "Shared Visions – Shared Ideals".

Since then IFOAM has continuously supported the development of PGS and are now advocating for their recognition by governments as valid local certification systems in cases where the organic sector is legally regulated. In parallel, other sectors have been looking into the concept to certify various products or processes.

Participatory Guarantee Systems (PGS) represent an alternative to third party certification, especially adapted to local markets and short supply chains. They can also complement third party certification with a private label that brings additional guarantees and transparency. PGS enable the direct participation of producers, consumers and other stakeholders in:

- the choice and definition of the standards
- the development and implementation of certification procedures
- the certification decisions

IFOAM defined (2008) Participatory Guarantee Systems as locally focused quality assurance systems. They certify producers based on active participation of stakeholders and are built on a foundation of trust, social networks and knowledge exchange.

The organic movement has been a pioneer in the implementation and definition of Participatory Guarantee Systems (PGS). Organic certification started in various parts of the world in the 70s and 80s based on associative systems that were very close to what is now called PGS. Some of these associations are still doing participatory certification today, such as for example Nature & Progrès in France. Even though third party certification (following ISO 65 requirements) has become the dominant form of certification in the food sector, as well as many other sectors, alternative certification systems have never ceased to exist.

In India, PGS has been accepted for domestic organic market and it is being implemented through active participation of non-governmental organizations.

6.6 STEPS IN ORGANIC CERTIFICATION

At a farm level, the certification is awarded to **whole farm** (and not to particular crop/produce) looking to various considerations regarding-

1. The source of seed material (seed must have been produced under organic conditions).
2. The source of manure or other organic inputs (the materials used to produce manure must have been originated from organic farm).
3. The feed given to animals, whose dung/urine is used to produce manures.
4. The soil and its physical, chemical and biological characters.

Following steps are followed in certifying a farm as organic and to allow the products of such farm to carry the tag `organic'.

- The farmer approaches a certifying agency with the prescribed fees for certification.
- Depending on the standards followed by each certifying agency, they visit the farm and collect all basic information regarding soil, crops and cropping pattern, cultivation practices and chemicals/fertilizers used.
- Usually soil, water, manure and produce samples are collected and analyzed.
- Then, the farmer is advised about "conversion to organic". Depending on the standards, they insist on the farmer to follow specify cropping pattern, specific set of inputs (obviously precluding the fertilizers and pesticides) as well as specific pattern of cultivation.
- The farmer must follow these methods till end of second year from the commencement of conversion process. Third year is reserved for production of organic goods, which are rigorously tested.

- Then the farm is awarded certification and is allowed to sell its products under `organic' tag-usually carrying logo of certifying agency/standard used.
- If processing is involved, the products are sent for processing, which again should be certified under similar standards. Otherwise, the product is packed and transported according to stipulated standards.

Certification in Indian context

The organic agriculture was promoted in India by Department of commerce under APEDA scheme. At later stages, Ministry of Agriculture and Cooperation initiated and supported number of programmes to promote organic farming. In India, the certification programme was developed and promoted mainly for export purposes like many other developing countries. Only in recent years, domestic market for organic goods has been recognized and comprehensive administrative steps are necessary to make use of large domestic market potential. Participatory Guaranteee Scheme (PGS) has been recently launched throughout the nation through net work of non-governmental organizations

IFOAM standards are adopted as national standards for organic certification in India. However, when organic consignments are exported to some countries, who accept their own standards, the growers should comply the required standards of importing countries

6.7. REFERENCES

www.ifoam.org/en/ifoam-norms

www.goma-organic.org

www.ifoam.org/.../coros_201101_forconsultation_ifoamversion_4.doc

www.ifoam.org/.../ifoam_norms_version_august_2012_with_cover.pdf

http://en.wikipedia.org/.../International_Federation_of_Organic_Agriculture_.

Organic agriculture and the Law: FAO Legislative case study 107. Pub: FAO, 2012

http://en.wikipedia.org/wiki/Demeter_International

http://en.wikipedia.org/wiki/Participatory_Guarantee_Systems_PGS

Chapter - 7

SCENARIO OF ORGANIC FARMING IN INDIA AND WORLD

7.1 BACKGROUND

Often it is held that organic farming originated in India. Prior to introduction of chemicals in agriculture, the Indian farmers were following the conventional practices that are now recognized as organic practices. Most of such practices were evolved over centuries of practical experience. Since such practices were eco-friendly in nature and were principally evolved to use the natural biological wastes, it is generally held that India was the origin of organic farming. Chinese also hold similar view to infer that organic farming was originated in China. The reasons for such a view are almost similar to Indian reasons. In fact, all the ancient civilizations supporting the agriculture can claim that some form of organic agriculture was practiced in the past. The recent publications of American National Library have attempted to trace the history of organic farming in Europe and USA from 50 A.D.

Although the origin of organic farming practices could be traced to ancient civilizations, the epicenter of practicing the holistic model of modern organic farming was central and Southern Europe. The European countries like Germany, Italy, Netherlands and France were pioneers to adopt the organic farming in its new expanded meaning. Most other countries including USA, India and China commenced their modern organic farming at later date. Obviously, the global scenario of organic farming must be studied in this perspective. Predictably, in India, China and few other countries organic farming is practiced by default or by the restriction of capital or poor knowledge about agrochemicals. Although no-chemical farming was adopted as a tradition for centuries, they were certainly short of all the modern concepts of organic farming.

The chemical farming, popularly called as modern scientific farming till recently, was evolved gradually after Leibig's discovery of fundamentals of plant nutrition. There are many fundamental differences between the ways in which modern chemical farming is globally adopted in contrast with the manner in which modern organic farming is being spread. Although few early adopters were voluntary, by and large the chemical farming was popularized and patronized by respective governments in most countries of world. Because, the inputs like fertilizers and pesticides were either manufactured or imported in most countries under the patronage and support of respective governments. Thereby, the adoption and spread of chemical agriculture was found in global literature, as the practices and the inputs were common to all the countries. But, the modern organic farming movement was started by few innovative thinkers, monks, philosophers and farmers, obviously taking clues from early practiceners of ancient agriculture. The government policies, government led research and endorsement of organic practices as means of organic agriculture at national level were all added in later years in respective countries. Besides, the practices and the inputs used in organic agriculture vastly differed between countries.

This differentiation of chemical farming and modern organic agriculture has led to few peculiarities:

- The organic practices and scenario of organic agriculture in one part of world are even now not easily perceived and understood in other parts of world. In the early era of modern organic farming, related information was not reported in global literature, although at domestic levels of Germany and other early adopting countries, the initial literature on organic farming was published. In recent 5-6 years global level information on organic agriculture is being published, as organic agriculture is taking a shape of global movement.
- Depending on the evolution and spread of modern organic agricultural knowledge, the adoption of organic agriculture is more concentrated in European countries like Germany, France, Switzerland and Netherlands, but insignificant in many parts of world.
- The spread and adoption of modern organic agriculture was driven by demand for pollution free food created initially in Europe, Japan and USA, presumably due to very high level of use of agro-chemicals in these countries. Many developing countries including India promoted organic agriculture in their respective countries to facilitate export of organic goods to these countries after stringent certification procedures.

To add to confusion, in many countries including India the farmers who adopted organic practices by default or by capital constraint were also to be regarded as organic growers.

In the light of these developments a clear scenario of modern organic farming in world is necessary to be studied. Focus of Indian developments in adoption of modern organic agriculture also need a clear understanding both due to its ancient tradition of organic practices as well as due to present position of India in world. An attempt is made to present global and Indian scenario of organic agriculture in this chapter in the light of global pollution, mounting pressures of international demands for organic products, public awareness of organic production as well as undisputed perilous effect of chemical on soil, animal and human health.

7.2 GLOBAL SCENARIO

Although many traditional Indian and Chinese and even African practices of agriculture are part of modern organic agriculture, the epicenter of global movement of organic agriculture is undoubtedly central and southern Europe. Falling in line with historical developments to seek an alternate form of agriculture, organic farming was conceived and adopted on modern scientific basis in many European countries as early as 1960's. India and most other developing countries had not even started using agro-chemicals in a large way in 1960. In 1990's, it spread to Australia and USA. In the last 8-10 years organic farming has fully been transformed into a global environment and spread to many developing countries and poor economies too.

Organic farming is now practiced in 162 countries covering all the continents. It is estimated by the latest survey of SOEL – FiBL that around 37.2 million ha of global organic area is reported from various continents (SOEL – FiBL, 2011), covering 0.86 per cent of global cultivated area this indicates improvement of organic area by 5 million has compared to 2009 statistics. Australia (12 million ha) and Argentina (6.8 million ha) are leading countries for organic area. In addition, 32.5 million hectare wild collection area is managed under organic production. Leading countries with high wild collection area are Finland (7 million ha) and Zambia (6 million ha). (Wild collection area refers to forest area, where the forest products are directly collected and does not involve any management). Nearly 1.8 million organic producers are recorded, with India and Uganda leading (5, 47,000 and 1, 89000 respectively).

The size of organic market is in the order of 63 billion US dollars, with US as leading market (29 billion $) followed by Germany (9.2

billion US$). Nearly 86 countries have organic regulation in place and others are in the process of enacting them.

The present scenario of global organic agriculture can be understood in following broad developments:

A. Continent wise spread of organic area-continent wise
B. Global demand for organic products
C. Global certifying agencies and need for certification
D. Endorsement and promotion of organic agriculture by global organizations.
E. Trading processing and marketing arrangements of organic products across the world
F. Government support and promotion of organic farming in different countries

A. Continent wise spread of organic area

Oceania (covering Australia, New Zealand etc.) itself covers 33 per cent of total organic area followed by Europe sharing 29 per cent of area (Fig 7.1) Asia supporting more than half of the global population has around 10 per cent of the global organic area. North American continent patronizing the biggest organic market has only 7 per cent of global area. Africa has smallest share of organic area (3%). It is vivid from Fig 7.1 that major organic area is confined to Europe and Oceania while Asia and Africa account for only 13 per cent of global area despite having biggest chunk of population. This also indicates that awareness to cultivate organic food is more in Oceania and Europe than other developing counties.

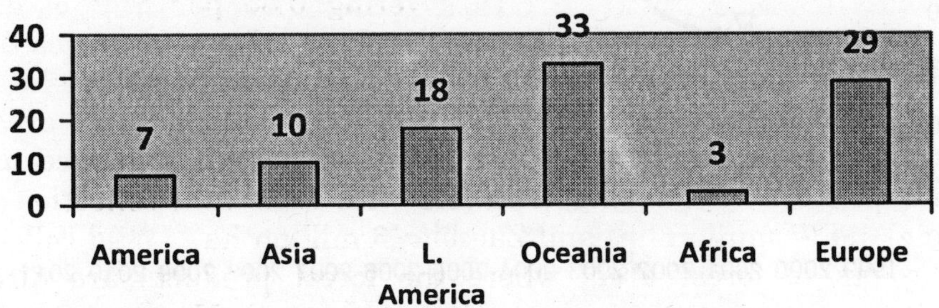

Fig. 7.1. Distribution of organic area in world (in % global organic area)

It is evident from the above statistics that major share of organic agricultural production is from Europe and Australia, African organic area is as little as 0.4 million ha, and its share in global organic area is as small as 3%. This reveals that spread of organic area and production are, at present, more dominant in developed countries with population enjoying higher per capita income. They are dismally poor in poorer continents like Asia and Africa.

On a continent level, the share or organic land in proportion to all agricultural land is highest in Australia/Oceania (2.9 per cent), followed by Europe (2.2%). It should be noted, though, that some countries in each continent exhibit a much higher percentage of organic land than continental average (ex: Austria (20%), Sweden (15%), France (17%) and Switzerland (11%). Ten countries have more than 10% share of their total agricultural land as organic, while 15 countries have their share between 5-10% and 35 countries between 1-5% of their respective agricultural land under organic management.

These global statistics consider only certified organic area in each continent, while a sizable organic area exists outside the ambit of certification in many continents like Asia, Africa and Latin America. Similarly, the global statistics excludes the area, where the production is organic by default in forest area (It is popularly referred as wild collection area – indicating that the products are collected directly from these areas and no conventional cultivation takes place). Both certified organic area and wild collection areas have spread fast in the last one decade.

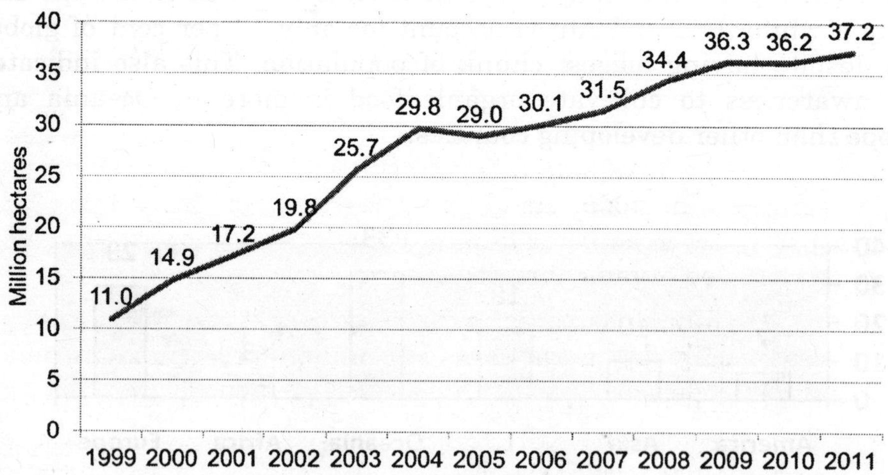

Fig. 7.2. Development of Certified Organic land (m.ha) worldwide 1999 to 2011

In the last 12 years, global organic area has increased by 27.2 million ha at a mean growth rate of 2.3 million ha /year (Fig 7.2).

However, the growth of organic area was highly dissimilar across the continents, with Europe witnessing 4 million ha from 2005 to 2011 (Fig. 7.3), while in the same period area in Oceania grew by only 0.38 million ha. Similarly, the growth of organic area was impressive in Latin America (1.8 million ha) as compared to North America (0.57 million ha) during the same period. Asian and African growth were of the order of 1.03 and 0.58 million ha. India and China have added 3.0 and 5.1 lakh ha new organic area during this period.

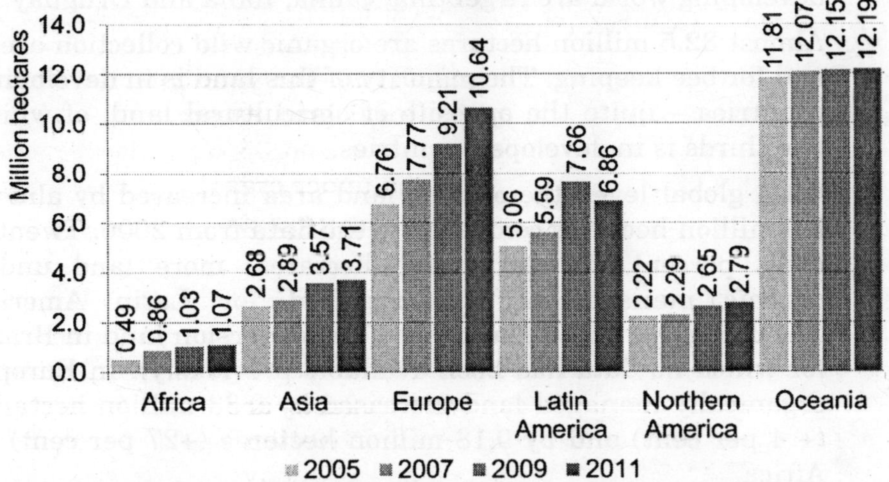

Fig. 7.3. Growth of organic area in different continents between 2005 and 2011

The global scenario of organic agriculture is appropriately summarized by FiBL as follows:

Organic agriculture is developing rapidly, and statistical information is now available from 162 countries of the world, instead of 139 countries in 2005. Its share of agricultural land and farms continues to grow in many countries. The main results of the global survey on certified organic farming show:

- 37.2 million hectares of agricultural land are managed organically by more than 1.8 million producers, including smallholders.
- The regions with the largest areas of organically managed agricultural lands are Oceania, Europe and Latin America. Australia, Argentina and China are the countries with the largest organically managed land areas.
- The highest shares of organically managed land are in Europe: Falkland Islands, Liechtenstein, Austria and France.

- The countries with the highest numbers of producers are India (5.47 lakh), Uganda (1.88 lakh), and Mexico (1.80 lakh). 1/3rd of the world's organic producers are in India.
- About one third of the world's organically managed land – almost 11.6 million hectares - is located in developing countries. Most of this land is in Latin American countries, with Asia and Africa is in second and third places. Countries with the largest area under organic management are in developing world are Argentina, China, India and Uruguay.
- Almost 32.5 million hectares are organic wild collection areas and for bee keeping. The majority of this land is in developing countries – quite the opposite of agricultural land, of which two thirds is in developed countries.
- On a global level, the organic land area increased by almost 5.0 million hectares compared to the data from 2006. Twenty-eight percent (or 1.4 million hectares) more land under organic management was reported for Latin America (including 0.9 million hectares of in-conversion land in Brazil for which no data had been available previously). In Europe, organically managed land increased by 0.33 million hectares (+ 4 per cent) and by 0.18 million hectares (+27 per cent) in Africa.

The global spread of organic agriculture is simultaneously supported by two important developments witnessed at global level.

(1) Global market for organic products.

(2) Standards and Certification.

B. Global market for organic products

The successful spread of organic agriculture across different countries is essentially supported by consistent growth in demand for organic products. The global demand, as estimated by Organic Monitor in terms of international sales has been to the tune of 63 billion US dollars in 2011. In a span of 6 years from 200, 85 per cent growth in international sales of organic products is witnessed. However, the consumers demand for organic products is mainly concentrated in North America and Europe, jointly accounting for 97 per cent of global revenues. The organic market grew by 21 billion Euros in USA, while it grew by 6.6 and 3.8 billion Euros in Germany and France. Denmark, Switzerland and Austria have highest share of organic food sales out of total food sale in their respective countries. They are also known for high per capita consumption of organic food.

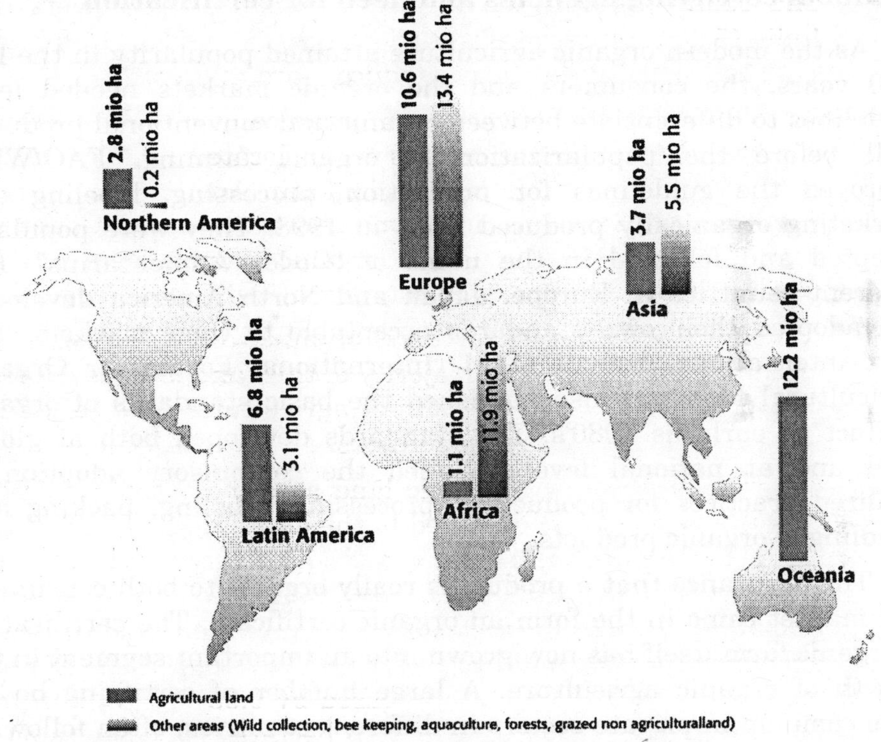

Fig. 7.4. Continent wise distribution of certified organic area

In general, 44% of global organic market is shared by USA, while Europe has a share of 41%.Canada and Japan have share of 4% and 2% respectively.

There is a clear vertical division between consuming and producing countries of organic products. Europe and North America consume a major share of organic products, well beyond their own domestic production. But Asia, Latin America and Oceania are the main producers and exporters of organic products. The impetus for faster spread and popularization of organic agriculture in developing countries is the incentive of export revenues and more remunerative prices offered to their exportable organic products. The main organic products being internationally traded are fruits, vegetables, beverages, grains, herbs, spices and seeds.

While the producing and exporting continents have mainly adopted organic farming for export purpose, the domestic demand and domestic sales of organic products in producing countries are not accounted in computing the global demand.

C. Global certifying agencies and need for certification

As the modern organic agriculture attained popularity in the last 8-10 years, the consumers and the organic markets needed legal guidelines to differentiate between organic and conventional products. Well before the popularization of organic farming, FAO/WHO approved the guidelines for production, processing, labeling and marketing organically produced foods in 1998. They were popularly accepted and followed in the name of "Codex Alimentarius". But different countries in Europe, Japan and North America developed and adopted their own standards acceptable to their markets. The apex international body IFOAM (International Forum for Organic Agricultural Moment) also published the basic standards of organic product as early as 1980's. The standards developed both at global level and at national level included the compulsory adoption of required practices for production, processing, labeling, packing and handling of organic products.

The assurance that a product is really organic to both consumers and market came in the form an organic certificate. The certification of organic farm itself has now grown into an important segment in the growth of organic agriculture. A large number of certifying bodies have come up in private sectors in different countries, often following their domestically accepted national or international standards. The acceptance of standard by certifying bodies became an integral part of certification and thereby facilitated the easy acceptance by importing countries. IFOAM has also emerged as accreditation body to award and recognize the certifying bodies about the truthfulness of standards being adopted.

As more than 576 organizations offer organic certification services adopting diversified standards, a great deal of confusion was experienced in production and export of organic product both for producing and importing countries. To resolve this, an international task force on harmonization and equivalence in organic agriculture (ITF) was launched in 2003 jointly by FAO, IFOAM and UNCTAD. The objective of task force included the evolution of general consensus on harmonizing the standards and regulation evolved by private and government agencies in different countries.

After the advent of certification process the producing countries like India have experienced the simultaneous growth of certified and uncertified organic area, as the domestic certification process has not yet started. The organic agriculture continues either for certification and export or for domestic consumption. On the similar lines many countries have experienced the difficulties of tagging some products from forest areas as organic because of constraints in avoiding certificate to such areas. The large area of wild collection has expanded over years and even exceeded the certified organic area in 2006 as depicted in Fig. 7.2.

D Endorsement and promotion of organic agriculture by global organizations

Organic farming has no more remained as mere venture of a few innovative farmers; it is not the story of few successful adopters; it has not remained confined to growing selected crops for attractive export prices. But, the knowledge and adoption of organic agriculture have grown by leaps and bounds globally, well beyond the academic interest. The cultivation and consumption of organic crops has been a worldwide phenomenon.

Such an exponential growth in organic agriculture in a short span of thc last 6 8 years is sponsored and endorsed by many global organizations. The important global organizations involved in various facets of promoting organic agriculture and bringing it as global movement are Food and Agricultural Organization (FAO), International Forum for Organic Agricultural Movement (IFOAM) and International Trade Centre (ITC).

Food and Agricultural Organization (FAO), Rome, Italy

It is an apex global organization and wing of United Nations on all the matters of promoting agricultural production with a specific interest on global hunger and its alleviation. Food security and sustainable agriculture are its main themes for global development. Although it started in 1949 and worked for improving the overall global food production through all the improved agricultural practices, its strategies and programs included organic agriculture since 1998. From 2002 onwards its declared policy to promote organic agriculture as a long term strategy to achieve the sustainability and food security is well pronounced. Some evidences of its policies supporting the organic agriculture are as follows:

- FAO and WHO jointly developed international organic standards for food products in the name and style of Codex Alimentarius, which are now popularly used internationally as Codex standards for organic certification. The first such standards was evolved and published in 1998. Later, number of products has been standardized. The Codex standards are used and acceptable in as many as 35 countries and now

regarded as acceptable standards for as many as 16 certifying agencies globally

- **FAO has sponsored organic agriculture as a global strategy to attain sustainability in various countries. This is evidenced by FAO publications like**
  - **Evaluating the potential contribution of organic agriculture to sustainability (1998)**
  - **Opportunities and constraints of organic agriculture- a socio ecological analysis (2000)**
  - **Organic agriculture, sustainable agriculture and rural development (2007)**
- **In its portal www.fao.org/organicag, the long term objective of FAO- organic agriculture program is stated as "to enhance food security, rural development, sustainable livelihood and environmental integrity by building capacities in organic production, processing, certification and marketing"**
- **FAO has drawn up number of programs to promote organic agriculture as means to attain food security in different countries with the focus of producing pollution free food. This is evidenced in following FAO publications:**
  - **Green revolution revisited: New needs, New strategies (1995)**
  - **Organic agriculture, environment and food security (2002)**
  - **Key lessons learned (1994- 2006) in special program for food security (2006)**
  - **FAO views on organic agriculture and food security (2006)**
  - **Organic agriculture and food utilization (2007)**
  - **Organic agriculture and food security in Africa (2007)**
  - **Organic agriculture and food availability (2007)**
  - **Organic agriculture and food security (2007)**
  - **Organic agriculture and access to food (2007)**
  - **Organic agriculture and environmental stability of food supply (2007)**

International Federation for Organic Agricultural Movement (IFOAM), Germany

IFOAM is a now considered as leading international organization steering the organic agricultural movement in different parts of world, spreading its activities in more than 130-150 countries. But, before it grew to the present status, it made a humble start in 1972 as a small group of German farmers to practice and promote organic agriculture. From the voluntary status, it was graduated into a national organization to offer an efficient platform to disseminate technical information, to carry out market surveys, to arrange trade fairs besides serving as certifying agency. However, over a period of time, it was further expanded to the global level after United Nations recognized it as steering body to promote organic agriculture in entire world.

The designated goals of IFOAM include

- IFOAM builds the global platform for the organic movement.
- IFOAM develops, communicates and defends the principles of organic agriculture.
- IFOAM advocates and facilitates the adoption of organic agriculture.
- IFOAM promotes the development of organic markets.
- IFOAM ensures an effectively managed organization with sufficient and sustainable resources.

Its present functions include:

- In the status of federation, its principal function include liasioning and coordinating all the activities of national level organizations of different countries, who take up the promotional and developmental works related to organic agriculture. This also includes the liasioning with respective governments in respects of organic standards and other promotional policies.
- Accreditation of certifying agencies working in different countries to instill trust and standards in the system of certifying and thereby promoting the exports of organic products from producing country.
- IFOAM actively participates in international agricultural and environmental negotiations with the United Nations and multilateral institutions to further the interests of the organic agricultural movement worldwide.

- IFOAM provides a market guarantee for integrity of organic claims. The Organic Guarantee System (OGS) unites the organic world through a common system of standards, verification and market identity. It fosters equivalence among participating IFOAM accredited certifiers, paving the way for more orderly and reliable trade whilst acknowledging consumer trust in the organic 'brand'.
- Through IFOAM programs, conferences and events, IFOAM is laying the groundwork for the further development of organic agriculture and its markets worldwide. Through IFOAM World Congresses, International Trade conferences, commodity specific organizations (coffee, seeds, wild products, etc.) and other events, IFOAM brings together the key players from all over the planet to facilitate trade in organic products.

Under the leadership of its new World Board, elected at the general assembly in Vignola, Italy, in June 2008, the International Federation of Organic Agriculture Movements (IFOAM) will continue to work on further enhancing organic growth in 2009, through advocacy, the facilitation of trade, and capacity building. In particular, IFOAM will be working on a new leadership program: Education and training, both vocational and academic, play an important role in disseminating the benefits of organic agriculture at all levels. The 1st International IFOAM Conference on Animal and Plant Breeding 'Breeding Biodiversity' will bring both animal and plant breeding together with the aim of explicitly highlighting the important their inter dependences and holistic approaches of organic agriculture

JOINT STRATEGIES OF FAO AND IFOAM

The organic market is confronted with hundreds of private sector standards and governmental regulations, two international standards for organic agriculture (Codex Alimentarius and IFOAM) and many certification and accreditation systems. Mutual recognition and equivalency among the systems is extremely limited. The multitude of standards, certification requirements and regulations are considered to be a major obstacle to the continuous development of the organic sector, especially for producers in developing countries. IFOAM, FAO and UNCTAD joined forces to search for solutions to this problem and formed the International Task Force on Harmonization (ITF) in 2003. The Task Force consists of representatives of governments, inter governmental agencies, and stakeholders from the private sector.

From 2003 to 2008 the International Task Force worked together to complete a series of proposals and tools aimed at its objective of helping to reduce organic trade barriers. Two practical tools have been developed to streamline acceptance of products that are traded internationally. One tool is for recognizing organic certification bodies and the other is for determining the equivalency of production and processing standards. The ITF has put a great deal of effort into these tools because it believes that they can effectively address and reduce organic trade barriers. The first tool, International Requirements for Organic Certification Bodies (IROCB), is an international reference norm that can be used by governments and private accreditation and certification bodies as a means of accepting certification of organic products outside of their own system. The second tool, the Equip-Tool is a set of guidelines, which include both procedures and criteria that can be applied for deciding when a standard applicable in one region of the world is equivalent to a standard applicable in another region.

INTERNATIONAL TRADE CENTRE (ITC)

International Trade Centre is a joint agency of World Trade Organization (WTO) and United Nations (UN) with following goals:

- Strengthen the international competitiveness of enterprises
- Develop the capacity of trade service providers to support businesses.
- Support policymakers in integrating the business sector into the global economy.

As the development partner for small business export success, ITC's goal is to help developing and transition countries achieve sustainable human development through exports.

It partners with trade support institutions to deliver integrated solutions for "Export Impact for Good". Its services are delivered through five complementary business services: business and trade policy, export strategy, strengthening trade support institutions, trade intelligence and exporter competitiveness.

Through strategic development and capacity building based on these business services, ITC connects opportunities to markets. As a result, it achieves long-term, tangible benefits at both national and community levels.

ITC has great role in promotion of organic farming in the entire world. Because, organic farming was popularized outside Europe mainly as an avenue to export and most developing countries encash upon this opportunity to adopt and promote organic farming at domestic level with consistent and assured market support in developed countries. Thus assistance of ITC in promoting organic farming through market research, market information has immensely helped in spreading the organic farming area in most developed countries.

7.3. CONTINENT WISE DEVELOPMENT OF ORGANIC AGRICULTURE

7.3.1 AFRICA

In Africa, there are almost than 1.1 million hectares of certified organic agricultural land. This constitutes about 3 per cent of the world's organic agricultural land. 6,00,000 organic producers were reported. The countries with the most organic land are Uganda (2,96,203 hectares), Tunisia (1,54,793 hectares), and Ethiopia (1,40,308 hectares). The highest shares of organic land are in Sao Tome and Prince (5 per cent), Uganda (2.3 per cent) and Tunisia (1.6 per cent).

Three countries have organic regulation and seven countries are in the process of drafting their regulations. It is difficult to get a clear data on organic production – referring to both certified and non-certified organic production – in Africa. Data related to organics in Africa has only begun to be collected in recent years, and as such, figures are sometimes approximate and incomplete. Organic agriculture appears to be a viable and sustainable development option, particularly for (groups of) smallholder farmers in Africa. Local NGOs and farmers' groups and development agencies are increasingly adopting organic techniques as a method of improving productivity and addressing the pressing problems of food security and climate change. Organic farming fits in a package that addresses a number of concerns.

They resonate with, and are being used in, initiatives designed to:

- maintain and enhance soil fertility;
- mitigate the effects of climate change and reverse desertification;
- promote tree-planting and agro forestry;
- maintain and improve biodiversity;
- promote the use of local seed varieties and develop strategies for combating pests based on locally available inputs;
- strengthen group formation, promote joint marketing;

- support the most vulnerable social groups such as female headed households;
- improve human health by avoiding chemical use and diversifying diets;
- ensure food security; and
- contribute to the eradication of poverty.

Most African countries do not have data collection systems for organic farming. Data on the numbers of farms are not available for every country.

African organic area

According to available sources, 4,17,059 hectares are currently managed and certified organic by at least 1,75,266 farms; the countries with the largest organically managed land are Tunisia (1,54,793 hectares), Uganda (88,439 hectares), South Africa (50,000 hectares) and Tanzania (23,732 hectares). Additionally, 8.3 million hectares are certified as forest and 'wild harvested' areas. The largest wild collection areas are in Zambia (7.2 million hectares), Sudan (4,90,000 hectares), Kenya (1,86,000 hectares) and Uganda (158'328 hectares).

African organic Markets

Farmers in Africa produce a diversity of organic crops. The range of certified organic products currently being produced in Africa is listed below.

Table 7.1. Organic Produce from Africa (by type and country)

Product group	Countries
Bananas	Cameroon, Ghana, Rwanda, Senegal, Uganda, Mali
Cereals incl. Rice	Ethiopia, Madagascar, Mozambique, Sudan, Egypt
Citrus Fruits, Grapes	Egypt, Morocco, South Africa
Cocoa	Cameroon, Ghana, Madagascar, Tanzania, Uganda, Sao Tome
Coconut Oil	Mozambique
Coffee	Cameroon, Ethiopia, Kenya, Madagascar, Rwanda, Tanzania, Uganda
Cotton	Benin, Burkina Faso, Egypt, Mali, Senegal, Sudan, Tanzania, Uganda, Zambia

Product group	Countries
Dried Fruits	Algeria, Benin, Burkina Faso, Cameroon, Egypt, Ghana, Madagascar, Morocco, Tanzania, Togo, Tunisia, Uganda
Essential Oils	Ghana, Kenya, Madagascar, Tanzania, Uganda, Zimbabwe
Fresh Vegetables & culinary herbs	Cameroon, Gambia, Egypt, Kenya, Madagascar, Malawi, Mali, Morocco, Rwanda, Sao Tome and Prince, South Africa, Tunisia, Zambia, Ethiopia, Ghana, Mozambique, Tunisia and Uganda
Ground Nuts	Cameroon, Mozambique, Tanzania, Zambia
Honey	Algeria, Malawi, Tanzania, Tunisia, Uganda, Zambia
Medicinal / Therapeutic Herbs and Spices	Egypt, Morocco, Namibia, Tunisia, Zambia
Olive Oil	Tunisia
tropical fresh fruits	Cameroon, Egypt, Ghana, Madagascar, Senegal, South Africa, Tanzania, Uganda
Palm Oil	Ghana, Madagascar
Processed fruits	Ghana, Tanzania, Uganda
Sesame	Burkina Faso, Mali, Mozambique, Senegal, Uganda, Tanzania
Spices (culinary)	Cameroon, Egypt, Ethiopia, Madagascar, Malawi, Mozambique, South Africa, Tanzania, Uganda, Zimbabwe
Sugar	Cameroon, Madagascar, Mauritius
Tea	Kenya, Tanzania, Rwanda
Tree Nuts	Burkina Faso, Ghana, Kenya, Malawi, Mali, Morocco, Tanzania, Togo, Uganda

Source: updates from the IFOAM /FiBL Survey 2008

African organic Exports

The majority of certified organic produce is destined for export markets, with the large majority being exported to the European Union, which is Africa's largest market for agricultural produce. Export of organic produce from Egypt totalled to 12,542 tons in 2004; this figure represented some 90 percent of organic production. Export of organic produce from Uganda, as summarized in the table below, totaled 7'877 tons in 2006

Source: NOGAMU/IFOAM-FiBL Survey 2008

The domestic market

The African market for organic products is still small. Certified organic products are currently recognized in only a few domestic markets, including Egypt, South Africa, Uganda, Kenya and Tanzania. There are a number of factors leading to expansion of the domestic market in these countries. To begin, there is a growing middle class – most notably in Egypt and South Africa – who share some similar values to European organic consumers. As a result, there are growing domestic market opportunities for diverse organic products in these locations, including organic fresh fruit and vegetables, dairy products, meat, wine, herbs and beauty products. In Egypt, mainly in Cairo, specialized shops and a number of supermarket chains (Metro and Carrefour) have organic sections, and sell mostly fruits and vegetables. Similarly, organic shops in South Africa and Uganda have also raised the profile of organic produce. In South Africa, the major supermarkets stock organic products. The NOGAMU shop in Kampala, Uganda, is directly adjacent to the NOGAMU office. Visitors to the shop have direct access to information about organic agriculture and related issues.

State support, standards and legislation

Organic agriculture is not yet integrated into most African State's agricultural policies. In some countries, mostly in East Africa, policy development is being undertaken, and the national organic movements are strongly involved in the process.

For exports, most African countries rely upon foreign standards. To date, the majority of organic production that is certified in Africa according to the EU regulation for organic products. Some producers are also certified to the US National Organic Program or the organic standards of the Japan Agriculture Standards (JAS) and to numerous private-sector organic standards, such as Soil Association, KRAV and Naturland. Certification services have been offered by foreign-based certification bodies. Originally, all work, including inspection, was carried out by foreigners flying in and out. Gradually, more and more work (in particular the inspections) is done by local staff. Some certification bodies have established with regional representation or developed closer cooperation with national bodies. Costs of certification are generally considered to be high. In East Africa, for example, an individual farm will pay 500 to 3'000 US Dollars for a foreign certification. However, the cost per individual farmer in an Internal Control System (ICS) can be as low as a few US Dollars for very big groups. For a typical ICS with 500 farmers, the cost is likely to be in the range of 10 US Dollars per farmer, and for very small ICS groups maybe as high as 100 US Dollars per farm. There are also substantial costs involved in the operation of the ICS itself.

Launch of the East African Organic Products Standard and the East African Organic Mark

The East African Organic Products Standard (EAOPS) - the second regional organic standard in the world, following that developed by the European Union - and the associated East African Organic Mark (EAOM) were developed by a public-private sector partnership in East Africa, supported by IFOAM, and the UNCTAD-UNEP Capacity Building Task Force on Trade, Environment and Development (CBTF). Both the mark and the standard were officially launched by the Prime Minister of Tanzania at the East African Organic Conference in May 2007. The consumer awareness-raising activities were implemented to increase consumer awareness in East Africa and support the introduction of the organic mark. The EAOPS serves to unite organic regulations and markets in Kenya, Tanzania, Uganda, Rwanda, and Burundi. It is expected that the organic mark and the consumer awareness campaign will contribute to the development of the regional market.

The NGO sector

In several African countries, organic agriculture has reached a critical stage of development, and networks are established to represent the organic sector both at national and international levels. These umbrella organizations serve to link the stakeholders of national movements, strengthen the sector and enhances its impact (Rundgren, 2007). The following national movements have been established:

- The National Organic Agricultural Movement of Uganda (NOGAMU);
- Organic Producers and Processors Association of Zambia (OPPAZ);
- Organics South Africa (OSA);
- The Organic Movement of Madagascar;
- The Kenyan Organic Agriculture Network (KOAN);
- The Organic Movement of Mali (MOBIOM);
- The Ghana Organic Agriculture Network (GOAN);
- The Ethiopian Association of Organic Agriculture (EAOA);
- The Tanzania Organic Agriculture Movement (TOAM); and
- The Zimbabwe Organic Producers' and Processors' Association (ZOPPA).

Research, extension and training

Agricultural research in Africa is fragmented between the international research centers, universities, national research institutes, and formal or informal field level research. There are some outstanding examples of innovative organic research at all of these levels. Pioneering research on organic farming techniques has emerged from the World Agro forestry Centre (formerly ICRAF) and the International Centre for Insect Physiology and Ecology (ICIPE). Other centers, such as the International Institute for Tropical Agriculture (IITA) and the International Livestock Research Institute (ILRI) could potentially contribute to finding solutions to the problems facing organic farmers. National initiatives, like that of the Uganda Martyrs University, the Organic Agriculture Project in Tertiary Institutions in Nigeria (OAPTIN), and the Olusegun Obasanjo Center for Organic Research and Development (OOCORD), are also growing. The OOCORD was launched in 2007 and is based in the city of Ibadan. The Center aims to develop research and knowledge exchange on sustainable, organic agricultural systems to address the dual needs of food security and incomes in Africa. Collaboration and information exchange with other institutions is expected to strengthen OOCORD's functions (OOCORD, 2007).

Outlook

The fact that traditional African agriculture is essentially low external input type provides a potential basis for organic agriculture as a development option for Africa. Organic farming practices deliberately integrate traditional farming methods and make use of locally available resources. As such, they are highly relevant to a majority of African farmers.

There is undoubtedly a large potential for a substantial increase in certified organic production in Africa. Smallholders engaged in it often derive significant benefits, improving their incomes, nutritional status and livelihoods as a result. Yet, there are also significant constraints on the potential for development. Some of them cost of certification, problems of infrastructure, maintaining links with distant markets and the vagaries of world markets. There are internal constraints as well. The overarching priority for African agriculture is that of achieving sustainable food security. Organic agriculture has a huge potential in helping meet this aim. In addition to expanding international market access, there is a need to develop local and regional markets for organic produce in Africa. Increased consumer awareness, cooperation among stakeholders and producers in the supply chain and the development of conformity assessment mechanisms for local marketing that are accessible for smallholders, such as Participatory Guarantee Systems (PGS), are key elements to achieving long-term sustainability of organic production systems.

The IFOAM Africa Office

The International Federation of Organic Agriculture Movements (IFOAM) established the IFOAM Africa Office in 2004 to help the growth of organic agriculture in the continent. The Africa Office is presently based at the IFOAM Head Office in Bonn, Germany. The IFOAM Africa Office coordinates IFOAM's advocacy network in Africa to take a proactive approach to the development of the organic sector. For the entire African continent, the IFOAM Africa Office empowers organic agricultural advocates - IFOAM Contact Points and other member organizations - to make concerted efforts to promote organic agriculture among farmer groups, NGOs, governments, and developmint organizations. The Africa Office works through IFOAM Contact Points across Sub-Saharan Africa. Each of them represents or is the coordinating office for a national organic agriculture movement or of a regional organic network.

- Ethiopian Association of Organic Agriculture (EAOA), Ethiopia
- Ghana Organic Agriculture Movement (GOAN), Ghana
- Kenya Organic Agriculture Network (KOAN), Kenya
- Laulani· Green University and Association (LGU and LGA), Madagascar
- National Organic Agriculture Movement of Uganda (NOGAMU), Uganda
- Organic Producers and Processors Association of Zambia (OPPAZ), Zambia
- Organic Agriculture Project for Tertiary Institutions in Nigeria (OAPTIN), Nigeria
- Participatory Ecological Land Use Management (PELUM) Regional desk, Zambia
- Sustainable Agriculture Development Network (REDAD), Benin
- Tanzania Organic Agriculture Movement (TOAM), Tanzania
- Zimbabwe Organic Producers' and Processors' Association (ZOPPA), Zimbabwe

The Africa Office publishes the Africa Organic News, an electronic newsletter, on a monthly basis. The newsletter features news about organic agriculture in Africa, and it is distributed freely in English and French to a wide audience in and outside Africa in a format that can be printed and distributed locally. The newsletters are available at the Africa Office homepage at www.ifoam.org/newsletter/newsletter_africa/AOSC_Newsletter_Archive.html or www.ifoam.org/about_ifoam/around_world/africa.html.

7.3.2 ASIA

The total organic area in Asia is nearly 3.7 million hectares. This constitutes ten percent of the world's organic agricultural land. 6,12,000 producers were reported. The leading countries are China (1.9 million hectares) and India (1.1 million hectares). The highest shares of organic land of all agricultural land are in Timor Leste (seven percent). Organic wild collection areas play a major role in India and China. Production of final processed products is growing, although a majority of production is still fresh produce and field crops with low value-added processing, such as dry or processed raw ingredients. Aquaculture (shrimp and fish) on the other hand, is emerging sector in China, Indonesia, Vietnam, Thailand, Malaysia and Myanmar. Textiles are another important sector showing positive growth. Sectoral growth is now also driven by local markets in many of the big cities in the South and Eastern part of region besides Japan, South Korea, Taiwan and Singapore. Kualalumpur, Manila, Bangkok, Beijing, Shanghai, Jakarta, Delhi, Bangalore and other cities are increasing internal consumption of organic products. Nine organic regulations are in place. In seven countries work on national standards and regulations is in progress.

7.3.3 EUROPE

As of the end of 2011, 10.6 million hectares in Europe were managed organically by more than 2,88,000 farms. In the European Union, 7.2 million hectares were under organic management, with more than 1,80,000 organic farms. 2.2 per cent of the European agricultural area and 6 per cent of the agricultural area in the European Union is organic. 28 per cent of the world's organic land is in Europe. The countries with the largest organic area are Italy (1,150,253 hectares), Spain (1.6 million hectares) and Germany (1.0 million hectares). The highest percentage share in respective cultivated area is in Liechtenstein (29 per cent), Austria (20 per cent) and Sweden (15.4 per cent). Compared to 2006, organic land increased by more than 0.3 million hectares. Sales of organic products were approximately 16 billion Euros in 2007. The largest market for organic products in 2007 was Germany with a turnover of 5.3 billion Euros (2008: 5.8 billion Euros), followed by the UK (2.6 billion Euros), France and Italy (both 1.9 billion Euros). As a portion of the total market share, the highest levels have been reached in Austria, Denmark and Switzerland, with around five percent for organic products. The highest per capita spending is also in these countries. Support for organic farming in the European Union and the neighboring countries include grants under rural development programs, legal protection and European as well as national action plans. One of the key instruments of the European Action Plan on organic food and farming, an information campaign, was launched during 2008, with the aim of increasing awareness of organic farming throughout the European Union. Furthermore, most EU member states have national action plans. In order to boost organic farming research, a technology platform joining the efforts of industry and civil society in defining organic research priorities and defending them vis-a-vis the policy-makers was launched in December 2008. The platform's vision paper reveals the potential of organic food production to mitigate some of the major global problems from climate change and food security, to the whole range of socio-economic challenges in the rural areas.

7.3.4 LATIN AMERICA

In Latin America, 3,24,000 producers managed 6.86 million hectares of agricultural land organically in 2011. This constitutes 18.4 percent of the world's organic land. The leading countries are Argentina (3.8 million ha), Brazil (1'765'793 hectares) and Uruguay (930'965 hectares). The highest shares of organic agricultural land are in the Dominican Republic and Uruguay with more than six percent and in Mexico and Argentina with more than two percent. Most organic production in Latin America is for export. Important crops are tropical fruits, grains and cereals, coffee and cocoa, sugar and meats. Most organic food sales in the domestic markets of the countries occur in major cities, such as Buenos Aires and Sao Paulo. Fifteen countries have legislation on organic farming, and four additional countries are currently developing organic regulations. Costa Rica and Argentina have both attained third country status according to the EU regulation on organic farming. In recognition of the growing importance of the organic sector to Latin America's agricultural economy, governmental institutions have begun to take steps towards increasing involvement; governments are beginning to play a central role in the promotion of organic agriculture. The types of support in Latin American countries range from organic agriculture promotion programs to market access support by export agencies. In a few countries, limited financial support is being given to pay certification cost during the conversion period. An important process underway in many Latin America countries is the establishment of regulations and standards for the organic sector.

7.3.5 NORTH AMERICA

In North America, almost 2.8 million hectares are managed organically, representing approximately a 0.7 per cent share of the total agricultural area. Currently, the number of farms is 18,000. The major part of the North American organic land is in the US (1.6 million hectares in 2005). Seven percent of the world's organic agricultural land is in North America. Valued at more than 25 billion US Dollars in 2011 (Organic Monitor), the North American market accounted for 44 per cent of global revenues. Growing consumer demand for healthy & nutritious foods and increasing distribution in conventional grocery channels are the major drivers of market growth. The U.S. organic industry grew 21 per cent in sales in 2006, and was forecast to experience 18 per cent sales growth each year on average from 2007 through 2010. Whether this rate will actually be realized is uncertain due to the economic downturn and reduction in consumer spending in the last quarter of 2008. Likewise, a downturn is expected in Canada, even though the market growth in Canada, paired with the introduction of the new organic regulations, should provide a good outlook over the coming years. In the United States, the National Organic Program has been in force since 2002. Canada has had an improved organic standard since 1999; this had been, however, voluntary and not supported by regulation. Canada's Organic Product Regulation will be fully implemented on June 30, 2009. Canadian labeling requirements will be similar to those of the US and the EU. In 2008, the new Farm Bill was passed by the US Congress. Increasing expenditures on organic agriculture and programs to approximately 112 million US Dollars over the course of its five-year life, the 2008 Farm Bill provides a five-fold increase for the organic sector compared with federal funding in the previous bill.

7.3.6 OCEANIA

This region includes Australia, New Zealand, and island states like Fiji, Papua New Guinea, Tonga and Vanuatu. Altogether, there are 18000 producers, managing almost 12.2 million hectares. This constitutes 2.9 per cent of the agricultural land in the area and 33 percent of the world's organic land. Ninety-nine percent of the organically managed land in Oceania is in Australia (12 million hectares, 97 per cent extensive grazing land), followed by New Zealand (65'000 hectares) and Vanuatu (8'996 hectares). The highest shares of all agricultural land are in Vanuatu (6.1 per cent), Samoa (5.5 per cent) and the Solomon Islands (3.1 per cent). Growth of the organic industry in Australia, New Zealand and the Pacific Islands has been strongly influenced by rapidly growing overseas demand; domestic markets are, also growing. In New Zealand, a key issue is lack of production to meet growing demand. Australia has had national standards for organic and biodynamic products in place since 1992, and like New Zealand, it is on the third country list of the European Union. It is expected that the Australian Standard, based on the National Standard employed since the early 1990s for the export market, will be adopted in 2009. In New Zealand, a National Organic Standard was launched in 2003. There is little government support to encourage organic agriculture in Australia. However, over the recent past, governments have been supportive of the Australian Organic Standards issue. Furthermore, funding is made available to promote an understanding among consumers. In New Zealand, through the establishment of the sector umbrella organization like Organics Aotearoa New Zealand and the Organic Advisory Programme as well as other initiatives, there is political recognition of the benefits of organic agriculture. In the Pacific Islands work on a regional strategy and national plans to lay the foundation of sustainable organic agriculture development in the region is in progress. The Regional Organic Task Force, a technical group representing all sectors and countries involved in organics, was charged with developing the Pacific Standard and will be responsible for implementing the Regional Action Plan. Pacific High Level Organics Group consists of Pacific leaders who have shown a commitment to the development of organic agriculture in the region and provide high level political support and advocacy. The first Pacific Organic Standard was endor-sed by Pacific Leaders in September 2008. This provides a platform for further regional policy development around organic agriculture.

7.3.7 WILD COLLECTION AREA

A major share of organic area, not accounted as regular organic but playing an important role in organic movement is "wild collection area". This includes all area, where the products are collected/ harvested in region, where no management (including application of fertilizers/manures/pesticides and no cultivation is practiced) is involved and these products are considered as organic. Such areas include all areas- where forest products are collected/harvested as well as bee keeping activities. The collection of wild harvested crops is defined in the IFOAM Basic Standards (IFOAM 2006), and wild collection activities are regulated in organic laws.

The global wild collection area (31.7 million ha) is almost equivalent to regular organic area. Nearly 35 per cent of wild collection area is located in Europe and 34 percent in Africa, while it is 18% in Asia and 10% in Latin America. It is significant to note that North America and Oceania have negligible or no wild area. Country wise, maximum wild collection area is in Finland (7.0 million ha), while it is Zambia and India respectively have 5.91 and 4.48 million ha under wild collection.

7.3.8 GLOBAL ORGANIC MARKET

In 2011, the global market for certified organic food and drink was estimated to be 63 billion US Dollars according to Organic Monitor (Sahota 2013). The turnover with organic products has more than fourfold since 1999, when it was estimated to be 15 billion US Dollars. More than ninety percent of the organic product revenues are made in the Northern hemisphere.

In 2011, the by far largest market was the United States with 21 billion Euros of organic food sales, followed by Germany with 6.6 billion Euros and France with 3.8 billion Euros (FiBL-AMI-IFOAM 2013). The highest annual per capita consumption was in Switzerland (177 Euros) and in Denmark (162 Euros). Denmark, Switzerland and Austria have the highest share of organic food sales.

Country wise, USA and European Union have respectively 44 and 41 per cent of global organic market- indicating that countries in other continents are net exporters while, European countries and USA are net importers of organic products. Countries with largest domestic markets. The countries with largest domestic market for organic products are USA (21 billion Euros),Germany (6.5 billion Euros) and France (3.7 billion Euros)

7.4 COUNTRY-SPECIFIC DEVELOPMENTS IN ORGANIC AGRICULTURE

Organic agriculture has gained the momentum globally mainly because of sustained market demand for organic products like organic fruits, organic vegetables, organic grains, organic cotton etc. The public interest in organic products is two folds.

1. *Pollution free food:* As the organic food products preludes all chemicals of persistence value, the organic agriculture leads to production of pollution free food.
2. *Concern about environment:* Use of chemicals in agriculture has led to large number of ecological hazards like eutrophication, nitrate pollution, imbalances in microbial population etc. Organic agriculture seeks to reverse these processes and bring them to natural mode.

Both the concerns were prominently and discernibly exhibited by consumers of Europe and US than any other part of world way back in 1990s. Although innovative practiceners of Germany started the modern organic farming in 1970s, it gained movement in Europe in later half 1990s and quickly spread to USA. The public Awareness of consuming pollution free food was the main driving force to steer such a movement. Such awareness was poor in other countries. But slowly countries like Australia, Japan, Canada and South Africa joined the about these issues due to growing concern. But, countries mainly comprising developing and under developed countries did not support and adopt organic farming till 2003.

The spread of modern organic farming movement from developed nations to developing nations was mainly supported by strong impetus of export avenues. The demand for organic products in developed countries of Europe and USA very soon fell short of domestic organic production necessitating the popularization of organic movement in developing countries to harness the export the avenue to these countries.

Hence, the spread of organic agricultural movement to different countries of world is characterized by organic exporters and organic importers. This is evidenced by developments such as

a. Most certification programs/organizations were started in Europe/ USA and the standards of certification were evolved to domestic necessities of the countries. These certifying agencies opened their officers in developing countries after 2000 mainly to promote organic production and their export to Europe/ USA due to consistent demand for organic production in these countries.
b. Many developing countries including India and China promoted the organic movement in their countries as promotion of export trade to earn both foreign exchanges as well as to offer premium prices to organic growers. For example in India

organic agriculture was popularized at the initiatives of APEDA working as a wing of ministry of commerce in 2004.

In later part of 2000, FAO sponsored the movement to spread organic agriculture as a tool achieve food security and sustainability in agriculture. In most developing countries these necessities are yet to be percolated to stimulate the growth of organic agricultural area. But most developing countries have realized the scope of adopting modern organic farming beyond the scope of export potential in recent years. The respective governments have offered number of supportive measures like subsidies and incentives to promote organic agriculture. Many countries have started developing the domestic market for organic products, although such measures are extremely limited. In India, the domestic organic demand is greatly supported by urban markets. But, lack of certification policies for domestic consumption as well as other policy initiatives are still lacking.

Based on share of organic area in total cultivated area (Table 7.2), the organic agricultural movement is not fully spread even in Europe/ USA, where most organic products are in demand. Although organic agricultural movement is now spread 139 countries, in most countries organic area is less than five percent of their respective of cultivated area, except Australia. Even in Australia, this percentage is small but appears to be high because of inclusion of wild forest/ pastures areas inorganic area. In most developing countries the share of organic area is meager indicating the poor status of organic movement. Looking at vast cultivated areas of India and China, there is a great scope to expand the organic movement not only to harness export potential but also to provide pollution free food to domestic population.

Table 7.2 Area under organic agriculture –country wise (2007)

Sl No	Country	Organic area (ha)	% organic area in respective country	No. of organic producers
1	Afghanistan	22	0.00%	233
2	Albania	77	0.01%	100
3	Algeria	1,550	0.00%	No data
4	Argentina	2,777,959	2.15%	1,578
5	Armenia	336	0.02%	35
6	Australia	12,023,135	2.70%	1,438
7	Austria	372,026	13.36%	19,997
8	Azerbaijan	21,240	0.45%	312
9	Bangladesh	177,700	. 2.00%	852
10	Belgium	32,628	2.37%	821

Sl No	Country	Organic area (ha)	% organic area in respective country	No. of organic producers
11	Belize	1,810	1.19%	68
12	Benin	1,488	0.04%	2,354
13	Bhutan	57	0.01%	323
14	Bolivia	41,004	0.11%	11,743
15	Bosnia Herzegovina	691	0.03%	304
16	Brazil	1,765,793	0.67%	7,250
17	Bulgaria	13,646	0.26%	240
18	Burkina Faso	7,267	0.07%	5,808
19	Cambodia	11,350	0.21%	9,350
20	Cameroon	336	0.00%	92
21	Canada	556,273	0.82%	3,782
22	Chile	13,564	0.09%	550
23	China	1,553,000	0.28%	1,600
24	Colombia	38,587	0.09%	No data
25	Congo (Democr. Rep.)	6,068	0.03%	1,053
26	Costa Rica	7,860	0.27%	2,921
27	Croatia	7,647	0.28%	483
28	Cuba	14,314	0.22%	2,954
29	Cyprus	2,322	1.53%	305
30	Czech Rep.	312,890	8.89%	1,318
31	Denmark	145,393	5.46%	2,835
32	Dominican Rep.	123,089	6.33%	14,992
33	Ecuador	49,196	0.65%	221
34	Egypt	14,165	0.40%	460
35	El Salvador	7,478	0.44%	2,000
36	Estonia	79,530	8.77%	1,220
37	Ethiopia	140,305	0.41%	165,560
38	Faroer Islands	12	0.40%	No data
39	Fiji	100	0.02%	No data
40	Finland	148,760	6.49%	4,406
41	France	557,133	1.88%	11,978
42	Gambia	86	0.01%	No data
43	Georgia	251	0.01%	49
44	Germany	865,336	5.11%	18,703
45	Ghana	24,449	0.17%	3,900
46	Greece	278,397	3.33%	23,769
47	Guatemala	7,285	0.16%	11
48	Guinea-Bissau	5,600	0.34%	401

Sl No	Country	Organic area (ha)	% organic area in respective country	No. of organic producers
49	Guyana	109	0.01%	28
50	Honduras	8,448	0.29%	1,825
51	Hungary	1,22,270	2.89%	1,242
52	Iceland	6,229	0.27%	36
53	India	12,31,311	0.57%	195,741
54	Indonesia	66,184	0.14%	No data
55	Iran	913	0.00%	5
56	Ireland	41,122	0.99%	1,134
57	Israel	5,693	1.10%	283
58	Italy	1,150,253	9.05%	45,231
59	Ivory Coast	943	0.00%	No data
60	Jamaica	437	0.09%	11
61	Japan	6,626	0.14%	2,463
62	Jordan	1,047	0.10%	13
63	Kazakhstan	2,393	0.00%	No data
64	Kenya	4,636	0.02%	1,811
65	Korea, Republic of	9,729	0.52%	7,507
66	Kyrgyzstan	14,400	0.13%	No data
67	Laos	1,537	0.08%	811
68	Latvia	173,463	9.78%	4,108
69	Lebanon	1,946	0.50%	190
70	Liechtenstein	1,048	29.68%	39
71	Lithuania	120,418	4.55%	2,855
72	Luxemburg	3,380	2.58%	81
73	Macedonia, FYR	1,333	0.11%	127
74	Madagascar	9,456	0.02%	5,455
75	Malawi	325	0.01%	13
76	Malaysia	1,540	0.02%	21
77	Mali	3,402	0.01%	7,526
78	Malta	12	0.12%	30
79	Mauritius	175	0.15%	5
80	Mexico	393,461	2.86%	128,819
81	Moldova	11,695	0.46%	121
82	Montenegro	25,051	4.84%	13
83	Morocco	3,590	0.01%	No data
84	Mozambique	728	0.00%	1,928

Sl No	Country	Organic area (ha)	% organic area in respective country	No. of organic producers
85	Namibia	80	0.00%	6,000
86	Nepal	8,187	0.19%	1,424
87	Netherlands	47,019	2.46%	1,374
88	New Zealand	65,000	0.38%	600
89	Nicaragua	70,972	1.33%	7,407
90	Niger	131	0.00%	No data
91	Nigeria	3,154	0.00%	No data
92	Niue	159	1.99%	61
93	Norway	48,863	4.72%	2,611
94	Pakistan	25,001	0.09%	28
95	Palestine	541	0.15%	303
96	Panama	5,244	0.24%	7
97	Papua New Guinea	2,497	0.23%	4,558
98	Paraguay	17,705	0.07%	3,490
99	Peru	124,714	0.59%	36,093
100	Philippines	15,344	0.13%	No data
101	Poland	285,878	1.85%	11,887
102	Portugal	233,475	6.35%	1,949
103	Romania	131,401	0.89%	2,238
104	Russian Federation, European Part	3,577	0.00%	12
105	Rwanda	13,356	0.69%	2,565
106	Samoa	7,243	7.79%	213
107	Sao Tome and Prince	2,862	5.02%	1,179
108	Senegal	1,589	0.02%	1,306
109	Serbia	920	0.02%	35
110	Slovak Republic	117,906	6.27%	280
111	Slovenia	29,322	6.00%	2,000
112	Solomon Islands	3,628	4.27%	352
113	South Africa	50,012	0.05%	500
114	Spain	988,323	3.93%	18,226
115	Sri Lanka	17,000	0.72%	4,216
116	Sudan	56,324	0.04%	No data

Sl No	Country	Organic area (ha)	% organic area in respective country	No. of organic producers
117	Suriname	40	0.04%	1
118	Swaziland	3	0.00%	No data
119	Sweden	248,104	7.96%	3,028
120	Switzerland	116,641	11.00%	6,199
121	Syria	28,461	0.20%	3,256
122	Taiwan	1,746	0.21%	905
123	Tanzania	62,180	0.18%	90,222
124	Thailand	19,123	0.10%	3,924
125	Timor Leste	23,790	7.00%	No data
126	Togo	2,545	0.07%	4,183
127	Tunisia	154,793	1.58%	862
128	Turkey	124,263	0.49%	16,276
129	Uganda	296,203	2.33%	206,803
130	UK	660,200	4.15%	5,506
131	Ukraine	249,872	0.60%	92
132	United Arab Emirates	5	0.00%	1
133	Uruguay	930,965	6.23%	630
134	USA	1,640,804	0.51%	8,493
135	Uzbekistan	1,854	0.01%	302
136	Vanuatu	8,996	6.12%	No data
137	Venezuela	2,441	0.01%	No data
138	Vietnam	12,120	0.13%	No data
139	Zambia	2,530	0.01%	20,000
	Total	**32,220,912**	**0.81%**	**1,219,438**

[Country wise data is available as on 2007, while continent wise data is available as on 2011. Hence, number of countries practicing in organic farming differ]

Across the globe, number of organic farms in each country is not proportional to the organic area in respective country. This may due to vast differences between countries in respect of cultivable area available to each farmer. In most developing countries including India, the number of farmers practicing organic agriculture is very high, as compared to the area occupied in organic agriculture. But, in country like Australia and USA, the number of organic farmers is less, although the organic area in these countries is substantially high.

Although global organic area as meager as 0.81 percentage of cultivable area, in countries like Liechtenstein (29.68%), UK (4.15%),

Switzerland (11%), Austria (13.36%), Australia (2.70%), Argentina (2.15%), Uruguay (6.23%) and Estonia (8.77%) the organic agriculture is more popular and occupied a sizable part of the cultivable area. There is clear evidence at global level that organic farming is gaining popularity and is expected to become a mainstream agriculture in the near future in many countries. For countries with vast cultivable area, even small percentages (Brazil with 0.67%, India with 0.57% and China with 0.28%) of organic area results into vast absolute areas. An overview of such a scenario suggests that organic agriculture has definitely not remained as a mere theoretical possibility but has been a way of life for millions. Further, looking to its spread and rate of growth, organic agriculture is justifiably expected to be major part of world agriculture in the next decade. In terms of absolute organic area, the leading countries are Argentina (2.78 million ha), Australia (12.02 million ha), Brazil (1.77 million ha), India (1.03 million ha), Italy (1.15 million ha), USA (1.65 million ha) and China (1.55 million ha). They have 69 percentage of global organic area revealing where the organic agriculture is widely practiced. This data explicitly excludes the wild harvested forests naturally managed under organic conditions.

The scenario of organic farming in selected countries is presented below briefly:

7.4.1 ASIA

7.4.1.1 BANGLADESH

Situated in South Asia, Bangladesh has a total area of 144,000 km^2. Bangladesh is bordered in the West, North, and East by India, on the South East by Myanmar (Burma), and on the South by the Bay of Bengal, with a total boundary length of 4,246 km^2. Bangladesh has a tropical monsoon climate

Agriculture is one of the largest sectors of the economy. It comprises 20% of the country's GDP, whilst employing 60% of the total labour force. Major agricultural products include cotton, rice, jute, tea, wheat, cane, oilseeds, potatoes, beef, milk, poultry, tobacco, pulses, spices, and fruit (FAOSTAT, 2005-2006). Rice is produced on 60% of all cropped land in Bangladesh. Jute is the main cash crop of Bangladesh; it produces about one-quarter of the total jute supply of the world. The combined total exports of jute and jute products represent 13–15% of Bangladesh's annual export earnings. The fishery sector has also an important role in the economy of Bangladesh, and it has grown notably in the last years. The land is fertile, but yields are usually low due to a lack of capital for input. The land use is divided as follows: arable land 55%, permanent crops 3%, others 42%.

Overview of organic farming

The organic sector in Bangladesh emerged in 1988. By 2006, Bangladesh became the second country in Asia with large area of under organic management. The total land under organic cultivation is estimated to be about 177,700 hectares, accounting for approximately 2% of the total agricultural area. In 2002, 2000 organic farms were operating in Bangladesh (IFOAM & FiBL, 2006). There is also the practice of Organic shrimp farming in Bangladesh.

Brief overview of key organic products

Certified organic produce from Bangladesh includes tea, shrimps, cotton, and rice. These products are mainly exported. So far, there is no evidence of a domestic market for organic products.

Organizational support

A grassroots initiative, called Nayakrishi Andolon (New Agriculture Movement) began in 1988 with the support of the UBINIG (Policy Research for Development Alternatives). The aim of the movement was to create a community-based work, which was organic in nature. They intended to incorporate traditional knowledge with newer scientific innovations suitable for both farmers and the environment. According to the IDRC, there are presently more than 2000 farmers all over Bangladesh engaged in Nayakrishi agriculture.

7.4.1.2 CHINA

The People's Republic of China claims an area of 9,596,960 km^2. The mainland's 5,774 km coastline is extending from the mouth of the Yalu River in the northeast to the Gulf of Tonkin in the south. China is bordered on the North by Mongolia and Russia; on the North East by the Democratic People's Republic of Korea; on the East by the Yellow and the East China seas; along the southern border are Hong Kong, Macau, the South China Sea, the Gulf of Tonkin, Vietnam, and Laos; on the South West by Myanmar, India, Bhutan, and Nepal; on the West by India, Jammu and Kashmir (disputed areas), Pakistan (west of the Karakoram Pass), and Afghanistan; and on the North West by Tajikistan, Kyrgyzstan, and Kazakhstan. The climate in China varies greatly with topography. The Minimum winter temperature ranges from –27°C in the northern region to 16°C in the southern region.

Although agriculture accounts for only 12.5% of the GDP, the sector employs 49% of the population. China's cereal production (mainly rice) is the largest in the world. China is also the world's largest producer of garlic, ginger, wheat and potatoes. Its main export crops include tea, rice, carrots, soybeans, cotton, sesame seeds, cinnamon, citrus fruits, maize, apples and pears. It also exports pork and fish (FAOSTAT, 2005-06). The majority of agricultural production is sold domestically. A greater portion of China's soil is acidic red clay, but irrigation and heavy fertilizer use - both organic and chemical - have provided Chinese farmers with high yields. The land use is divided as follows: permanent pastures 43%, forests and woodlands 14%, arable land 10%, permanent crops 0%, other 33%.

Overview of organic farming

The organic sector in China has rapidly expanded in the last decade. China is the first country in Asia and the second in the world for area under organic cultivation. With 3,466,570 hectares, accounting for 0.60% of the total agricultural area, approximately 1,560 Chinese farmers use organic production methods (IFOAM & FiBL, 2006). This growth in organic production has been driven not only by the increasing international demand, but also by an expanding domestic market in urbanized areas, which showed market growth up to 30% (IFOAM & FiBL, 2006). The increasing international demand and the expansion of the internal market for organic products indicate that the sector has potential to develop further in the future.

Brief overview of key organic products

Certified organic produce from China includes tea, pumpkin, sunflower seeds, soy products, various mushroom, medicinal plants, and kidney and black beans. Most of these products are sold in semi-processed or raw forms (IFOAM & FiBL, 2006). Most existing organic fruit farms presently produce for domestic market, but aim at international market in the longer run (FAO, 2003). Chinese organic products are exported mostly to Europe, U.S and Japan. China's organic food exports totalled 142 million US$ in 2003 and 200 million US$ in 2004.

Organizational support

Organic production has attracted over the past years an increasing political recognition by the Chinese government for its role in protecting the environment. The China Greenfood Development Centre (CGFDC) was established in 1992 under the jurisdiction of the

Ministry of Agriculture. Its major activities include inspection, certification, teaching, and training. The CGFDC's aims at the protection of the agro-ecological environment and the extension of sustainable food production methods. CGFDC formulates programme, policies and plans for developing Green Food; it also administers the use of the Green Food label and organizes the formulation of the various standards for Green Food.

The Organic Food Development Centre of China (OFDC) is another governmental organization directly under the jurisdiction of the State Environmental Protection Administration (SEPA). OFDC has been involved in organic inspection and certification since 1994. The main tasks of OFDC are to control agro-environmental pollution caused by pesticides and fertilizers, to promote the development of organic agriculture, and to provide certified quality organic food for consumers. It also administers the use of the OFDC organic logo, which is approved by SEPA.

Established in 1998 with the aim of increasing the productivity and sustainability of agricultural production systems in China, the Intercontinental Center for Agro ecological Industry Research & Development (ICAIRD) promotes the development of agro ecological industry and organic agriculture.

The Jiangsu Ruikang Organic Food Trade (JROFT) was established in 1995 with the aim of stimulating organic farming in China. Over the last years, JROFT has cultivated organic tea, organic liquorice and organic star anise.

Notably, in 2006 BioFach China took place for the first time in Shanghai. BioFach is a public-private partnership between Nurnberg Global Fair and the German Development and Investment Society. The BioFach China was organized together with CGFDC and SEPA.

7.4.1.3 INDONESIA

The Republic of Indonesia consists of five large islands and 13,677 smaller islands (about 6,000 of which are inhabited) forming an arc between Asia and Australia. With a total area of 1,919,440 km^2. The five principal islands are Sumatra; Java; Borneo of which the 72% belonging to Indonesia is known as Kalimantan; Sulawesi, formerly called Celebes; and Irian Jaya (West Irian), the western portion of the island of New Guinea. Indonesia has land boundaries with Malaysia (on Borneo), Papua New Guinea (on New Guinea), and East Timor (on Timor). It is bounded on the North by the South China Sea, on the North and East by the Pacific Ocean, and on the South and West by the Indian Ocean. Indonesia has a tropical climate characterized by heavy rainfall, high humidity, high temperature, and low winds.

The agricultural sector in Indonesia accounts for 13% of the GDP and involves 47% of the total labour force (CIA, 2007). Three decades of steady progress in agricultural development were abruptly interrupted by the financial and environmental shocks at the end of the 1990s and the country's agricultural sector is just recovering from those shocks. Indonesia is one of the biggest producers and exporter of oil palm and oil palm kernels. Other major export crops are: pepper, cashew nuts, rubber, cocoa beans, coffee and tea (FAOSTAT, 2005-06). The arable land accounts to 11.03% of the land in use, permanent crops account to 7.04%, while 81.93% is destined to other uses (CIA, 2007).

Brief overview of organic farming

A large proportion of Indonesia's farmers are organic farmers simply because they were not targeted or did not participate in the "green revolution" and are continuing traditional methods of farming. In other areas, farmers could no longer afford pesticides and fertilizers when prices went up as a result of the economic crisis (Down to Earth No. 49, May 2001). In recent years, the government of Indonesia has realized the importance of sustainable agricultural development and became aware that sustainable agricultural development requires integrating natural resource management strategies into agricultural development policies. Since 1986, the Indonesian government banned several types of pesticide, and established the integrated pest management (IPM) program. This programme involved controlling pests through the use of the pests' natural predators. According to the Food and Agricultural Organization (FAO, 2006), pesticide use had declined by 90%, while average rice yields had risen from 6.1 tons per hectare to 7.4 tons per hectare.

Brief overview of key organic products

Certified organic produce from Indonesia includes: palm oil, coconuts, coffee, cocoa beans, rice, vegetables, red ginger, mushrooms, cinnamon, pepper, honey, cashew nuts, pineapple and shrimps (BIOCert). Most of the products are sold in semi-processed or raw forms. Organic farms produce mainly for the international markets. The IDEP Foundation (IDEP) reports that consumer demand for organic produce is very low in Indonesia, whilst IFOAM & FiBL (2006) observe an organic boom is taking place in Indonesia. This includes an increase of organic food imports in the major cities. Specialty health food stores can be found in larger cities in Indonesia, catering mostly to an expatriate population and the increasingly aware educated Indonesians with strong interest in organic food.

Organizational support

Although no official government programme is implemented in support of organic farming in Indonesia (Food and Fertilizer Technology Center), many NGOs like Pesticides Action Network (PAN) Indonesia, SPTN-HPS, ELSPPAT (Bogor), BITRA, and Sintesa in North Sumatra are trying to promote the debate about agriculture in public whilst organizing practical projects with farmers' groups. These NGOs are members of the national network of organic farmers (Jaringan Kerja Pertanian Organik), which includes both NGOs and farmers' groups. Although not a member of IFOAM, the national network, works together with IFOAM on some activities. The Indonesian Organic Alliance (an alliance of 41 members) has set up a national certification centre called BIOCert (Board of Indonesian Organic Certification).

7.4.1.4 JAPAN

Situated off the eastern edge of the Asian continent, the Japanese archipelago is bounded on the North by the Sea of Okhotsk, on the East and South by the Pacific Ocean, on the South West by the East China Sea, and on the West by the Sea of Japan. The total area of Japan is 377,835 km^2. The five districts are Honshu, Hokkaido, Kyushu, Shikoku, and Okinawa. Each of the five districts consists of a main island of the same name and hundreds of surrounding islands. Japan is located at the northeastern edge of the Asian monsoon climate belt, which brings much rain to the country.

The economic importance of agriculture in Japan has rapidly declined since 1950, with the sector constituting 1.7% of national GDP and employing 4.6% of the total labour force. A striking feature of Japanese agriculture is the shortage of farmland. Approximately 11.64% of the total land area in 2005 is cultivated (CIA, 2007). Japan is thus a major importer of food. According to FAOSTAT (2005-06), Japan is the largest importer of pork, maize and canned chicken, the second largest importer of beef, soybeans and wheat, and a major importer fruit and vegetables. Rice is the principal domestically produced crop. Non rice farmland is planted with wheat and barley in the autumn and with sweet potatoes, vegetables, and dry rice in the summer. Intercropping is common: such crops are alternated with beans and peas. Although production is kept at high level thanks to use of technically advanced fertilizers and farm machinery, the agricultural sector in Japan suffers from many constraints such as the rapidly diminishing availability of arable land (FAO, 2004).

Brief overview of organic farming

According to IFOAM & FiBL (2006), the area under organic management in Japan is 29,150 hectares, which constitutes 0.56% of the total agricultural land. There are currently 4.539 organic farms registered in Japan. The same constraints affecting conventional agriculture also have an impact in the production of organic crops. Furthermore, the Japanese hot and wet climate conditions make it even more difficult to cultivate crops without chemical fertilizers. Production trends for organic food and vegetables appear to be moving towards value added product lines, such as pre-packaged salad mixes. It is expected that although conventional agriculture is still in decline, organically certified products will increase both in area of production and in the range of products (FAO/ITC/CTA, 2001). After some years of confusion between green food (low use of chemical pesticides) and organic food, the Japanese government implemented in 2001 a new law (JAS Standards for Organic Agricultural Products and Organic Agricultural Processed Foods) with a clear definition of organic food and stringent rules for certification and imports.

Brief overview of key organic products

Certified organic produce from Japan includes: rice, green vegetables, green tea, sweet potato, taro, pumpkin, and potatoes, citrus and other fruits. Potentially, Japan can be an enormous organic food market. Currently, consumer demand exceeds domestic supply and most organic products are imported. The demand for organic food is growing rapidly in Japan and it is expected that this will continue in the future.

Organizational support

The development of organic agriculture in Japan is mainly promoted by the farmers NGOs, and by the private sectors. There is no specific government support for organic agriculture. However, there are two national organizations that promote the development of organic agriculture: the Japan Organic Agriculture Association and the Japan Organic & Natural Foods Associations.

7.4.1.5 THAILAND

Comprising an area of 514,000 km^2 in Southeast Asia, Thailand (formerly known as Siam) extends almost two-thirds down the Malay Peninsula. It is bordered on the North East and East by Laos, on the South East by Cambodia and the Gulf of Thailand (formerly the Gulf of Siam), on the South by Malaysia, on the South West by the Andaman Sea, and on the West and North West by Myanmar, with a total boundary length of 8,082 km. Thailand has a tropical climate. For much of the country there are three distinct seasons: the hot season, from March through May; the rainy or wet monsoon, June to October; and the cool season, November through February.

The agricultural sector in Thailand accounts for 9.9% of the GDP and involves 49% of the total labour force. The economic growth recorded by the country in the 1970s and the early 1980s owed mainly to the steady expansion of the agricultural sector. The sector not only provided for domestic food demand but also produced substantial surpluses of some commodities for export. With 20.4 million hectares of farmland, Thailand relies heavily on agricultural sector (ESCAP). Rice is the major crop grown and Thailand is the world's biggest rice exporter. Other crops grown in the country include: rubber, sugarcane, cassava, fruit, cashew nuts, corn, tobacco, cotton, cocoa, peanuts, soybeans, medical plants, dairy, and fishery products. Fresh flowers, especially orchids, are important exports. The land use is divided as follows: arable land 27%, permanent crops 7%, other 65% (FAO, 2006).

Brief overview of organic farming

According to IFOAM & FiBL (2006), there are 13.900 hectares of land under organic management, with a share of total agricultural land of 0.07% and about 2,500 organic farms operating in the country in 2003. Some successful initiatives promoted the expansion of organic agriculture. Although both the domestic market and exports have been growing slowly in the last few years, Thai organic agriculture expanded rapidly and seems to reach a take-off stage. Most organic products are exported to Europe and the USA, but growing concerns on healthy food among Thai consumers further contributes to an expanding demand on organic foods. In 2005, the Thai organic market was valued at US$ 23 million, up from US$ 9.4 in 2002 (ITC, 2006).

Brief overview of key organic products

The main certified organic products from Thailand are rice, and vegetables. There is also production of cotton, soybeans and shrimps. Almost all certified organic products are currently exported with only a small amount sold in domestic market. The domestic market for certified organic products is estimated to be slightly below US$ 1 million, the non-certified and health food market is estimated to US$ 75 million (UNEP/UNCTAD, 2006).

Organizational support

Thai farmers have practiced traditional farming for hundreds of years and enriched it through farmers' knowledge of local agro-ecology and environmentally sustainable way of farming. Local indigenous knowledge of sustainable farming is still widely spread, favoring policies aimed at promoting organic agriculture (FAO, 2006). In 1980, the Thai organic agricultural movement was initiated by farmers and local NGOS and in 1984 the Alternative Agriculture Network (AAN) was established in 1984 as a national network for sustainable agriculture and organic farming. Producer organizations, private companies, exporters and NGOs have initiated many organic production projects. Government policies towards organic farming have generally been favorable. Many organic farming projects aimed at setting national standards and certification services have been supported by local authorities . The Thai Department of Agriculture (DOA) established the Organic Crop Institute (OCI) with the aim of setting up an organic crop production standards based on the Codex Alimentarius Guideline of Organically Produced Food. The Department of Export Promotion (DEP) has also funded some projects supporting organic production and exports. Also the Department of Livestock Development (DLD) and the Department of Land Development have both funded projects aimed at increasing organic production.

As organic agriculture became more popular in Thailand, several organizations working on organic production have emerged. Green Net and Earth Net Foundation were founded in 1993 and is now one of the leading organizations with an instrumental role in organic conversion. A national private certification body, the Organic Agriculture Certification Thailand (ACT) founded in 1995 was also set up to provide professional organic certification services for all farm production as well as processing and handling operations.

7.4.1.6 VIETNAM

Situated on the eastern coast of mainland Southeast Asia, the Socialist Republic of Vietnam (SRV) has an area of 329,560 km^2. The nation is bordered on the North by China, on the East by the Gulf of Tonkin, on the East and South by the South China Sea, on the South West by the Gulf of Thailand, and on the West by Cambodia and Laos, with a total land boundary of 4,639 km and a coastline of 3,444 km. Before unification, which was proclaimed on 3 July 1976, Vietnam was divided in two by the 17th parallel. To the south was the Republic of Vietnam (RVN), also known as South Vietnam; to the north, the Democratic Republic of Vietnam (DRV), also known as North Vietnam. Vietnam is entirely located in the tropical belt lying between the equator and the Tropic of Cancer. While there are slight variations in temperature, depending on the season and the altitude, the primary seasonal changes are marked by variations in rainfall.

According to the Ministry for Agriculture and Rural Development of Viet Nam, agricultural growth rate has reached rapid and stable level with an average growth of 4.5% per year over the last 10 years. Today, agriculture contributes about 30% of the total export turnover and 25% to total GDP (MARD). In the last decade, Viet Nam has surged as one of the biggest coffee producer of the world, becoming in few years the third longest coffee exporter in the world. Viet Nam is also the first exporter of pepper, the second exporter of cashew nuts and the third exporter of milled paddy rice. Other major export crops include: rubber, tea, groundnuts (shelled), honey, tobacco, maize, fruit and verge-tables (FAOSTAT, 2005-06). Of the total land in use, 20% is arable land, 7% is permanent crops, and 73% is for other uses (CIA, 2007).

Brief overview of organic farming

According to IFOAM & FiBL (2006), there are 6,475 hectares of land under organic management in Vietnam. There are currently 1,022 organic farms registered in Vietnam. The use of chemical fertilizer has increased dramatically in this country since 1980. Government policies have favored the use of chemical fertilizers with the goal of promoting food security in the country. Organic agriculture in the country is limited, and there is no official government programme supporting the sector. Organic agriculture is at its initial phase and efforts for simplifying organic certification in the country are underway (IGCI).

Brief overview of key organic products

The organic products grown are mainly rice, tea, maize, wheat, legumes, and shrimps. There is no evidence of a domestic market for organic products, and production is mainly exported.

Organizational support

By December 2006, there were no evidences of government-funded programmes aimed at the promotion of the organic sector. The Swiss Import Promotion Programme (SIPPO) is currently assisting Vietnam growers to breed certified organic prawns. National agricultural research, although very advanced in some respects, is focused on conventional agriculture. The International Global Change Institute (IGCI) has funded three projects to develop and test technologies and processes for adopting organic production methods for tea and other crops in selected communes.

7.4.1.7 SOUTH KOREA

Occupying the southern 45% of the Korean Peninsula in East Asia, the Republic of Korea (ROK), also known as South Korea, has an area of 98,480 km². Bounded on the North by the Democratic People's Republic of Korea (DPRK), on the East by the Sea of Japan (known in Korea as the East Sea), on the South by the Korea Strait, and on the West by the Yellow Sea.

The rapid development of the economy of the Republic of Korea of the past 40 years has resulted in a declining role for the agricultural sector. The share of the agricultural sector fell from 27% of GDP and 50% of the labour force in 1970, to 3.3 per cent and 6.4 per cent in 1999, respectively (CIA, 2007). In other words, agriculture in the Republic of Korea is characterized by a small family farm structure cultivating rice as a basic crop. Barley is the second most important crop produced in the country. Other crops produced in South Korea are millet, corn, sorghum, buckwheat, soybeans, and potatoes, fruits and vegetables which include pears, grapes, mandarin oranges, apples, peaches, Welsh onions, Chinese cabbage, red peppers, persimmons, cabbage, peaches, and radishes. Other important cash crops include cotton, hemp, sesame, tobacco, and ginseng. Of the total land in use, 16.58% is arable land, 2% is crop land, and 81.40% is for other uses (Ministry of Agriculture and Forestry of Korea).

Brief overview of organic farming

According to IFOAM & FiBL (2006), in the Republic of South Korea there are 28,218 hectares of land under organic management, with a share of total agricultural land of 1.46% and a presence of around 28,951 organic farms. According to UNESCAP, Korea is one of the major organic markets in Asia. In the early 1990s, the government decided to adopt a sustainable agriculture development policy as an important agricultural policy and the Direct Payment Programme for sustainable farming practices was adopted in 1999. The main purpose of this programme is to pay farmers directly for the additional costs or farm income loss as a result of practicing sustainable agriculture. Consequently, the number of farms and cultivated areas of organic farming has been increasing very rapidly since the mid-1990s. Nevertheless, almost no organic agricultural products and foods are exported to foreign countries because the quantity is not enough to process for export. For this reason, no foreign donor agency is working in the Republic of Korea in the organic sector. Some Korean food companies import processed foods such as orange juice for baby food production (UNESCAP, 2006).

Brief overview of key organic products

The local organic products grown are: rice and vegetables. Reportedly, they produce 500 metric tons of organic rice a year. There are also productions of organic grapes, strawberries, and small oranges. Because most crops are produced in costly greenhouses, production of these goods is limited. The organic sector in the Republic of Korea is mostly domestic-oriented. Although the domestic demand of organic products has increased in South Korea, production has not been able to satisfy the growing domestic demand and therefore, most of the organic food sold in South Korea is imported. According to the largest Korean retailers of processed organic products, this segment has shown the most growth in the sector and is anticipated to continue to do so in the future (OTA).

Organizational support

As already mentioned, there is no NGO or international organization working on the development of the organic sector in South Korea. Due to a high level of cancer caused by the indiscriminate use of chemical during in the past decades, the local population is aware of the advantages of sustainable agriculture and organic food. However, initiatives in the development of this sector are left to the government and the private sector. It is expected that the increasing amounts of organic products demanded, will be imported because of the limited organic production in the Republic of Korea. The organic sector in South Korea offers many opportunities for foreign imported organic products (OTA).

7.4.1.8 PHILIPPINES

The Republic of the Philippines consists of an archipelago of 7,107 islands situated South East of mainland Asia and separated from it by the South China Sea. The total land area is approximately 300,000 km^2. The Philippines is separated from Taiwan on the North by the Bashi Channel (forming part of the Luzon Strait) and from Sabah, Malaysia (northern Borneo), on the South West by the Balabac Strait (off Palawan) and the Sibutu Passage (off the Sulu Archipelago). Bordering seas include the Philippine Sea and the Pacific Ocean on the East, the Celebes Sea on the South, the Sulu Sea on the South West and the South China Sea on the West. The Philippine Islands, in general, have a maritime tropical climate and, except in the higher mountains, temperatures remain warm, the annual average temperature ranging from about 23° to 32°C throughout the archipelago. Annual normal relative humidity averages 80%. Rainfall and seasonality differ markedly throughout the islands.

The agricultural sector in the Philippines accounts for 14% of the GDP and involves 36% of the total labour force (CIA, 2007). Palay (unhusked rice) and corn are the two cereals widely grown in the Philippines. Coconut is a major export earner. Other agricultural crops are sugarcane, pineapples, Cavendish bananas (a dwarf variety), mango and coffee, which are also important earners of foreign exchange. The Philippines is the leading supplier of Eucheuma seaweed accounting for about 80% of total world supply. Low productivity and low incomes from agriculture and fisheries are consistent with the prevalence of rural poverty. The situation is further aggravated by low farm gate prices of produce and high retail prices of food, which are among the highest in the region. In terms of area, about a third of country's 30 million hectares is used for agriculture (food crops 52%; food grains 31%; non food (pasture and cut flowers) 17%).

Brief overview of organic farming

According to IFOAM & FiBL (2006), there are 35,000 organic farms on 14,140 hectares of land under organic management, with a share of total agricultural land of 0.12%. Although organic agriculture in the Philippines is still in its emergent or incipient phase, the organic agriculture is fast gaining ground in the country. Organic agricultural production is limited though steadily growing, reportedly between 10-20 percent annually (FAS/USDA, 2000), but reliable statistical data are hardly available. The organic market in the country has been described as a "niche market". A number of organic products are increasingly being sold in major supermarkets, with a price premium reportedly ranging from 20-30 to 30-50% over non-organic products (Yussefi & Willer, 2003). In 2005, the President of Philippines approved a document on the "Promotion and Developmint of Organic Agriculture in the Philippines", recognizing the potential of organic agriculture in the country and providing government support to the development of the sector.

Brief overview of key organic products

Locally grown organic products include rice, fruits and vegetables (both fresh and processed), herbs and spices, along with, soybean, and honey. Some livestock and poultry, fish, dairy products, and fertilizers are also sold as organic products. The main organic export products include muscovado sugar, fresh bananas, banana chips, and coconut oil, with Japan, Western Europe, and the U.S. as the primary destinations. Most of the organic production in the Philippines is exported to international markets. Most recently, a niche domestic market has developed and domestic demand increased (PCARRD, 2006)

Organizational support

The organic movement in the Philippines was initiated in the 1980s by a series of uncoordinated initiatives promoted by some NGOs. With no support by the government, for the rest of the 1980s a number of other projects emerged and introduced organic farming in the Philippines. Today there are many private companies and NGOs involved in the production of organic food. The Organic Producers and Traders Association (OPTA) were formed in 1995. The Organic Farming Information Network (Phil-Organic) is an information service that provides accessible data/information to various stakeholders in the organic farming industry. MASIPAG is a farmer-led network of people's organizations, non-government organizations and scientists working towards the sustainable use and management of biodiversity through farmers' control of genetic and biological resources, agricultural production and associated knowledge. The OCCP is an independent, private, membership-based, organic-standard setting and organic certification body. The PDAP is a national network of Philippine NGOs working on the promotion of rural enterprises for poverty reduction and as tool for peace building in conflict-affected areas and also active in the promotion of organic agriculture.

7.4.1.9 MALAYSIA

Situated in Southeast Asia, Malaysia, with an area of 329,750 km^2, consists of two non-contiguous areas: Peninsular Malaysia (formerly West Malaysia), on the Asian mainland, and the states of Sarawak and Sabah, known together as East Malaysia, on the island of Borneo. Peninsular Malaysia, protruding southward from the mainland of Asia, comprises an area of 131,587 km^2. It is bordered on the North by Thailand, on the East by the South China Sea, on the South by the Strait of Johore, and on the West by the Strait of Malacca and the Andaman Sea. The climate of Peninsular Malaysia is equatorial, characterized by fairly high but uniform temperatures (ranging from 23° to 31°C throughout the year), high humidity, and copious rainfall (averaging about 250 cm/ in 100 days annually).

The agricultural sector in Malaysia accounts for about 8.4% of the GDP and employs 14.5% of the total labour force (CIA, 2007). Malaysian agricultural production consists of tree crops (mainly for export), rice and livestock (mainly for domestic consumption), and fruits and vegetables (both export and domestic consumption). Main export crops include oil palm, rubber, cocoa, pineapple and pepper and cover over 75% of cultivated land. The government is encouraging a shift of production to higher value crops. A minimum area will remain under paddy because of its strategic importance. The domestic rice self sufficiency production target is set at 65 per cent (UNEP/UNCTAD, 2006). The land use is divided as follows: arable land 5.46%, permanent crops 17.54%, other 77% (FAOSTAT, 2005-06).

Brief overview of organic farming

According to IFOAM & FiBL (2003), there are 600 hectares of land under organic management, with a share of total agricultural land of 0.01%. In 2001, the Department of Agriculture (DoA) reported there were 27 organic producers in the country with a total area of 131 hectares, a fivefold increase in the number of organic initiatives listed in an earlier NGO country report on sustainable agriculture in 1996. Domestic production is largely limited to vegetables and fruit with one organic poultry operator in East Malaysia. Whilst organic sector is growing, its total estimated contribution is still statistically insignificant (UNEP/UNCTAD, 2006). The National Agriculture Policy (NAP) recognized that organic agriculture can offer export opportunities in the organic niche market, particularly for fruit and vegetables. The government encourages small-scale producers to venture into organic farming as part of the strategy to raise producers' income, overcome problems of chemical residues in food production, protect the environment, reduce food imports as well as enhance the country's export of high quality safe food (DoA). Under the 9th Malaysia Plan (2006-2010), the government is targeting the organic farming industry to be worth RM800 million in 5 years' time. The Ministry of Agriculture plans to have 20,000 hectares under organic farming methods by year 2010, increasing local production by 4,000 hectares per year. Organic consumption is expected to grow by 20% per annum (DoA).

Brief overview of key organic products

Most organic production is sold domestically with some being exported to Singapore. Sales channels include dedicated organic/ health food shops and supermarket chains. So far, there is no reliable market statistics available. The main certified organic products from Malaysia are vegetables, fruit and rice.

Organizational support

The first of several efforts to form a national network was the Malaysian Organic Farm Network (MOFAN) initiative in 1990 but it is currently not active. In 2001 Organic Alliance Malaysia (OAM) was founded as a membership based private sector association to fill the gap. It has currently over 30 members, mostly from the trade (importers and retailers). OAM hosts a monthly lunch meeting for organic operators in Kuala Lumpur and publishes the Organic Directory. It is also considering a private label scheme.

7.4.1.10 IRAN

Situated in southwestern Asia, Iran covers an area of 1,648,000 km^2. Iran is bounded on the North by Armenia, Azerbaijan, Turkmen-is tan, and the Caspian Sea, on the East by Afghanistan and Pakistan, on the South by the Gulf of Oman and the Persian Gulf, on the West by Iraq, and on the North West by Turkey, with a total land boundary length of 5,440 km. The coastline is 2,440 km. The shoreline on the Caspian Sea is 740 km. Iran's territory includes several islands in the Persian Gulf. Iran has a continental type of climate, with cold winters and hot summers prevalent across the plateau. On the plateau, the annual rainfall does not exceed 30 cm, with the deserts and the Persian Gulf littoral receiving less than 13 cm. Snow falls heavily on the mountain peaks and is the principal source of water for irrigation in spring and early summer. The Caspian littoral is warm and humid throughout the year, and the annual rainfall is from about 100 to 150 cm.

Iran's agricultural sector contributed 12 percent of the GDP in 2005 and employed 30% (2001) of the labour force (CIA). Some northern and western areas support rain-fed agriculture, while other areas require irrigation for successful crop production. Wheat, rice, and barley are the country's major crops. Total wheat and rice production fails to meet domestic food requirements, however, making substantial imports necessary. Other principal crops include potatoes, legumes, vegetables, fruits, fodder plants, oil seeds, nuts (pistachios, almonds, and walnuts), cotton, sugarcane, sugar beets, herbs, spices, tea, and tobacco (FAOSTAT). Honey is collected from beehives, and silk is harvested from silkworm cocoons. Livestock products include lamb, beef, goat meat, poultry, eggs, milk, butter, cheese, wool, and leather. Major agricultural exports include fresh and dried fruits, nuts, animal hides, processed foods, and spices (Food and Fertilizer Technology Center).

Brief overview of organic farming

Traditional small-scale farming was the main structure of farming communities for centuries and this has caused a tremendous accumulation of indigenous knowledge in farming practices and food production. Therefore, natural farming used to be a widespread practice not more than half a century ago and is still appreciated to some extent by the farmers. According to IFOAM & FiBL (2006), in the Islamic Republic of Iran there are 100 hectares of land under organic management and only one registered organic farm. Although the remains of traditional systems of food production, completely organic both in technical and social terms, are still in operation in remote areas, these systems have not been regarded as organic in the present context of organic production and in fact they could be considered as ignored organic farmers (IFOAM).

Brief overview of key organic products

The organic products grown are rice, maize, wheat ,legumes and nuts. There is no evidence of a domestic market for organic products, and certified production is mainly exported.

Organizational support

At December 2006, there is no organic organization within Iran registered with IFOAM, nor are there any national organic standards. A few public and private initiatives in the country do state their intent to develop organic production: some non-certified production for local consumption, and some certified for export, mainly jasmine and rice. There is no evidence of a domestic organic market.

7.4.1.11 INDIA

India fills the major part of the South Asian subcontinent (which it shares with Pakistan, Nepal, Bhutan, and Bangladesh) and includes the Andaman and Nicobar Islands in the Bay of Bengal and Lakshadweep (formerly the Laccadive, Minicoy, and Amindivi Islands) in the Arabian Sea. The total area is 3,287,590 km^2. India is bordered on the North by the disputed area of Jammu and Kashmir (west of the Karakoram Pass), China, Nepal, and Bhutan; on the East by Myanmar, Bangladesh, and the Bay of Bengal; on the South by the Indian Ocean; on the West by the Arabian Sea; and on the North West by Pakistan. The monsoon is the predominant feature of India's climate and helps to divide the year into four seasons: rainy, the southwest monsoon, June–September; moist, the retreating monsoon, October–November; dry cool, the northeast monsoon, December–March; hot, April–May. The southwest monsoon brings from the Indian Ocean the moisture on which Indian agriculture relies.

The agricultural sector in India accounts for 18.6% of the GDP and involves 60% of the total labour force. During the past five years the agricultural sector has witnessed a huge increase in the production and productivity of food grains, oilseeds, commercial crops, fruits, vegetables, poultry and dairy. India has emerged as the second largest producer of fruits and vegetables in the world. In addition, it is also the largest overseas exporter of cashews and spices. Other agricultural products include: cotton, jute, tea, sugarcane, cattle, water buffalo, sheep, goats, poultry, and fish (FAOSTAT, 2005-06). The land use is divided as follows: arable land 49%, permanent crops 3%, other 47% (CIA, 2007).

Brief overview of organic farming

According to the International Fund for Agricultural Development (IFAD), organic production in India has been growing steadily in the last years. In India, there has been a remarkable growth in organic farming. The Research Institute of Organic Agriculture (FiBL) reported a total 2 lakh organic farms operating in the country in 2009. The total area reported under organic agriculture is 1.2 million ha of certified and around 8.0 million uncertified organic area. Recently, an increasing number of companies, NGOs, farmer organizations, and government agencies have been promoting organic agriculture in India. The growth in organic production has been driven mainly by the increasing international demand, but the domestic market is also strengthening due to a large population and increasing wealth (IFAD).

Brief overview of key organic products

Certified organic produce from India includes tea, rice, bananas, cotton, wheat, spices (mainly pepper and ginger), coffee, nuts, pulses, and herbal products. Most of these products are sold in semi-processed or raw forms (IFOAM & FiBL, 2006). Most organic farm produce are for the international markets, but there is an emerging robust domestic market.

Organizational support

The Agricultural and Processed Food Products Export Development Authority (APEDA) has set up a Programme for Organic Production (NPOP) in order to promote and support the organic sector in India. INDOCERT is one of the national certification body, accredited by NPOP, which has technical collaboration with FiBL and bio.inspecta and is financially supported by SECO. Organic production has attracted, over the past years, an increasing political recognition by the Indian government both at central and local level. In November 2006, the second edition of India Organic 2006 took place in Lalbagh, Bangalore. This organic business trade fair was organized by the International Competence Center for Organic Agriculture (ICCOA) with the active support of IFOAM and the Government of Switzerland (SECO). Both ICCOA and IFOAM set up a joint initiative in Bangalore. The centre aims at promoting nationwide adoption the principles of organic agriculture, taking advantage of the opportunities provided by the large development of organic stakeholders. Government of India has adopted a strong promotional policy to promote organic farming. It has adopted IFOAM standards in 2003 for certification. Simultaneously, 12 different national and international certification organizations assist in certification. National centre for organic farming (NCOF) at Ghazibad and 5 Regional Centres for Organic Farming (RCOFs) are government sponsored agencies to promote organic farming. In addition a network of NGOs and farmers organization also helps in promotion of organic farming.

7.4.2 AFRICA

7.4.2.1 ZAMBIA

Zambia is a landlocked country in south central Africa. It shares borders with the Democratic Republic of the Congo and Tanzania to the north, with Malawi to the east, Mozambique to the southeast, Zimbabwe, Botswana and Namibia to the South and Angola to the West. The country extends at its maximum about 821 km from north to south and 1,385 km from east to west, covering a total area of 7,52,614 km². Although lying within the tropical zone, much of Zambia enjoys a pleasant subtropical climate because of the high altitude. The rainy season runs from November to April.

About 70 percent of Zambia's working population is engaged in agriculture, largely subsistence farming. The sector accounts for 20% of GDP. Principal crops are maize, cassava, fresh vegetables, sugar cane, wheat and groundnuts. Beef and dairy cattle are raised for domestic use. Major export crops include cotton, tobacco, maize, sugar, green coffee, peas, beans and fresh vegetables (FAO, 2006). The agricultural sector remains underdeveloped and vulnerable to weather fluctuations. High transportation costs resulting from long distances to seaports pose a constraint to agricultural exports.

Overview of organic farming

A recent survey reports significant increases in organic production in Zambia. Approximately 2,500 Zambian farmers cultivate 187,694 hectares of land using organic methods. This accounts for 0.53% of the total agricultural area (IFOAM & FiBL 2006). The data include fully converted land as well as "in conversion" land area. Having a lot of virgin arable land, Zambia continues to show bright prospects for further developing its organic sector due to low labour cost and favorable climate. However, in order to compete with surrounding countries, more investment in local infrastructure is required.

Overview of organic products

The bulk of the area under organic production in Zambia is virgin land for wild harvesting of products such as honey, mushrooms and indigenous tree seed oil. The remainder is commercial land, which is used for the production of export vegetables, herbs, spices, medical plants, groundnuts, sesame, green manures, soybeans and maize (IFOAM 2003). Basically, all organic exports from Zambia are destined for Europe.

Organizational support

The Organic Producers and Processors Association of Zambia (OPPAZ), which was set up in 1999 as a body affiliated to the Zambian National Farmers Union, promote the organic movement in Zambia. It also developed a technical advisory service for its members. In general, it provides marketing, certification and technical support to producers from the smallholder sector as well as to large-scale commercial enterprises.

The main certifiers are Ecocert and the Soil association. OPPAZ has set up an initiative to develop a certification label for the national market (IFOAM 2003).

7.4.2.2 UGANDA

Uganda is a landlocked country on the equator in East Africa. It is bordered by Kenya to the east; Sudan to the north; Democratic Republic of the Congo to the west; and Rwanda, Tanzania, and Lake Victoria to the south. The country has a total area of 241,038 km^2 measuring 625 km east to west and 638 km north to south. Although generally tropical in nature, the climate differs between parts of the country. However all regions, except for the northeastern border area and small areas in the southwest, usually receive sufficient rain to permit crops to grow once or even twice a year.

Fertile soils and regular rainfalls are Uganda's most important natural resources. Agriculture forms the heart of its economy, accounting for 45% of GDP and employing over 80% of the work force, with coffee accounting for the bulk of export revenues (IFOAM 2003). Besides, plantains which dominate Uganda's internal market, cassava, sweet potatoes, dry beans and maize are among the country's major crops (FAO, 2006). In general, most farms are small-scale.

Overview of organic farming

Uganda has the most developed sector of certified organic production in Africa. About 33,900 farmers manage 122,000 hectares of land using organic methods, an area that accounts for 1% of Uganda's arable land (IFOAM & FiBL, 2006). Although still small and far below the increasing global demand, the country's export of organic agricultural produce has been growing substantially in recent years. In Uganda, which has one of the lowest agro-chemical usages in Africa, the majority of farmers practice de facto organic agriculture without being certified yet (ACODE, 2006). Since no significant domestic market exists, certified organic agriculture targets mainly export markets in Europe and North America.

Overview of organic products

Major organic export products include coffee, cotton, dried bananas, paw paws, pineapples, passion fruits, chillies, ginger and sesame. While traditional cash crops remain as the backbone of certified organic exports from Uganda, two sectors of particular focus in recent years have been fresh and dried fruits and vegetables, and –

most recently – vanilla (Gibbons, 2006). The exceptionally high growth rates in some markets reflect a combination of factors including growing awareness of opportunities in the sector and the availability of donor support, especially for certification.

Organizational support

The two key organizations making up Uganda's wider institutional environment for organic farming are the National Organic Agricultural Movement of Uganda (NOGAMU) and the national certification and inspection body, UgoCert. NOGAMU was set up in 2001 as a membership organization, which unites farmers, producers, exporters and other stakeholders in the promotion of organic agriculture. It represents over 35,000 farmers and works with partner organizations throughout the country.

NOGAMU is also the owner of UgoCert, whose formation has mitigated the impact of the worldwide rise in organic certification costs by providing local inspectors for some international certification bodies.

7.4.2.3 SOUTH AFRICA

South Africa occupies the southern tip of the African continent, its long coastline stretching more than 2,500 km and across two oceans, the Atlantic and the Indian. It borders the countries of Namibia, Botswana, Zimbabwe, Mozambique, and Swaziland. The nation of Lesotho forms an enclave in the eastern part of the country. South Africa extends some 1,500 km from east to west and 1,000 km from north to south, comprising a territory of 1,219,090 km^2. South Africa enjoys a generally warm, temperate climate. Most of the country experiences light rainfall and long hours of sunshine.

South Africa is not only self-sufficient in virtually all major agricultural products, but in a normal year it is also a net food exporter. However, with very low average rainfall and high variability within and between seasons, its agriculture is vulnerable to the effects of drought. The agricultural sector contributes 3.4% of GDP, accounts about 7% of exports and employs 10% of the population. Production includes Maize, cattle and chicken meat, grapes, dairy produce, sugarcane, wheat, eggs, fresh fruits and vegetables. South Africa has a growing reputation as a producer of quality wines, which represent the country's main export commodity by now (FAO, 2006).

Overview of organic farming

South Africa has had an organic sector for many years, although it has grown in 'fits and starts'. It comprises 250 certified farms and 45,000 hectares of certified organic land, which account for 0.05% of the country's total agricultural area (IFOAM & FiBL 2006). South Africa is one of the two countries – the other is Egypt – with a robust domestic market, although it is underdeveloped and there are few, price premiums for organic products. Most organic growers therefore look to Europe as their outlet area (IFOAM 2003). South Africa is also a market destination for organic produce from nearby southern African states, particularly Mozambique.

Overview of organic products

Certified organic production started with rooibos tea, mangoes, avocados, herbs, spices, and vegetables (ITC 1999). It has expanded to include a much wider range of products: organic wines, olive oil, specialty vegetables and berries, citrus, sub-tropical fruit, and dairy products (IFOAM 2003). Different sorts of vegetables and culinary herbs form the core of the organic produce sold in South Africa itself (EPOPA 2006).

Organizational support

Two key organizations promoting organic agriculture in South Africa are the Organic Agricultural Association of South Africa (OAASA) and the Biodynamic Agricultural Association of South Africa (BDAASA). Both non-profit membership organizations intend to further the organic movement by providing a network for all stakeholders. Their activities comprise public relations, teaching and training, publishing and other forms of information dissemination.

Two local certification agencies exist. Afrisco, is in the process of becoming ISO accredited, so that it will soon be able to provide EU certification. Other reputed organization to offer certifications is ECOCERT.

7.4.2.4 MADAGASKAR

Madagascar is an island nation in the Indian Ocean, separated from the southeastern coast of Africa by the Mozambique Channel. The country extends 1,600 km from north to south and up to 570 km wide. Its area totals 587,041 km^2, which makes it the fourth largest island in the world. The climate varies by region: coastal areas are warm, humid and tropical, while the inland region enjoys temperate conditions; the south is almost always dry.

Agriculture is a mainstay of Madagascar's economy, accounting for more than one fourth of GDP, contributing 70% of export earnings and employing about 80% of the population. However, because of the mountainous terrain, only 5% of Madagascar is farmed. The chief food crop is rice, which is grown on about one half of the agricultural land. Other important food crops are cassava, sweet potatoes, fresh vegetables, bananas, maize and beans. Leading export crops are vanilla, cloves, fruits, cocoa, sugarcane, coffee, sisal and cotton (FAO, 2006).

Overview of organic farming

Madagascar's organic sector remains very small. Only about 129 hectares are under organic cultivation, which accounts for less than 0.005% of the total agricultural area (IFOAM & FiBL 2006). Several attempts to develop a commercial organic sector in Madagascar have failed because of poor quality produce, poor communication between buyers and sellers, non-competitive prices and high transport costs. Another obstacle, that continues to inhibit growth, is the general lack of awareness about certification and logistical requirements (IFOAM 2003). Improving the situation of Madagascar organic agriculture requires further investment in training, market information and certification capacity.

Overview of organic products

Main organic exports from Madagascar include cocoa, sugar, coffee, palm oil, processed fruits, pepper and vanilla. In addition, farmers grow a variety of organic spices and herbs on a small scale such as cinnamon, turmeric, ginger, cloves, chillies, nutmeg and essential oils (IFOAM 2003). A number of companies have made several attempts to produce and export organic fresh fruits and vegetables from Madagascar. However, until now, virtually all failed to succeed (FAO/ITC/CTA 2001). Small-scale farmers produce most organic products, while commercial companies are active in processing, grading and exporting.

Organizational support

The main organic organization in Madagascar is PROMABIO. Founded in 1993 by three entrepreneurs under the name PROBIOMAD, it has now expanded to include 15 members (IFOAM 2003). PROMABIO is a member of IFOAM. The association aims to represent the interests of its members by following its main objectives of promoting organic agriculture, supporting producers, processors and exporters of organic products and developing the brand name "PROMABIO ".

7.4.2.5 GHANA

Ghana is situated on the west coast of Africa with a total area of 238,540 km^2. The country has a north-south extent of about 670 km and a maximum east-west extent of about 560 km. It borders the Ivory Coast to the west, Burkina Faso in the north and Togo to the east. Ghana has a tropical climate. The eastern coastal belt is warm and comparatively dry; the southwest corner, hot and humid; and the north, hot and dry.

Agriculture is Ghana's most important economic sector, employing almost two thirds of the population and contributing half of GDP and export earnings (IFOAM 2003). The country produces a variety of crops, including cocoa, maize, yam, palm oil, coffee, rubber, and timber. Smallholder farms dominate the agricultural sector, accounting for about 80 percent of total production.

Overview of organic farming

As in most West African countries, the organic sector in Ghana is still relatively underdeveloped. About 19,132 hectares of land are under organic cultivation, which accounts for 0.13% of the total agricultural area in Ghana (IFOAM & FiBL 2006). These numbers including fully converted land as well as "in conversion" land area have been growing substantially in recent years. The whole region has a potential for developing a formal certified sector, especially with regard to tropical fruit and palm oil. Since Ghana does not have any significant domestic market for organic produce, the future development relies on organic trading links being established to export markets, particularly in Europe.

Overview of organic products

Ghana's main organic export commodities are palm oil and fresh fruits. The certification of farms already using organic methods makes progress in these markets. Other key organic products include cocoa, bananas, cashews, culinary herbs, cereals, vegetables, cotton and shea butter (IFOAM 2003). Ghanaian NGOs and farmers groups promote the expansion of organic production in the existing product range as well as in new sectors. These indigenous groups are active in developing and disseminating improved organic farming methods.

Organizational support

The Ghana Organic Agriculture Network (GOAN) is the main grouping of organic NGOs and trade associations, which is working actively with concerned organizations such as ITC, HDRA, DfiD and PAN-UK in developing the organic sector in the country. It has about 150 member groups and organizations, representing a thousand individual members.

With the assistance of HDRA and PAN-UK, GOAN in 1995 established an agriculture centre to provide information, training and advice on organic agriculture practices. It also has links with research institutes to examine alternative methods of pest control, particularly for cocoa, oil palm cotton, cereals and fruit vegetables.

7.4.2.6 EGYPT

Located in northeastern Africa, Egypt is bounded on the north by the Mediterranean Sea; on the east by the Gaza Strip, Israel, and the Red Sea; on the south by Sudan; and on the west by Libya. Covering a total area of 997,739 km^2, the country stretches 1,105 km from north to south and up to 1,129 km from east to west. More than 90 percent of Egypt consists of desert areas. Its climate is characterized by hot, dry summers and moderate winters.

The agricultural sector contributes about 15 percent of Egypt's GDP. Approximately one-third of Egyptian labour is engaged directly in farming, and many others work in the processing or trading of agricultural products. Food production for domestic consumption includes tomatoes, rice, wheat, milk, maize and fruits. Livestock and fish also make an important contribution to the national diet (IFOAM 2003). Major export crops are cotton, rice, oranges, vegetables and molasses (FAO, 2006). Egypt's area of cultivable land is very small but highly fertile. It is located for the most part along the Nile and in the Nile Delta.

Overview of organic farming

In Egypt, intensive cultivation on a relatively small agricultural area corresponded with a traditionally high rate of pesticide and fertilizer use. Problems with this have been stimulating a rising interest in organic production (IFOAM 2003). Today Egypt has a well developed and still rapidly growing organic sector. About 24,548 hectares of land are under organic management, accounting for 0.72% of the country's total agricultural area. Organic production concentrates on about 500 farming enterprises (IFOAM & FiBL 2006). Many of these farms are 'desert' farms, using irrigation from the Nile.

Overview of organic products

Egyptian organic farmers grow a variety of crops, including fruits (notably grapes, citrus, dates mangoes and strawberries), vegetables, cereals and spices as well as non-food crops as cotton and medical plants (IFOAM 2003, IFOAM & FiBL 2006). While much produce is exported, primarily to Europe, Egypt is also one of the few African countries that enjoy domestic demand for a number of organic products. The ongoing expansion of the domestic market continues to diminish the local producers' reliance on export sales, thereby encouraging buoyant investment activity in the sector.

Organizational support

NGOs play a significant role in supporting the organic movement in Egypt. The Egyptian Biodynamic Association (EBA) has been established in 1990 aiming to promote organic agriculture in Egypt. The organization providing training, research, and advisory services in field of organic farming is a subsidiary initiative of the SEKEM group, which has been responsible for most of the early development of the organic sector in Egypt.

Organic certification in Egypt is mainly provided by two local organizations: the Egyptian Center for Organic Agriculture (ECOA) and the Center of Organic Agriculture in Egypt (COAE). Both companies are members of IFOAM and accredited to certify for EUREPGAP.

7.4.2.7 CAMEROON

Cameroon forms a bridge between West Africa and Central Africa. The triangle-shaped country is bounded on the east by Chad and the Central African Republic; on the south by the Republic of the Congo, Gabon, and Equatorial Guinea; and on the west by the Atlantic Ocean and Nigeria. Measuring about 700 km east to west and 1,200 km from north to south, Cameroon covers an area of 475,442 km2. The country's climate varies with terrain, from tropical along coast to semiarid and hot in north.

Agriculture is the main pillar of Cameroon's economy. It accounts for 43% of GDP, employs 70% of the working population and generates more than one third of total export earnings. Main agricultural products include plantains, cattle meat, cocoa beans, taro, bananas, maize, fresh vegetables and groundnuts. Cameroon's primary export crops are cocoa, cotton, coffee, bananas, rubber and palm oil (FAO, 2006). The country's main export partner is the European Union. However, the bigger part of agricultural activity is confined to subsistence level farming.

Overview of organic farming

As in most West African countries, the organic sector in Cameroon is still underdeveloped. Estimates of certified organic production suggest that 7,000 hectares of land are under organic management accounting for 0.08% of the total agricultural area (IFOAM & FiBL 2006). The country shows a strong potential for further developing the organic sector to meet export opportunities and create a more sustainable form of agriculture. High premiums for organic produce, the increasing costs of synthetic farm inputs and a lack of support for ordinary farming have led to some entire areas spontaneously converting to organic farming (IFOAM 2003).

Overview of organic products

Cameroon is well known for its climatic, geographic and ecological diversity, which enables farmers to grow a wide variety of crops. The most frequently cultivated export crops on Cameroonian organic farms are bananas, pineapples, avocadoes, mangoes, papayas, coffee and cocoa. Other certified products include herbs, spices, tubers and medical plants. There is a growing emphasis on processing some of the fruit crops through drying, pulping and juicing (IFOAM 2003).

Organizational support

In Cameroon, there are two organizations gathering operators of the organic sector: ASPABIC (Association for the Promotion of Organic Agriculture in Cameroon) in the French-speaking part and AVEGRO in the English-speaking area. Both groups concentrate on promoting the sector and raising public awareness. They provide their members with information services, technical assistance and training.

Two external bodies, IMO and Ecocert, offer organic certification services in Cameroon (IFOAM 2003).

7.4.2.8 ETHIOPIA

Ethiopia is a land locked country situated in the Horn of Africa. It shares frontiers with Eritrea to the north, Djibouti to the northeast, Somalia to the east, Kenya to the south, and Sudan to the west.

Ethiopia covers an area of 1,133,380 km^2 measuring about 1,200 km from north to south and approximately 1,600 km from east to west. Although the entire country lies within the tropics, only the lowlands have a hot climate. In the highlands, however, the proximity of the equator is counterbalanced by the elevation of the land, resulting in temperate conditions.

Agriculture is the main economic activity in Ethiopia, employing more than three quarters of the country's workforce. Half of this is subsistence level farming. State farms are responsible for much of the cash crop production (IFOAM 2003). Coffee is the most important crop, accounting for nearly two thirds of exports. Sesame, animal hides, oilseeds and cotton make a significant contribution, as well. The main staple foods are cattle meat, roots and tubers, milk, pimento, maize, cereals and sorghum (FAO, 2006). Ethiopia is also recognized as one of the centers of agricultural biodiversity, particularly for unexploited varieties of coffee, sorghum, chickpeas, wheat, barley and several other grains (IFOAM 2003).

Overview of organic farming

Certified organic agriculture was nonexistent in Ethiopia only a few years ago. However, the international coffee price crisis, which has seen prices at a 35-year low around the world, is creating severe hardships for conventional coffee farmers. Organic coffee planting as a measure of product differentiation holds a possible way out of the crisis: the international trade in organic products is an expanding niche market that Ethiopia is geographically well situated to exploit. Only very recently, some producers started successfully to tap the potential that organic production offers in this market. Growth prospects are bright due to the fact that estimated 80% of the Ethiopian coffee produced is de facto organic without being certified as such (IFOAM 2003).

Overview of organic products

In Ethiopia, green coffee is the prevailing export commodity. This applies for organic produce even more strictly than for conventional agriculture. Nonetheless, there have been various small initiatives promoting organic farming in other sectors, not only to build up export capacities but also to ensure food security and soil conservation (IFOAM 2003). These new crop varieties include culinary spices and herbs as well as bananas, plantains and potatoes.

Organizational support

An Ethiopian organization addressing the promotion of organic agriculture in particular does not exist yet. However, Agri Service Ethiopia (ASE), an indigenous NGO mainly targeting rural poverty reduction, is a member of IFOAM. Its strategic objectives are not only to improve food security and to facilitate access to basic social services but also to ensure sustainable management of the environment. In this context, the organization has been executing different programs that de facto promote the use of organic production methods.

Ethiopia still lacks local certification capacities. The establishment of such facilities could make a valuable contribution to a favorable development of the organic sector.

7.4.2.9 KENYA

Kenya lies astride the equator on the eastern coast of Africa. It is bordered by Ethiopia to the north, Somalia to the east, Tanzania to the south, Uganda to the west, and Sudan to the northwest, with the Indian Ocean running down the southeast border. Kenya covers an area of 582,646 km^2 measuring about 890 km east to west and 1,030 km from north to south. The country enjoys a tropical climate. It is hot and humid at the coast, temperate inland and very dry in the north and northeast parts of the country.

Agriculture is a cornerstone of Kenya's economy employing over three quarters of the population (IFOAM 2003). More than 50% of export earnings and 24% of GDP are attributed to agricultural products. Tea, coffee, tobacco, palm oil, cashew nuts, sisal and pyrethrum create the largest profit for Kenya. Exports of fresh fruits and vegetables are also attracting an increasing amount of foreign attention and money. The primary food crops are maize, potatoes, beans, cabbages and fruits (FAO, 2006). Although subsistence farming still represents half of agricultural output, the sector includes many large-scale commercial farms, plantations and specialist horticultural units.

Overview of organic farming

The organic sector in Kenya is still relatively small but growing fast. Over 182,000 hectares of land are under organic management, which accounts for 0.69% of the total agricultural area in Kenya. About 30,000 farms have changed over to organic cultivation methods so far (IFOAM & FiBL 2006). Many of the exporters of organic produce are large-scale farmers, already engaged in the export agriculture and horticulture, which are diversifying into organic production to meet demand from their established customers (IFOAM 2003).

Overview of organic products

Certified organic produce from Kenya includes fresh vegetables, fruits, coffee, tea, beans and nuts (macadamia, cashew, shea). Since no significant domestic market exists, Kenyan organic agriculture principally targets export markets, predominantly in the European Union and Japan (IFOAM 2003). In recent years, organic trade has developed beyond traditional export crops to include other products such as essential oils, dried herbs and spices, as well as products for the cosmetic and pharmaceutical industries. In these new sectors, smallholder farms possess a relatively higher market share (UNEP/UNCTAD 2006).

Organizational support

Small-scale organic farmers in Kenya have formed a national representative organization, the Kenya Organic Farmers Association (KOFA). Larger companies and commercial farmers who are already in the export market have organized themselves into the Kenya Organic Producers Association (KOPA). In 2004, organic agriculture stakeholders in Kenya, including KOPA and KOFA, formed the umbrella network KOAN to support the successful growth of the sector (UNEP/UNCTAD 2006).

The Kenya Organic Agriculture Network (KOAN) aims at developing and promoting local and export markets, supporting development of affordable local certification capacity and creating awareness of market opportunities in the organic sector.

7.4.2.10 TUNISIA

Tunisia is a small country by North African standards, sandwiched between the much larger countries of Algeria to the west and Libya to the southeast. The northernmost country in Africa, Tunisia is bounded on the north and east by the Mediterranean sea. It covers an area of 163,610 km², measuring 1200 km from north to south and an average of 280 km from east to west. Northern Tunisia has a Mediterranean climate, with mild, rainy winters and hot, dry summers. Toward the south the climate becomes progressively hotter and drier.

Even though Tunisia is on its way to become a modern service economy, the agricultural sector is still of vital importance, contributing 12.6% of GDP and employing almost one quarter of the country's labour force. Agricultural produce represents about 6% of the country's export earnings. The main cereal crops are wheat and barley. Tunisian farmers grow olives, dates and fresh fruits for both export and domestic consumption. Meat (chicken, sheep and cattle) and vegetables (tomatoes and potatoes) represent important comestible goods as well. The main agricultural export is olive oil (FAO, 2006). The European Union is the country's principal trading partner (ITC, 10.09.2006).

Overview of organic farming

Organic agriculture is relatively new in Tunisia. However, in the last ten years, organic land area, number of farmers, and crop diversification increased rapidly. With 155,323 hectares under organic management, that represent 1.59% of total agricultural area, Tunisia has now one of the most developed organic sectors in Africa (IFOAM & FiBL 2006). This favorable trend is largely due to an active government policy in the promotion of organic agriculture (IFOAM 2003). Since there is not yet a strong domestic demand market for organic products, most of the production is directed to the export market. Some of the farmers are producers and exporters at the same time.

Overview of organic products

About three quarters of organic land in Tunisia is dedicated to growing olives, many of which are processed into oil. Organic olive farmers receive a price premium ranging from 10 to 20% relative to non-organic products (Twarog, 2006). Other crops include dates, jojoba, almonds, fruits and vegetables, honey and aromatic plants (IFOAM, 2003). In recent years, organic livestock husbandry in Tunisia has expanded significantly.

Organizational support

Organic agriculture in Tunisia started in the eighties with private initiatives. However, the breakthrough came in the mid-nineties when the Tunisian government established a national strategy for the development of the organic sector. This program included the establishment of a national commission for organic agriculture (CTAB) and the provision of financial support. Over the first five years after conversion, organic farmers are eligible for financial help covering up to 30% of investment costs for equipment and 70% of certification costs (GTZ, 2003).

Four certification bodies licensed by the Ministry of Agriculture are operating in Tunisia: ECOCERT, IMO, LACON and BCS. They comply with the IFOAM standards, EU regulations and Tunisian legislation.

CTAB is acronym of Centre Technique de l'Agriculture Biologique.

7.4.3 LATIN AMERICA

7.4.3.1 ARGENTINA

Argentina is located in the southern part of South America, and is bordered by Chile to the west; Bolivia and Paraguay to the north and Brazil, Uruguay and the Atlantic Ocean to the west. The country has a total area of 2,780,400 sq km and national climate is mostly temperate, with variations between regions. The country has an Antarctic climate in the south and a more tropical jungle climate in its northern parts.

Argentina's main crops include soybeans, indigenous cattle meat, wheat, maize, whole fresh cow milk, grapes, indigenous chicken meat, sunflower seed, sugar cane and apples.

Brief overview of organic farming

There are 2,777,959 hectares of organic land in Argentina and approximately 1578 organic farms operating in the country. The vast majority of these farms (90%) are small to medium sized. Organic land makes up 2.2% of the country's arable land. Argentina has the highest area of organic farming land in South America, and the second highest in the world (FILB 2009), and yet the country's mean level of production is relatively low. In addition, there is much disparity between the land quality and farming land area of each organic farm. Most of Argentina's organic farmers work in dryer areas and possess more organic land, but only 5% of the organic farms. In the warmer and more humid areas, farmers possess less land and yet hold over 90% of Argentina's organic farms. Argentina's organic crops include sugar cane, raw wool, fruits, vegetables and beans.

Brief overview of organic products

Major organic exports include cereals and oilseeds: corn, wheat, soybean and sunflower. Fruits and vegetables also make up a part of Argentina's major organic exports: pears, apples, oranges, lemons, garlic, onions and beans. Exports of aromatic and medicinal plants are increasing. Argentina's also exports processed organic products such as olive oil, sugar, concentrated juices, honey and wine. Meat and organic wool continue to be exported. The national organic product market includes various different products of which vegetables fruits, oils, teas, breads, eggs, jams and organic milk are most produced (2007).

Organizational support

Several NGOs and organizations play vital roles in strengthening Argentina's organic sector. The Argentine movement for Organic Agriculture (MAPO), the Chamber of Commerce (CAPOC) and the Certifiers' Chamber (CA.CER) are all key parts of the country's Organic network. These three organizations work towards attaining agreements between public authorities in South America, and strategic planning. MAPO acts as a hub for all the activities being undertaken with this movement.

7.4.3.2 BOLIVIA

Bolivia is a landlocked country located in the heart of South America. It is surrounded by Peru to the west; Brazil to the north/east; Paraguay to the south-east; Argentina to the south and Chile to the south-west. The country has a total area of 1,098,581 sq km and its climate varies between regions of different altitudes; the country has both cold and tropical temperatures.

Bolivia's main crops include soybeans, indigenous cattle meat, indigenous chicken meat, potatoes, indigenous pig meat, sugar cane, bananas, maize, paddy rice, whole fresh cow milk.

Brief overview of organic farming

There are 41,004 hectares of organic land in Bolivia and approximately 11,743 organic farms operating in the country. Organic land makes up 0.1% of the country's arable land. The most important organic products produced by Bolivia are coffee, quinoa, chestnut, cocoa, vegetables, tea and herbs. The country also produces a smaller scale of amaranth, dried fruits and beans.

Brief overview of organic products

Major traditional organic export products include coffee, quinoa, chestnut, cocoa, vegetables, tea and herbs, and lesser volumes of amaranth, dried fruits and beans. Bananas, sesame seeds, sugar cane, soybeans and honey have been increasingly produced and are rising in prominence (2008).

Organizational support

The organic network in Bolivia is comprised of research institutions and programmes, several organization and associations. The Ecological Producers Association (AOPEB) is a key actor in Bolivia's organic field. It links up to 56 different organizations, connecting producers- mainly small holders.

7.4.3.3 BRAZIL

Location

Brazil is the largest country in South America, located to the east of the continent. It is bordered by French Guiana, Guyana and Suriname to the north; Venezuela and Colombia to the northwest; Peru to the west, Bolivia, Paraguay and Argentina to the southwest; Uruguay to the south and the Atlantic Ocean to the East. Brazil has a total area of 8,514,877 sq km. The climate varies between regions but is mostly tropical and otherwise temperate.

Agricultural sector

Brazil's main agricultural products include indigenous cattle meat, soybeans, indigenous chicken meat, sugar cane, whole fresh cow milk, maize, indigenous pig meat, oranges, paddy and cassava.

Brief overview of organic farming

There are approximately 7,250 organic producers in Brazil. The country's total share of land under organic management is 1,765,793 hectares, making up 0.7% of national agricultural land. The country's organic movement also includes organic land areas which may or may not have been informally certified. Most of Brazil's organic exports are coffee, banana, soybeans, and corn.

Brief overview of organic products

Brazil's key organic products include both raw and processed goods: coffee, banana, soybeans, corn. Meat exports are increasing rapidly. Juices, sugar, processed soy, organic textiles and cosmetics.

The domestic market's products include fresh vegetables and fruits, teas, coffee, mate tea, jams, oils, breakfast cereals and dairy products.

Organizational support

Association of Organic Agriculture (AAO)

Empresa Brasileira de Pasquisa Agropecuária (EMBRAPA)

7.4.3.4 CHILE

Location

Chile is located to the east of South America. The country is bordered by Peru to the north; Bolivia to the north east; Argentina to the east and the Pacific Ocean to the west. Chile has a total land area of 6,339 sq km. Climate varies between desert, Mediterranean and cool weather, between the north, central and southern regions respectively

Agricultural sector

Chile's main agricultural products include grapes, whole fresh cow milk, indigenous chicken meat, indigenous cattle meat, indigenous pig meat, apples, tomatoes, wheat, maize, potatoes.

Brief overview of organic farming

There are 13,564 hectares of organic land in Chile and approximately 550 organic farms operating in the country. Organic land makes up 0.1% of the country's arable land. Most (80%) of the organic good produced in Chile are exported. The government is supporting the organic industry through its Ministry of Agriculture, research is being undertaken, and regulation implemented, in order to improve organic standards. The Swiss Ministry of Agriculture is working in collaboration with the Chilean government to transfer technical support and expertise for an improved organic sector in Chile.

Brief overview of organic products

Most of the country's organic products are exported. Major domestically sold organic products include fresh fruits and vegetables, olive oil and wine. The main products which are exported are fresh fruits and vegetables, olive oil and wine.

Organizational support

Huertos Orgánicos de Chile S.A.

7.4.3.5 COLUMBIA

Location

Colombia is located to the northwest of South America. The country is bordered by Venezuela to the north east, Brazil to the south east, Peru to the south; Ecuador to the southwest and Panama to the northwest. The country has a total land area of 1,138,914 sq km. Climate varies between tropical in the coastal areas, and cooler weather in areas with higher altitude.

Agricultural sector

38,587 hectares of Colombia's land is being managed for organic use. It makes up 0.1% of the country's amount of total arable land. Colombia's top 10 agricultural products are whole fresh cow milk, indigenous cattle meat, sugar cane, indigenous chicken meat, plantains, green coffee, paddy rice, hen eggs, potatoes and bananas.

Brief overview of organic farming

Columbia is in the process of promoting a new organic brand. The different actors in the country's organic market are relatively independent; they do not form one body. They have been involved in the promotion of organic farming for the longest time period are NGOs, certification agencies and producer associations. Other bodies who have recently become actors in the sector include private companies, governmental institutions and universities.

Brief overview of organic products

Major organic products include coffee, tropical fruits, sugar, palm oil, cacao, herbs, medicinal plants, vegetables and basic grains, eggs, milk, yoghurt and cheese.

Organizational support

The National Inter-institutional Committee of Organic Agriculture work on Colombia's organic policies and national programme. Environmental authorities work alongside the National Inter-institutional Committee for Organic Agriculture to support and reinforce the national organic market.

7.4.3.6 ECUADOR

Location

Ecuador is located in the north west of South America. The country is bordered by Colombia to the north, Peru to the east, Chile to the south and the North Pacific Ocean to the west. Ecuador has a total area of 110, 860 sq km. Climate varies with altitude, tends to be cooler in elevated areas, and more tropical along coastlines.

Agricultural sector

Ecuador's main agricultural products include bananas, whole, fresh cow milk, indigenous cattle meat, plantains, indigenous pig meat, sugar cane, cocoa beans and maize.

Brief overview of organic farming

There are 44,661 hectares of organic land in Ecuador and approximately 2,427 organic farms operating in the country. Organic land makes up 0.6% of the country's arable land. The organic market sector in Ecuador includes different groups of actors, namely small farmers and producers possessing large quantities of land. The Ministry of Agriculture also plays a big role as it is recognized as the national authority for organic agriculture.

Brief overview of organic products

The most prominent products of Ecuador's organic market are cocoa, shrimp, banana and coffee. Ecuador also produces products originating from Amazonia such as mushrooms under wild collection, quinoa and other Andean cereals, mango, sugar cane and vegetables.

Organizational support

Ministerio de Agricultura, Ganadería,
Acuacultura y Pesca (MAGAP)

7.4.3.7 MEXICO

Location

Mexico is forms a bridge between North America and South America, located between the United States of America to the North and Guatemala to the south. It also shares its borders with the Gulf of Mexico and the Caribbean Sea to the East; and the North Pacific Ocean to the west. The country has a total area of 1,964,375 sq km and has both topical and desert climate tendencies.

Agricultural sector

Mexico's main agricultural products are indigenous cattle meat, whole fresh cow milk, indigenous chicken, meat, maize, hen eggs, indigenous pig meat, sugar cane, sorghum, oranges and avocados.

Brief overview of organic farming

There are 393,461 hectares of organic land in Mexico and approximately 128,819 organic producers operating in the country.

Organic land makes up 2.86% of the country's arable land. Most farmers in the organic market are small-scale farmers, although the proportion of medium and large-scale farmers is increasing due to appealing export business incentives. Mexico's organic market sector has grown rapidly; approximately 30% per year. Most organic land is used for crop cultivation. Organic land is also used for livestock and beekeeping.

Brief overview of organic products

Mexico is the world's leading coffee producer. Organic products include cacao, avocado, agave, mango, coconut, aloe vera, corn, citrus fruits, honey and sesame. All these exports are mainly produced for exporting purposes.

The network

Cultura Organica Integral, S.A de C.V.
Manuel Nájera

7.4.3.8 PARAGUAY

Location

Paraguay is located towards the south of South America. It is surrounded by Brazil to the east; Argentina to the southwest; and Bolivia to the north. Paraguay has a total area of 406,752 sq km. Climate varies between subtropical and temperate with rainfall tendencies in the east and more semi arid weather in the west.

Agricultural sector

Paraguay's main agricultural products are soybeans, indigenous cattle meat, cassava, indigenous pig meat, whole fresh cow milk, wheat, maize, hen eggs, sugar cane and indigenous chicken meat.

Brief overview of organic farming

There are 17,705 hectares of organic land in Paraguay and approximately 3,490 organic farms operating in the country. Organic land makes up 0.1% of the country's arable land. 90% of organic production is carried out by small farmers. The domestic market for organic products remains small, nevertheless several supermarkets do sell organic products such as sugar and mate tea. The Ministry of Agriculture is working in collaboration with Arasy Organic to strengthen technical capacity in the organic sector.

Brief overview of organic products

Major organic products include sugar; this has helped Paraguay significantly as the sugar market share had been falling due to reduced prices of sugar. The domestic market for organic products is minute but includes vegetables, sugar, mate tea.

The network

Asociación de Productores Orgánicos (APRO)

7.4.3.9 PERU

Location

Peru is located in the north of South America. The country is bordered by Chile to the south; Bolivia to the southeast; Brazil to the west; Colombia to the north; Ecuador to the northwest and the Pacific Ocean to the west. The country has a total land area of 1,285,216 sq km and has a tropical climate in its eastern region, more desert-like weather in the west and temperate weather in regions of higher altitude.

Agricultural sector

Peru's main agricultural products are indigenous chicken meat, paddy, potatoes, plantains, whole fresh cow milk, indigenous cattle meat, asparagus, hen eggs, maize and sugar cane.

Brief overview of organic farming

There are 124,714 hectares of organic land in Peru and approximately 36,093 organic farms operating in the country. The total amount of certified organic hectares in Peru has significantly increased. Organic land makes up 0.6% of the country's arable land. Coffee and cocoa make up most (94%) of Peru's organic exports. Almost all (97%)of Peru's organically produced products are exported.

Brief overview of organic products

Major organic products include coffee and cocoa; the two most prominent organic exports. Banana, quinoa, cotton, pecans, Brazil nut, onions, asparagus, sesame seeds, amaranth and tomato are also organic products which are exported from Peru.

Organizational support

Bio Latina - Certificadora Latino America

7.4.3.10 VENEZUELA

Location

Venezuela is located to the north of South America. The country shares its borders with Guyana to the west; Brazil to the south; Colombia to the west; and the Caribbean Sea to the north. Venezuela has a total land area of 912,050 sq km. Climate is mostly tropical but less so in area of higher altitude.

Agricultural sector

Venezuela's top 10 agricultural products are whole fresh cow milk, indigenous cattle meat, wheat, tomatoes, fresh vegetables, grapes, watermelons, hen eggs, sheep milk and maize.

Brief overview of organic farming

There are 2,441 hectares of organic land in Venezuela and approximately 4 organic farms operating in the country. Organic land makes up 0.01% of the country's arable land. Most organic products which are consumed in the domestic market, are imported, as the country has not yet developed its processing standards for organic products. The country's organic products are mainly exported to the USA and the EU. Amongst the different actors in Venezuela's organic market are private organizations, national research and education organizations, as well as government institutions.

Brief overview of organic products

Organic products include coffee, cocoa, banana, fruits, medicinal plants, macadamia nut, cassava, honey, herbs and spices. Some biscuits are also produced organically.

Organizational support

National Agricultural Research Institute (INIA) is one of the main agencies responsible for the establishment of producer organizations.

7.4.3.11 URUGUAY

Location

Uruguay is located to the southeast of South America. The country shares its borders with Brazil to the northeast; the South Atlantic Ocean to the south and east; and Argentina to its west.

Uruguay has a total land area of 176,215 sq km. Climate is considered to be generally warm and temperate.

Agricultural sector

Uruguay's main agricultural products are indigenous cattle meat, whole fresh cow milk, paddy rice, wheat, soybeans, grapes, greasy wool, indigenous chicken meat, indigenous sheep meat, and sunflower seed.

Brief overview of organic farming

There are 759,000 hectares of organic land in Uruguay and approximately 500 organic farms operating in the country. Organic land makes up 5.1% of the country's arable land. Meat, honey and vegetable are most produced on Uruguay's organic farms. The country's domestic market remains small; organic products are sold via home delivery, farmers markets, on-farms sales and supermarkets.

Brief overview of organic products

Organic exports include meat, wines, honey, rice, milk and citrus fruits.

Associación de Productores Orgánicos del Uruguay (APODU) is a small organic producers association in Uruguay.

7.4.4 EUROPE

7.4.4.1 FRANCE

In contrast with other European countries, organic agriculture in France developed fastest in the nineteen-eighties, particularly in areas of quality level of the products, number of organic farms and level of information amongst the population.

The first standards for organic agriculture were published in 1972 by the producers' association Nature and Progrès.

Legislation on organic agriculture was passed in 1981. The state logo for organic products, the AB-Logo (Agriculture Biologique), was launched in 1985. This state recognition led to the wide acceptance of organic agriculture in France and across its borders and at the time France was the most important European supplier for organic products, and 40 per cent of the European organic land was located in France.

In 1993 the budget provided by the French government for conversion aids was still very modest in comparison with other EU-countries. In the year 1998, however, the French Ministry of Agriculture had allocated 60 million Francs (approximately 9 million Euros).

In December 1997, because of the growing domestic demand (20 percent annual increase), the former agricultural minister, Louis Le Pensec, announced an action plan - Plan Pluriannuel de Development et la Promotion de l'Agriculture Biologique -to support organic production.

The plan involved subsidies totalling 80 million Francs (12,3 million Euro) to stimulate and improve organic production, distribution and sales. The goal required converting one million hectares of farmland and increasing the number of organic producers to 25,000 by the year 2005.

In order to achieve this goal, financial support for farm conversion was increased, and support was extended to marketing initiatives as well as to training and research.

More than 60 percent of the organic land is located in the six regions of Basse Normandie, Bretagne, Pays de la Loire, Languedoc-Roussillon, Midi-Pyrénées and Rhône-Alpes, which account for less than 40 percent of the French agricultural area.

The majority of converted land is grassland, and within organic animal-production a big increase can be recorded.

Until 2000 organic arable land had been growing at a slower rate than grassland. Because of the discrepancy between the development of the animal and plant sectors and the growing demand for organic products on the part of the consumers, the import of organic products from other countries rose by 40 percent in 1998 over the previous year.

In 2000, however, this picture changed: Growth of organic animal production slowed down and the arable land increased faster than the grassland. This change was due to increased payments for cereals and oil seeds.

Organic Agriculture Organizations

The logos of the producer associations are not used for labeling organic products.

The organic agricultural organizations in France enjoy political patronage. Three prominent organizations promote organic agriculture

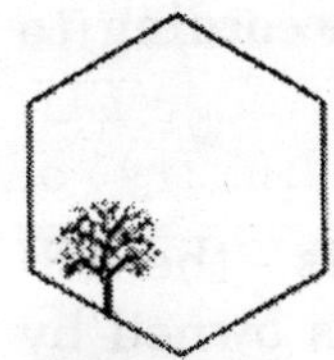

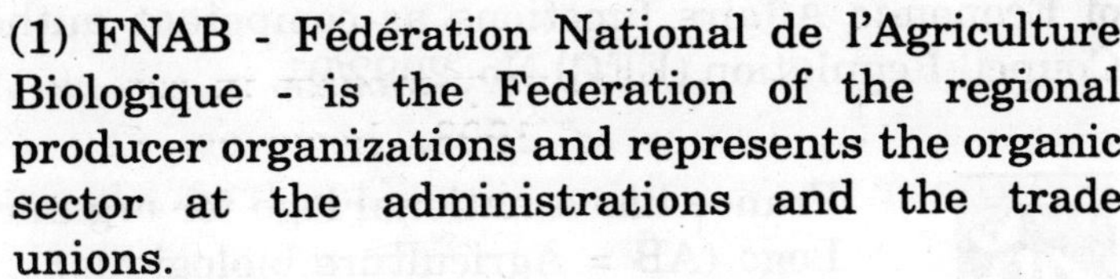

(1) FNAB - Fédération National de l'Agriculture Biologique - is the Federation of the regional producer organizations and represents the organic sector at the administrations and the trade unions.

(2) Bioconvergence and

(3) Syndicat Europèen des Transformateurs de Produits de l'agriculture Biologique (SETRAB). Bioconvergence and SETRAB represent the processors.

All three organizations are competent partners of the Ministry of Agriculture in questions relating to standards, inspection and certification.

The trademark for bio-dynamic products is the Demeter logo (Biodynamie).

Inspection and Certification

Six inspection bodies are accredited according to EN 45011 by the Ministry of Agriculture.

These are:

Ecocert SARL (http://www.ecocert.fr)

- Qualité France
- Ulase
- Agrocert
- Certipaq
- Aclave

These six organizations are authorised to certify organic plant and animal products. The Ministry of Agriculture as well as the Ministry of Economic Affairs functions as competent authorities according to Council Regulation (EEC) No. 2092/91.

France has a national logo for organic products - the AB-Logo (AB = Agriculture biologique) -, which is owned by the French state.

Organic products can be labeled with this logo when they contain more than 95 percent organic components, and were produced or processed within the EU, and were certified by one of the inspection bodies accredited according to EN 45011.

Marketing

The Market

"Organic" is still a niche market in France in terms of value, representing only 0.5 percent of total retail food sales in comparison with other European Union (EU) countries such as Denmark, where the share is three percent. However, the market has been growing at a rate of twelve percent per year, and that rate is expected to reach 20 percent per year in the future. In 1998, sales of organic food products in France were estimated at 700 million Euros. Current French production cannot meet demand (Piason, 1999).

Growing demand, coupled with the government action plan which aims at stimulating domestic production and improve distribution, is expected to boost organic food sales to about 2.6 billion Euro (Piason, 1999).

Product Range

The range of organic foods available in France is still relatively limited. Major domestically produced organic products include grains, prepared cereals, dairy products, fresh and processed fruit and vegetables, baby foods, meat, and poultry products (Piason, 1999).

In 2000, of the new products appearing on the shelves two thirds were of plant origin (baby food, juices, and soups). The new products of animal origin were mainly milk products and sausages (Source: Observatoire national de l'agriculture biologique, Resultats, 2000).

Consumers

Consumers of organic products fall into three main categories: the politically or ideologically motivated who are concerned about the environment and animal rights; the health conscious; and the switchers, who are easily swayed by the media and influenced by price and availability. About half of consumers of organic foods are managers or self-employed professionals between the ages of 25 and 49 (Piason, 1999).

Imports

In 1998, organic imports into France were valued at 50 million Euro. The main organic imports from non-EU countries included exotic fruits (avocado, mango, and bananas), coffee, tea, citrus fruits and cereals from African, South American and Asian countries as well as from North America.

From countries within the EU, fruit juices, dried fruit, cereal bars, muesli and other processed, new-to-market products such as ready-to-eat frozen foods and vegetarian meals are imported. Within the last few years, milk and meat came from Germany as the home production still was not high enough. Besides Germany, the Scandinavian countries were the predominant country sources, although 40 per cent of the imports came from non-European countries. Total imports accounted for about ten percent of the organic food market (Piason, 1999).

Exports

Exports represent about 17 per cent of all French organic products and consist mainly of value added grain products. The major French organic product exports are bulk cereals to Germany and Scandinavian countries. Most exports of dairy products, meat and poultry, and fruit and vegetables go to EU neighboring countries (GAIN-Report # FR 9070/ Piason 1999)).

Marketing Channels

Sales via small specialized natural food and health food shops were important until 1990. Today, however, supermarkets are gaining more and more importance for the marketing of organic products. Furthermore, the Superettes, small specialized bio-supermarkets with self-service and a surface of 200 to 500 square meters, partially replace, but also supplement the small specialized shops. Many of them and other sales persons are members of the association Biocoop which organizes the distribution of organic produce in France. It has 170 selling points in the whole country and gave itself special standards. Biocoop is the successor organization of Food-Coop since 1987.

Today, nearly half of the organic food is sold through supermarket chains, whereas the rest is sold through health food stores, direct sales, and open-air organic food markets. Retail prices for organic products are on average 25 to 35 per cent higher than conventional foods.

Most of the supermarket chains have their own organic food label, carrying an array of products from dry foods to dairy products to meat and fresh products.

7.4.4.2 DENMARK

Denmark is situated in the northern part of EU and is one of the Nordic and Scandinavian countries. It is a small country with a population of 5.3 million inhabitants, and the GNP per capita is approximately 35 000 USD.

Denmark is a flat country with rich agricultural land situated in a temperate climate. Summers are generally warm with an average temperature of 16.4 degrees centigrade and freezing temperatures are seldom experienced in winter for prolonged periods. In total, 664 mm of precipitation fall evenly spread over the year.

2.7 million hectares or sixty-three percent of Denmark's land area is cultivated farmland. Grain crops amount to over half of Denmark's agricultural production with wheat, barley and rye being the most widespread. Roughage - beets and grass - are also grown. Two thirds of production is utilized as fodder for animals.

23 million pigs are produced in Denmark each year, three-quarters of these for export. This constitutes 7 per cent of Denmark's total annual export. From the beginning of the 1980s the production of pigs has risen by almost 50 per cent to c. 1.8 million tonnes of pork in 2000. Over the same period milk production fell by 15 per cent to 4.7 million tonnes, partly as a result of the European Union's introduction of milk quotas.

Denmark is one of the top-ten countries in Europe as regards the organic share of the total cultivated area. There were 3714 organic farms cultivating approx. 180,000 hectares in 2002 corresponding to 6.7 percent of the total Danish farmland

Government support

Substantial government support has led to increased organic production. In 1998, the support accounted for 58 per cent of the total green farming grants, compared to only 7 per cent of grants in the United Kingdom. In addition to heavy financial support to organic farmers, the Danish government also discouraged conventional farming by levying high taxes on products such as insecticides and pesticides.

Table 7.3 Number of organic farms and farmland (1989-2002)

Year	Farms	% of all farms	Hectares	% of all agricultural land
1989	401		9554	
1990	523		11 581	
1991	672		17 963	
1992	675		18 653	
1993	640		20 090	
1994	677		21 145	
1995	1050	1.5	40 884	1.5
1996	1166	1.7	46 171	1.7
1997	1617	2.5	64 329	2.4
1998	2228	3.5	99 163	3.7
1999	3099	5.2	146 685	5.5
2000	3466	6.4	165 258	6,2
2001	3525	6.5	173 497	6.5
2002	3714	7.3	178 360	6.7

All EU countries are subject to the same rules for converting a conventional farm into an organic one. However, in Denmark the whole farm must be converted, whereas in other EU countries it is possible only to convert part of the farm. The system of funding to organic farmers is shown in the table. The financial assistance is restricted to per hectare 5,000 Danish kroner.

The Development of Organic Farming in Denmark

Organic farming has its roots in alternative farming systems, and these systems have existed for many years both in Denmark and other countries around the world, questioning whether intensive agriculture, which uses artificial fertilizers and sprays to provide the greatest possible yield, is the best way to produce foods that promote human health. Further, the impact of the production method on the surrounding environment is included as a parameter of quality.

Roughly, the alternative systems have been significant in two periods in modern times. The first period was 1920-1940, and the second period was from 1960 and onwards.

In the 1920´s Denmark experienced a widespread interest in natural living and natural foods, not least influenced by the biodynamic agricultural system from Germany. In 1936 The Biodynamic Association was established by influential landowners from the aristocracy

Highlights - Organic Production in Denmark

- 1936 The Biodynamic Association and Demeter Association were established.
- 1972 An umbrella organization, the International Federation of Organic Agriculture Movements (IFOAM), was set up for societies working towards alternative agriculture.
- 1981 The Danish Association of Organic Farming was founded and a number of specific rules were formulated. The association sets up its own inspectorate.
- 1982 The first Danish organic agricultural college was set up in northern Jutland.
- 1985 The first organic agricultural advisory service was set up.
- 1987 The Danish parliament adopted the world's first comprehensive legislation on organic farming. State inspection and certification scheme were introduced.
- 1988 The first litre of organic milk was bottled at a small dairy.
- 1990 Launch of a national campaign for organic agriculture.
- 1993 The largest Danish supermarket chain, Coop Denmark, reduced prices by 15 to 20 per cent on a large number of organic products. Boom in consumption.
- 1993 General economical support for organic farming was introduced
- 1995 The Danish Ministry of Food, Agriculture and Fisheries presented a comprehensive Action Plan-I designed to propel organic farming towards the year 2000.
- 1996 - The 11th IFOAM World Conference is held in Denmark
- The Danish Ministry of Food, Agriculture and Fisheries introduced an increased support to plant producers

- The Danish Research Centre for Organic Farming was established, and the Organic research station Rugballegaard was established.
- 1998 The number of organic farms tripled over a five-year period.
- 1999 Organic organizations establish co-operation in the Centre for Organic Agriculture. Action plan II was introduced.
- 2000 Organic e-commerce was launched by the company "Aarstiderne" – www.arstiderne.com.
- 2001 Denmark hosted the European Organic Food and Farming Conference

Inspection and Labeling

Denmark is exceptional in having an official set of regulations and a single unique symbol for organic products, and also in that the State undertakes inspections. In general, Danes have great confidence in the State as a serious and neutral body of inspection and labeling.

All farmers who practice organic farming must be authorised in organic management. Amongst other things, in collaboration with an organic agricultural adviser, the farmer must set out a plan for converting to organic management. Authorization is not granted until the farm has been inspected and the conversion plan accepted by the Plant Directorate.

In addition, a very comprehensive control of organic production is carried out at least once a year. To strengthen the control of organic products The Danish Veterinary and Food Administration also makes crosschecks. Each year a random selection of companies is made, and their accounts are compared with those of their suppliers and customers. This process establishes whether the amount of organic products bought and sold tallies between companies

The "Ø"-label is an inspection label launched in 1990. The regulations associated with the Ø label are based on EU legislation - although Danish rules still apply in a few areas because EU legislation still does not cover all aspects of organic activities.

Fundamentally, the red Ø label signifies that the Danish authorities have carried out a control on the farms and work places that produce, process, package or label the goods in Denmark. The red Ø-label shows that the latest preparation of the organic product has taken place in a Danish company under inspection of the public authorities. Therefore, the logo can be seen both on foods of Danish origin and on imported foods processed or packaged and labeled in Denmark

Recently a new EU label was introduced which signifies that an organic product is produced and controlled within the EU. It is the long-term objective of the Danish authorities to support the implementation of the EU label.

Advisory System

Denmark has a comprehensive and well-organized agriculture advisory system, which comprises approximately 60 independent local advisory centers and one joint advisory centre, The Danish Agricultural Advisory Service, National Centre.

Dansk Landbrugsrådgivning
Landscentret

Today approx. 150 advisers (the equivalent of 50 full-time jobs) provide advice on organic farming. Besides providing organic farmer with the latest information, the organic advisers supply information to conventional farmers who are gradually being inspired by organic methods of production, e.g. the increasing use of clover and grass on dairy and arable farms.

At the National Centre specialists co-ordinate the advisory development within organic farming systems. This is done by an internal group of 14 specialists representing the various professional fields. The farmers association, Danish Agriculture owns the National Centre.

Research

Danish research in organic farming has increased considerably since 1995, but there is still a need for targeted research activities to help promote and develop organic farming.

Many of the concerns of farming such as consideration of the environment and nature, animal welfare, product quality and health are all fundamental aspects of organic farming. The promotion of organic farming has been part of Danish government policy for several years. A major initiative in this respect has been the establishment of the Danish Research Centre for Organic Farming in 1996. The objective of DARCOF is to initiate and co-ordinate Danish research in organic farming. The centre synthesizes and communicates scientific information across traditional boundaries and disciplines.

Research concerning all aspects of organic farming and food processing has been initiated by and carried out by various Danish research institutions, e.g. The Danish Institute of Agricultural Sciences, National Environment Research Institute, Denmark, The Royal Veterinary and Agricultural University, Technical University of Denmark, Danish Institute of Agricultural and Fisheries Economics and Danish Agricultural Advisory Service, National Centre.

Under the auspices of DARCOF an open access archive, Organic Eprints, for electronic documents related to organic agriculture was established in 2002. Researchers and organizations are invited to join the archive.

Organisations

The successful development in production and sales of organic products is largely based on a consistent and fruitful co-operation between the many specialized organizations and agencies within Danish organic production and manufacturing.

GOVERNMENTAL ORGANISATIONS

The Organic Food Council

Appointed in 1987. The aim of the Council is to encourage, monitor and assess the opportunities to develop Danish organic food production, to assess the current advisory and research work, to formulate proposals for additional activities and to comment on standards for the control of production, marketing, storage, transport, labeling, distribution and retailing of organic goods.

The Danish Directorate for Food, Fisheries and Agro Business

Under the Danish Ministry of Food, Agriculture and Fisheries. It administers EU market schemes on farm products, fruit and horticultural products in addition to fish produce. The directorate is the sole Danish payment agency, which takes care of all payments financed by the European Agricultural Guidance and Guarantee Fund (EAGGF). Deals within organic agriculture with national strategy issues, product development issues, and equivalence agreements and pending negotiations with foreign certification bodies, Action Plans on organic farming and EU-related issues.

Danish Plant Directorate

Under the Danish Ministry of Food, Agriculture and Fisheries. The assignments of the Danish Plant Directorate cover legislation on seeds, feeding stuffs, Plants, potatoes, fruit and vegetables, use of fertilizers and the establishment of vegetation cover, EU agricultural schemes and organic farming. The Danish Plant Directorate lays down regulations, performs administrative functions, carries out inspections, e.g. on organic farms, prepares legislation, provides service to the authorities and prepares policies. It has several offices across Denmark. The inspections on organic farms take place when crops are still in the fields so that the inspectors can verify that everything is up to standard.

The Danish Veterinary and Food Administration

It is part of the Ministry of Food, Agriculture and Fisheries. The Food Administration accredits and checks the processing and trading of organic food products. Eleven regional veterinary and food control centers handle food control and veterinary inspection.

NON-GOVERNMENTAL ORGANISATIONS

Danish Agricultural Council

It is a joint committee for various professional farmers' associations and the food industry. Prime objective is to further the co-operation within the business in relation to the Government, the Danish Parliament, the central administration, the authorities, and other domestic and international trade organizations

Danish Agriculture

In 2003 the two previous central organizations, the Danish Farmers' Unions and the Danish Family Farmers' Association merged into one organization, Danish Agriculture. The aim is - in co-operation with the local farmers associations - to take care of Danish farmers business, political, social and cultural interests. Deals with organic farmers issues in a special committee for organic farmers.

The Danish Association of Organic Agriculture

Established in 1981. A merger between various organizations in 2002 led to formation of one association for organic farmers, consumers and companies. It publishes a magazine in Danish named Organic Farming. The association disseminates information about organic farming and its products and contributes in coordinated marketing efforts for organic products.

The Biodynamic Association and Demeter Association

Established in 1936, representing the Biodynamic farmers. Publishes a magazine in Danish named Biodynamisk Jordbrug.

The Danish Consumer Council

It represents the interests of consumers and is independent of public authorities and commercial interests. Founded in 1947, the Consumer Council is the spokesperson for consumers' interests, lobbying vis-à-vis the Government, the Danish Parliament, public authorities and the business community.

Network for Ecological Education and Practice

Founded in 1994, with the objective to establish a foundation for information, exchange of experience and dialogue about initiatives in promoting a sustainable development in Denmark. This is done by informing about meetings, courses etc., and by establishing connections between projects and people. The Network publishes the newsletter Eco-net Newsletter in Denmark.

The Danish Society of Practical Ecology

Association of organic gardening.

Research and Advisory Service

Danish Agricultural Advisory Service, National Centre

The centre provides the local advisers with the latest information, which forms the basis of the farmer advising. The specialist staff renders advice in nine subject areas. The Department for Organic Farming at the National Centre co-ordinates the advisory development within this field.

The Organic Agricultural School

Self-governing institution established in 1992. It offers agricultural education module, professional/technical education, manager course, and extended manager course. It also offers additional training courses and management consultant- and counseling functions related to organic farming projects.

The Royal Veterinary and Agricultural University

The Department of Agricultural Sciences

Founded in 1856. Research is carried out within the fields of natural science, veterinary medicine, animal science, environmental science, agriculture, horticulture, landscape architecture and landscape ecology, forestry, food sciences and human nutrition. The Department of Agricultural Sciences at The Agricultural University works mainly with research and teaching in areas related to plant breeding, crop science, plant nutrition, and the qualitative and quantitative aspects of agricultural and horticultural crop production in temperate and warmer climate zones. The Organic Farming Unit is responsible for the development of teaching activities and for the promotion of research activities within organic agriculture at the Department of Agricultural Sciences.

Technical University of Denmark

The Institute of Environment and Resources

E&R aims to develop technical and sustainable solutions to minimize the impact of society on the hydrologic and geologic environment. The institute is committed to research and teaching at an international level.

The Danish Research Centre for Organic Farming

DARCOF was established in 1995 as a so-called "centre without walls" where the actual research is performed in interdisciplinary collaboration between the participating research groups. The aim of DARCOF is to co-ordinate research for organic farming, with a view to achieving optimum benefit from the allocated resources. Its aim is also to elucidate the ideas and problems faced in organic farming through the promotion of high quality research of international standard.

Danish Institute of Agricultural Sciences

DIAS was established in 1997 and is a sector research institution under the Ministry of Food, Agriculture and Fisheries. Aims to conduct research and accumulate knowledge of importance to agriculture. Emphasis is on responsible resource utilization, environmental impacts both internal and external, animal welfare, and the quality and competitiveness of the products.

Danish Institute of Agricultural and Fisheries Economics

Conducts research in economics of organic farming.

7.4.4.3 BELGIUM

Belgium is one of the smaller countries of the European Union, surrounded by the Netherlands, Luxemburg, Germany and France and, with the Channel in between, the UK.

Belgium has a surface of 30,528 km² and 10.5 million habitants. This results in a very high population density (344/km² in 2006). Belgium is divided in two principal regions, Flanders and Wallonia, both with their own official language: Dutch and French. There is also a small German speaking minority.

Today agriculture is of minor importance: Only 1% of the gross national product GNP is realized through agriculture. The total agricultural area is about 13,71,955 hectares (in 2007) with 47'936 farms.

In 2004, Belgian families spent 15.8% of their budget on food, drinks and tobacco. That makes food, drinks and tobacco the second most important group. Belgium uses the Euro as currency.

Status of organic sector

The Belgian organic sector witnessed important growth in the late nineties until 2001. After five years of fall and stabilization, the sector has grown again slowly since 2006. This evolution should however not be seen apart from the general evolution in agriculture: the number of farms tends to decline every year (in 2007 by 4%) as well as the agricultural area (minus 0.8% in 2007). From that point of view, the growth in organic is rather remarkable.

The regional differences in the organic sector in Belgium are quite important: Flanders has only 0.6% of its cultivable area under organic (in 2007) whereas Wallonia comes up to 3.9% (2007). As a whole, Belgium achieves 2% which is comparable to the Netherlands, where the organic area has a share of 2,5%.

The weak growth in organic area in Belgium has several causes. The intensive character of the agriculture in Belgium, and especially in Flanders makes the conversion difficult. Belgian agriculture demands a lot of (expensive) work and capital; organic agriculture is even more demanding, and the conversion is financially difficult to bear.

Moreover, the country is very small, quite urbanized (especially Flanders) and import from the surrounding countries (like France, Germany, the Netherlands) is very common. The Belgian population is rather conservative and is not easily convinced to change its buying habits.

Finally, until 2007 the most important general farmers' organization was only marginally interested in organic production and did not stimulate conversion.

Since the early 1990 several non-profit organizations are working in the field to promote organic agriculture: Velt, Belbior, Probila-Unitrab, Unab, Nature & Progrès and others. In order to strengthen their actions, they professionalized and some of them united in Flanders into one organization - BioForum Vlaanderen - in 2008.

The demand for organic products seems to be growing. In 2007, domestic sales were evaluated at about 283 million Euros. The organic sector has a share of approximately 1.87% of food sales.

Politically, organic production gains more and more recognition and has been supported over the last few years by several Ministers of Agriculture. However it has been considered more as an experimental agriculture, interesting enough to try out techniques that can stimulate agriculture in general to become more environmental friendly, but not as a full alternative. In June 2008, the Flemish government started a new action plan to stimulate organic agriculture.

The most important farmers' organization finally seems to be interested in organic agriculture, which could make a big difference in the number of conversions. In Wallonia the growth of the organic area is obvious.

Legislation and government support

Before the government showed interest in the organic sector, a group of pioneers organized themselves. They wrote the standards for organic farming, developed a label 'Biogarantie' and organized inspection and certification.

In Belgium, the use of the word 'biologics' or 'biologique', which means 'organic', has been legally protected since 1991 (European regulation 2092/91) and 1994 (Belgian standards). One cannot use the word 'biologisch' (nor one of its translations) on food products without complying with the legal standards and being controlled by one of the two accredited control bodies. Only their certificate gives the authorization to place organic products on the market.

Companies that wish to market organic products have to bear the control cost themselves.

The Belgian organic sector has to respect legal standards that are equal for the whole European Union.

Information on the European legislation as well as the legislation itself can be found at the Organic Europe homepage. Information on the Belgian legislation can be found, in French, at: http://www.integra-bvba.be/Frans/f-framebio.html (look at 'notice explicative')

Labeling

Most organic products bear the Belgian label Biogarantie. This label is a private label existing since more than 25 years. Being private, it can only be used after having received an additional certificate and after payment of the royalties. The label is owned by the non-profit organization Biogarantie vzw. During 2008, BioForum will become the licence holder of the label.

However, this label is not compulsory. Some organic products only mention the word 'biologisch' or 'biologique', with additionally the name or code of the inspection body. With the words 'biologisch' and 'biologique' being legally protected, they guarantee the organic character of the concerned product.

There are quite a lot of foreign organic products on the Belgian market, with one or more labels. The current European label for organic production is not very common due to some additional conditions.

Most of the European countries have their own organic label. In Belgium, the most known are 'EKO' from the Netherlands and 'AB' from France.

Foreign organic products can obtain the Belgian label Biogarantie as soon as they have been certified as equivalent. This will facilitate recognition by the Belgian consumer.

Some products comply with the standards of the organic-dynamic movement. They are considered to be the pioneers of the organic sector and respect an additional range of standards. Their label 'Demeter' is the same in the whole world. The Demeter-label and standards are developed by Demeter International. In Belgium, only few farmers follow Demeter. Demeter products are only sold in specialized organic shops.

Organic area

The organic area has witnessed the growth in the late nineties (1998 - 2000), partly due to some food scandals. At that time organizations and a single minister had the ambition of reaching 10% organic area in 2010. However, after a short period of expansion (until the end of 2001), organic production lost much ground. By the end of 2007, the organic area had recovered and counted for 2.4% of the total agricultural area.

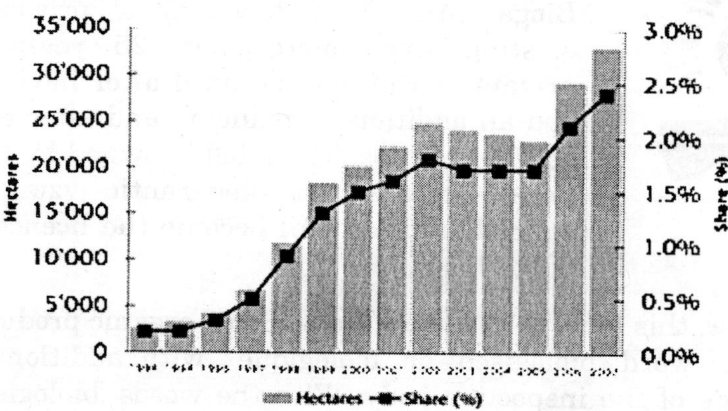

Market and product rage

According to BioForum Vlaanderen, organic sales reached an amount of 245 million Euros in 2006 and of 283 million Euros in 2007.

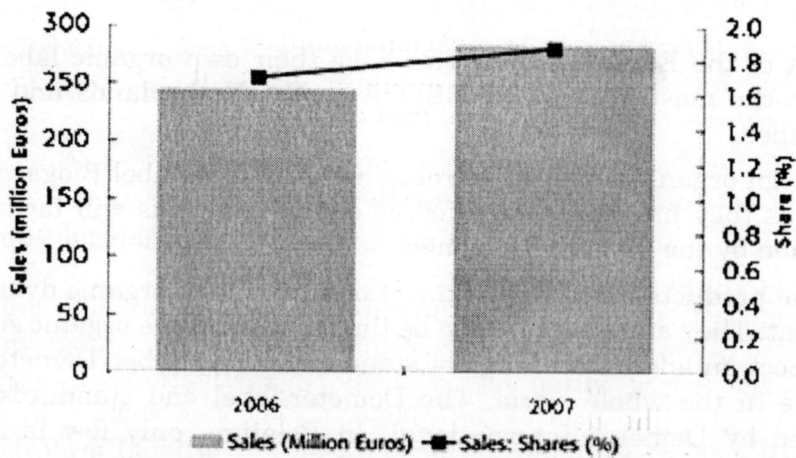

Table 7.4. Product range and their market share

Product group	Market share in 2006 (%)
Eggs	7
Rice	3.9
Marmalade	3.4
Flower	3.2
Vegetables	3.1
Pasta	2.7
Bread	2.3
Fruit	1.6
Dairy products	1.1
Potatoes	1
Poultry	1
Meat	0.7

Marketing channels

On the Belgian market, food is sold in supermarkets and discounter, groceries, bakeries, at the butcher's shop, on the market and at farms.

Organic food is nowadays available in many different ways. We can distinguish supermarkets, specialized organic shops, market sales and farm shops. Unlike the common food market, there are very few organic bakeries and butcher's shops.

Organizations

Politically, agriculture is a regional matter; this means that the politics on organic agriculture is quite different in the Flemish (northern) and Walloon (southern) part of Belgium.

The Flemish government has, since a few years, stimulated organic agriculture by means of an action plan.

This action plan includes among other things, financial support for organic farmers and for the umbrella organization BioForum Vlaanderen, actions to manage the supply chain, to promote organic products, to integrate organic agriculture in educational programs, to improve the coordination of scientific investigation on organic agriculture, etc. The new action plan for 2008 was written in close cooperation with the organic sector and the conventional professional organizations.

For organic processors or distributors, there is no specific support.

The Walloon government is favorable towards organic agriculture and they strongly hold on to the principles of organic agriculture. However, the support for the organic sector is rather minimal and is focused more on supporting the professional market channel. The interest group BioForum Wallonia receives less support than in Flanders.

Among the organizations that stimulate organic consumption and production, the following are worth consideration:

BioForum

BioForum is the umbrella organization of the whole organic sector and unites farmers, processors and retailers. BioForum is separated in two regional chambers that operate distinctively due to the different approach of the regional government.

Both offer information to all professionals as well as consumers, organize several promotional campaigns (for example 'Bioweek'), follow up the legislation etc. BioForum is member of the international group IFOAM.

Velt : Velt is one of the most known 'amateur organic gardeners' organizations that promote an ecological way of life. Eating (local) organics is one of their mottos. Their action radius is Flanders.

Nature and Progrèss : is a similar organization to Velt but operates mainly in Wallonia.

Belbior: used to be the Flemish organic farmers organization and Probila-Unitrab the national professional organization for organic processors. However, both organizations merged with BioForum Vlaanderen during spring 2008.

Unab : is the Walloon organic farmers organization.

Naredi : is the national federation for food supplements and dietary products. They support both organic and non-organic supplements.

Vredeseilanden : is an development organization that battles for sustainable and fair agriculture in the South and in Belgium. In

Belgium their actions mainly concentrate on organic agriculture. They organize organic fruit supply in elementary schools (www.biofruitopschool.be) and support the production of products that bear both fair trade and organic labels.

Bond Beter Leefmilieu : is the umbrella organization of the environment action groups. They support organic agriculture for being more environment-friendly.

VLAM : is a division of the Ministry of Agriculture and Fishery responsible for the promotion of all farm products. They also promote organic products.

7.4.4.4 ESTONIA

The development of Estonian organic farming began in 1989,when the Estonian Biodynamic Association (EBÜ) was founded. The Association used IFOAM (International Federation of Organic Agriculture Movements) standards to work out the first Estonian organic agriculture standards and started to use the trademark "ÖKO“ and also started to control the producers. In 1992 the regional organization South-East Estonian Bios was established. Later it acted also as control body. In the beginning of the 90's several local organic organizations were founded (in Võru, Saare, Lääne and Viljandi County).

The year of 1997, when the first Estonian Organic Farming Act came into force, can be seen as the beginning of a new phase of development. The Centre for Ecological Engineering started actively organizing educational events, published informational brochures and brought to life development projects. In 2000 the Estonian Organic Farming Foundation was created and has been very active in developing organic farming. The Agri-Environment Bureau was founded by the Ministry of Agriculture in 2000. The bureau was responsible for the oversight of organic farming. From 2004 the Organic Agriculture Bureau of Plant Health Department assumed oversight responsibilities. The first (and still the only) organic producers cooperative was founded in 2003, called Eesti Maheliha (Estonian Organic Meat). As the name suggests, the initial aim of the cooperative was to bring into the marketplace local organic meat products. This aim has been realized in 2008. In the course of time the union's vision widened and they started to include other food groups (grain products, vegetables and fruit). With a current membership of nearly 80, the organization has been renamed Eesti Mahe (Estonian Organic) in 2007. Recent years have seen the creation of several new local production organizations – in Saare, Hiiu, Harju and Pärnu County. In July 2006, eight organic farming organizations founded the

Platform of Estonian Organic Farming Organisations, the main aim of which is to develop organic farming sector. All the active Estonian organic farming organizations took part in founding the platform: the Estonian Biodynamic Association, the Estonian Organic Meat, the Estonian Organic Farming Foundation, Harju Organic Farmer's Association, Hiiu Organic, Läänemaa Organic Farmers' Society, Saare Organic and the Centre for Ecological Engineering. The newly founded Pärnu Organic joined the platform in 2007.

The Estonian Organic Farming Foundation and The Estonian Biodynamic Association are also members of international organic farming organization IFOAM. Information about organic farming can be found from many sources. With the support of the Ministry of Agriculture and the European Union several publications about organic production and marketing have been published, mainly put together by the Estonian Organic Farming Foundation and the Centre for Ecological Engineering. The latter organization has also been publishing a quarterly organic magazine "Mahepõllumajanduse leht" since 1996, recently with the support of the Ministry of Agriculture. Organic producers are offered free courses. For those who apply for organic farming support money it is mandatory to take part in the course. Producers can reach out for help from organic advisors, but unfortunately there are no advisers for processors. Very little scientific research has been conducted in the organic arena in Estonia. The few studies that exist have been carried out by the Agricultural Research Centre, the Estonian University of Life Sciences, the Jõgeva Plant Breeding Institute and the Estonian Research Institute of Agriculture. Organic production has grown rapidly, one of the reasons being the financial support given per organic hectare since the year 2000. By 2008 organic land (87 000 ha) was close to 10% of all agricultural land in use, with 1245 organic producers.

The size of organic farmland has expanded from year to year, to an average area of 70 hectares. Estonia's largest organic farm has over 1000 hectares of land. The largest number of organic producers is in Võru County, but the largest amount of organic land is in Saaremaa. Organic farming is also widespread in Viljandi, Hiiu and Lääne County. In Hiiu County nearly two third of all agricultural land is organic. In comparison to plant and animal production, the organic processing and marketing development has been modest. In 2008 the organic farming register had total of 32 organic processors. Organic products reach consumers mostly via direct sales and specialist organic shops, a few products are on sale in conventional stores. Consumer interest to purchase organic food has increased considerably. The areas of organic processing and marketing will undoubtedly develop soon, making organic food much more widely available, and this is a main aim of the Organic Farming Action Plan 2007-2013.

Organic production

In 2007 there were 81,528 ha of organic agricultural land in Estonia. In 2008 organic land has increased to 87,347 ha, out of which 71,849 ha (82%) had gone through the conversion period. As characteristic of organic agriculture, large areas were grassland (81%). This is not unique to Estonia and is found in many countries. 9265 ha were covered by organic grains, showing a growth of 2,5 times in four years (compared to 2004). The most popular crops were oats and barley (60% of total area covered by grain), which were mostly grown for animal feed. Spelt wheat (the most sought after grain by consumers, especially as flour or flakes) was grown in 89 ha. The area covered by industrial crops (mostly oil seed rape) enlarged almost eight times in four years. At the same time the amount of organic industrial crops grown is still very small - in 2008 it was only 336 ha. Unfortunately, organic vegetable growing has not expanded, and the current quantity and selection of vegetables is not enough to meet the consumer demand. Small yields and the large amount of handwork required are significant factors. In comparison to 2004 the area where potatoes were grown has even slightly decreased. The most popular fruits grown are apples. Plums and cherries are grown as well. It is worth noting that among berries the area of organic sea-buckthorn has nearly quadrupled in four years (366 ha in 2008). The first organic sea-buckthorn products are already for sale.

White mustard takes up the largest area of culinary and medicinal plants. Herbs are picked also from uncultivated organic land, 79 plant species from 247 ha of land in 2008.

Legislation

The basic rules of organic farming are uniform in all European Union member states. From January 1st, 2009 the new EU Organic Farming Regulations (EC) No 834/2007 and (EC) No 889/2008 came into effect. The main purpose for updating it was to verbalize the most important principles of organic farming more clearly and comprehensively while at the same time allowing greater flexibility than the current regulation for taking into account local conditions. On the national level organic farming is regulated by the Estonian Organic Farming Act and ordinances associated with it. Extensive EU organic regulation provides detailed rules for production, processing, marketing, control and labeling. Estonian legal acts mainly specify control and labeling. References to the EU Regulations and Estonian Legal Acts, and also general organic farming information can be found on the Ministry of Agriculture website www.agri.ee and from the Plant Production Inspectorate's website www.plant.agri.ee.

Organizational support

(1) Ministry of Agriculture, Organic Agriculture Bureau
(2) Plant Production Inspectorate Organic Agriculture Department
(3) Veterinary and Food Board, Office for retail sales, organic farming and food of non-animal origin
(4) Estonian Agricultural Registers and Information Board. Tel: 737 1200, e-mail: pria@pria.ee
(5) Estonian Organic Farming Foundation
(6) Centre for Ecological Engineering
(7) Estonian Biodynamic Association
(8) County Organic Farmers' Society
(9) Harju Organic Farmers' Association

7.5 SCENARIO OF ORGANIC FARMING IN INDIA

As compared to other developed and many developing countries, Indian agriculture is more focused on production of crops in arable land than production of animal products (except milk). This is due to vegetarian food habits of India and religious/political/social limitations on the use of meat, pork and other animal products. Loss of pasture lands and forest lands coupled with acute population pressure on arable land also laid focus on production of crops than animal products. In this background, the adoption and spread of organic cultivation are restricted to production of crops than animal products.

7.5.1 Back ground

It is commonly held by Indian agricultural scientists, farmers, policy makers and administrators that India is the origin of organic farming. The history of many organic farming practices like sequential cropping, green manuring and use of cattle manures is indeed deeply rooted in the history Indian agriculture dating back to 3500 B.C. Till late 1950s, conventional agriculture adopted by Indian farmers was organic by default, as it was free of chemicals, Even now, more than 50 percent farmers adopt chemical free agriculture not by option but by limitation of capital.

But, the conventional agriculture or ancient agriculture in India was fully organic neither in the sprit nor in the scope of modern organic farming. Many components of modern organic farming like use of biofertilizers, use of bio-control agents for pest control, preparation of vermicompost, modern methods of enriched compost preparation were completely absent. Moreover, the chemical free agriculture itself is not organic agriculture. The perception to adopt organic farming as a an effective means of improving the biological activities, as well as holistic nature of organic farming were never part of ancient/conventional agriculture. In this background, a clear distinctive knowledge between ancient/conventional chemical free agriculture and modern organic agriculture is absolutely essential before a scenario of organic farming in India is studied.

Although many present day practices of organic farming were evolved and followed in India since ancient times, they were not practiced in a module with specific aim either to improve the soil fertility or to produce pollution free food. In Europe and USA, organic farming was evolved as a module in the name of alternate agriculture (as an alternative to chemical farming), which later developed in to modern organic farming. In India, the large scale use of chemicals in agriculture was started in late 70s and early 80s, while in USA and Europe by then the chemical load on land and resultant pollution were already intensified. Hence, evolution of alternative agriculture was inevitable to escape out of pollution in USA and Europe. But such situation did not arise in India. Perhaps in this background, the necessity to conceive organic farming as a module to produce pollution free food and to save the eco-system from pollution did not arise. The conventional practices like green manuring, intercropping and use decomposed animal manure (Farm Yard Manure) remained as isolated practices adopted along with chemical farming practices. Many innovative indigenous practices followed in India up to 16th century (like use of plant extracts) inexplicably lost their place in Indian agriculture till 2000s.

For these reasons, the organic farming as is being conceived in global circles was never practiced in India before advent of chemical farming and even up to 2003. However, the practices like use of manures, green manures were exclusively adopted till 1970s and in later period till 2000, their importance gradually faded in the era of intensive use of fertilizers. In this context, the planning commission recommendations of 1974 and recommendations of state agricultural universities from 1970s to 2000 to use manure along with fertilizer (integrated nutrient management) were not adopted seriously on large scale. It is indeed ironical that focus of recommendations on INM intensified in these years only to synchronize with declining popularity of use of manures. In the light of such development in the last 40-50 years as well as in the light of pro-fertilizer/pro-chemical oriented agricultural research, most present day agricultural scientists believed that organic farming is equal to use of manures (or adopting conventional farming similar to non-chemical farming of the non chemical area) in variance from the holistic meaning of organic agriculture.

7.5.2 Origin of modern organic farming in India

The common feature of India and many other developed countries in respect of modern organic farming is that it originated at farmers' level and later the consumers, the respective governments and R & D organizations improved/refined the its shape at latter stages. Even in India, many innovative farmers deliberately shaped the modern organic farming by practical knowledge of natural laws in agriculture. These innovative organic farmers had obscure but profound ideas of how nature and natural laws are responsible to sustain the agriculture production. In this sense, they redefined the objectives of organic agriculture to improve the soil fertility and therefore sustained productivity.

Although such farmers practiced organic farming many decades prior to chemical farming, they remained unidentified and unrecognized by the modern world. From 1970 onwards, their efforts to adopt and propagate such farming practice were practically eroded due to government sponsored program to promote chemicals. Till late 1990s, the modern organic farming was not even perceived and adopted in India, although few innovative farmers resisted the use of chemical but continued age old farming devoid of latest practices of organic farming.

During 2000, the planning commission constituted a steering group on agriculture, who identified and recommended organic farming as a major thrust area for 10th plan. Originally, it was recommended as a practice suitable for selected areas, where the use of agro-chemicals was substantially less. Later, national agriculture policy adopted in 2000 recommended the promotion of traditional knowledge of agriculture relating to organic farming and it's scientific up gradation. The Department of Agriculture and Cooperation constituted, during 2000, a task force on organic farming under the chairmanship of Shri Kunwar J Bahi Yadav, which also recommended promotion of organic farming. But, neither of these steps really resulted in any nationwide action in propagating the organic farming.

By noticing immense opportunities to export organic products to developed countries, Ministry of commerce launched National Organic Programme and appointed Agricultural and Processed Food Products Export Development Authority (APEDA) as implementing authority of national organic programme. APEDA formulated and implemented National Programme on Organic Production (NPOP), where in standards of organic products, certification/accreditation criteria and modalities of certification were finalized. By this time, many European and American economies, whose demand for organic food was very high, were looking for organic imports from such countries where load of agro-chemical is relatively less. They had already established certification norms and many European/American certifying agencies opened their offices in India to certify the organic production for export. The synchronization of great export demand for organic products and programmes under NPOP achieved the spread of organic farming from 2001 onward. The major incentive for global market was less load of chemicals on Indian soils and thereby production of pollution free organic food was naturally easy.

Thus, modern organic farming entered India as a business opportunity to export their organic foods through APEDA and less as viable means to achieve sustainability. In this sense, early adopters of organic farming in India more stringently followed certification requirements than following its holistic approach to achieve improved biological activity in soil. In promoting organic farming as an avenue of export, NPOP programme developed standards for organic production and laid down certification standards besides facilitating the whole process of organic exports. They mainly adopted IFOAM standards for these purposes. In 2004, these standards were approved as equivalent to EU and American (MOP) standards, paving way to export Indian organic products to Europe and America. Under NPOP programme 12 international and national level certifying agencies were approved to facilitate certification. List of certifying agencies in India are :

- Natural Organic Certification Association, Pune.
- Bureau Veritas Certification India Pvt. Ltd., Mumbai.
- Ecocert, Aurangabad.**
- Control Union Certifications, Mumbai. **

- International Resources for Fairer Trade, Mumbai.**
- IMO Control Private Limited, Bangalore.**
- SKAL International, Bangalore.**
- Aditi Organics Certification Pvt. Ltd., Bangalore.
- APOF Organic Certification Agency, (AOCA), Bangalore.
- Indian Organic Certification Agency (INDOCERT), Ernakulum.
- Lacon Quality Certification Pvt. Ltd, Thiruvalla, Kerala.
- Rajasthan Organic Certification Agency (ROCA), Jaipur.
- One Cert Asia Agri Certification Private Limited, Jaipur.
- SGS India Pvt. Ltd., Gurgaon; Uttaranchal State Organic Certification Agency (USOCA), Dehra Dun.
- FoodCert India Pvt. Ltd, Hyderabad; Chhattisgarh Certification Society, India (CGCERT), Raipur.
- Vedic Organic Certification Agency, Hyderabad.
- Tamil Nadu Organic Certification Department (TNOCD), Coimbatore.
- ISCOP (Indian Society for Certification of Organic Products), Coimbatore.

But, different certifying agencies followed different standards (like Codex standard, IFOAM standards etc), to meet the requirements of respective countries. This resulted in greater confusion, as a farmer certified by an agency had to export his goods only to countries accepting such certification. The government of India initiatives encouraged the organic production in a slow and steady way but the support extended to certify and export these goods was limited and restricted to only few centers around periurban areas. These measures did not help to spread the organic farming to deep interiors of rural areas. Further, the export of organic goods was entirely left to the description exim traders, who dictated term on farmers. It must be noted that India has no Indian standard for organic products and has simply endorsed other international standards. This is obvious, because the Indian organic products were not aimed to be consumed in India and were aimed to be exported, therefore, the standards acceptable in importing countries were to be adopted and approved in India.

7.5.3 Establishment of National Project on Organic Farming

Department of agriculture and cooperation (DAC) under the Ministry of Agriculture launched an ambitious programme called National Project on Organic Farming (NPOF) in 2004 for various promotional activities such as

1. Establishment of NCOF/ RCOF
2. Capacity building
3. Setting up production units
   I. Fruit /vegetable compost units
   II. Biofertilizers units
   III. Vermicompost hatcheries
4. Training on certification, input production
5. Field demonstration
6. Market development and promotion

NCOF (National Center for Organic Farming) in Ghaziabad with 6 RCOFS (Regional Centre for Organic Farming) at Bangalore. Bhuvaneshwar, Hissar, Jabalpur, Imphal and Nagpur have taken up various promotional works to strengthen organic farming in India the concept of group certification was promoted by capacity building through an approved service provider. The service provider would help the organic farming by imparting training on record management, supervising international control system, assisting in certification and marketing access. One service provider would cater to 1500 organic farmers.

Other programmes in support of organic farming

Projects like 'Technology Mission on Horticulture' 'Cotton Mini Mission' taken up at national level promote organic farming on mass scale by providing support to adopt organic farming practices. The MOA also supports different state sponsored organic schemes under macro-management programmes.

Ministry of agriculture has also taken up the technical cooperation programme of FAO to overcome the knowledge gap by quickly producing basic information on organic farming suitable to different agro-ecological regions of India. Its aim is ultimately come out with production package for organic cultivation of different crop.

7.5.4 Role of NGOS

Large numbers of voluntary non-government organization are encouraged in the last 10-12 years thoroughout the country under central and state programme. The services of NGOs are utilized for activities such as

- Extending technical knowhow of organic farming..
- Identification of locally suitable practices.
- Management of result oriented subsidy schemes.
- Encouragement of adopting technologies by front line demonstration.

7.5.5 Organic production scenario in India

As a sequel to different promotional measures, the area under organic farming has gradually increased from a meager 4200 ha in 2003-04 to 1.2 million ha in 2008-09 (Table 7.5). The organic area in India has increased by 80 times in a span of 5 years from 2003-2004.

Table 7.5. Growth of organic area on India (source: NCOF, 2009)

Year	Certified organic area	Yearly increased (ha)
2003-04	4200 ha	—
2004-05	76000 ha	71800 ha
2005-06	173000 ha	97000 ha
2006-07	538000 ha	365000 ha
2007-08	865000 ha	327000 ha
2008-09	1207000 ha	342000 ha

Total production of some commodities under organic management in 2007-08 is depicted as follows:

Commodity	Quantity products (Mt)
Rice	44159
wheat	10161
other cereals/millets	28894
pulses	17225
soybean	20210
oilseeds	20151
cotton	77266
spices	26217
tea/coffee	19647
fruits/vegetables	369814
sugarcane	83480
herbal medicine	145943
Total	902169

(Source: NCOF, 2009)

Other details of organic farming in India are as follows (as of 2008-09):

1.	Area under fully converted organic production		640161
2.	Area under conversion		566803
3.	Number of farmers fully converted to organic		256518
4.	Number of farmers under conversion		457599
5.	Number of operating organizations		1812
6.	Number of processors		276
7.	Number of grower groups		886
8.	Number of exporters		299
9.	Total organic production	(Metric tons)	1811111
10.	Total quantity exported	(Metric tons)	53918
11.	Value of export	(US$)	119 million

Table 7.6. Details of organic products exported during 2007-08

S.N.	Name of the products	Export Qty (in M.T.)	Ex value (US $)	Destination
1	Cotton	16503.63	26357352	China, Japan, Pakistan, Bangladesh, Taiwan, Indonesia, Mexico, Italy, France
2	Basmati Rice	5923.19	12596530.32	USA, U.K., Germany, Italy, Denmark, Finland, France, The Netherland, Austria
3	Honey	4220.24	9874839	USA, U.K., Germany
4	Tea	3027.705	19602525	USA, U.K., Germany, Canada, Australia, Japan, France, New Zealand, Russia
5	Dry fruits	2479.38	17939248	Germany, France, Austria, U.K., USA, Holland, Scotland
6	Sesame	1395.64	1943169.83	Germany, Netherland, France, Austria
7	Spices	1181.748	4237600	UK, USA, Germany, South Africa
8	Processed Food	1534.21	3827730.49	UK, Germany, Netherland
9	Medicinal & Herbal Plants	614.08	2252361	USK, USA, Germany, Australia, Belgium, Netherland, Malaysia, Czech Republic,

				Korea, France, Australia, New Zealand, Russia
10	Misc. (Guar Gum, Saffron, Bea wax, Morels, Arrowroot)	249.87	1073960.07	Switzerland, Germany, USA, Netherland, Spain, Italy
11	Oil crops (excluding sesame)	243.21	165128.4	Germany, UK, Canada
12	Coffee	138.2	420346.07	France, Germany, Australia, UK
13	Fruits	15.73	77541	UK
14	Pulses	5	2850	Germany
15	Flower	0.928	6359	Germany
	Total	**37533**	**100413540.2**	
			(US$)	**IN INR 498 CRORE**

Source: APEDA and Annual Report of Certification Bodies.

7.6 SCENARIO OF ORGANIC FARMING IN DIFFERENT STATES

Although NPOP aimed originally at spreading organic farming mainly to NE Indian states (due to their low level of fertilizer use) and other rainfed area of poor fertilizer consumption, the organic area has spread to almost states of India. The Table 7.7 indicates the scenario of organic area and organic producers in different states.

Madhya Pradesh (463553 ha) with 167141 organic growers and Maharashtra (277780 ha) with 213456 organic growers are leading states for organic production. Other important states having substantial certified area under organic production are Orissa (81560 ha), with 43852 growers, Gujarat (50927 ha) with 17057 growers, Delhi (33289 ha with 14873 growers), Mizoram (34906 ha) with 41841 growers and Andhra Pradesh (32450 ha) with 34185 growers. Large area is also under conversion in the states like Andhra Pradesh, Karnataka, Delhi, Himachal Pradesh, Maharashtra, Madhya Pradesh, Orissa, Uttar Pradesh and Uttara Kannada.

In addition to regular organic area, India has vast wild collection area under organic management, wherein, different forest based products are prominently harvested, processed and exported. They include walnut kernels, black morels, amla, harad, bhera, vibhuti kai, haritaki, amalkai, arjun, kali musli, varieties of medicinal plants, dye plants and aromatic plants. Nearly, 8.47 million ha area is accounted as wild collection area. The major area under wild collection is in the states of Madhy Pradesh (5.0 million ha), Uttar Pradesh (2.4 million ha) and Rajasthan (1.47 million ha).

Table 7.7 Area under organic certification process (ha) and number of farmers in India (2008-09)

States	Total organic area	Organic growers
Andhra Pradesh	32450	34185
Arunachal Pradesh	311	613
Assam	4629	1985
Chhattisgarh	503	84
Delhi	33289	14873
Goa	10979	3281
Gujarat	50927	17057
Haryana	12220	6063
Himachal Pradesh	15435	5259
Jammu& Kashmir	419	112
Karnataka	22230	19243
Kerala	10507	8892
Manipur	10818	16650
Maharashtra	277780	213456
Madhya Pradesh	463553	167141
Mizoram	34906	41841
Meghalaya	1813	951
Nagaland	24042	28365
Orissa	81560	43852
Punjab	4192	2616
Rajasthan	29267	15214
Other states	85750	38654
Total	**1,20,7000**	**6,35,733**

Source: NCOF, 2009

7.7 FUTURE CHALLENGES

Although foregoing discussion reveals a great deal of achievement, the organic farming is still not a main stream agriculture in major part of India. It has remained a choice of select farmers either with interest to export to improve the soil fertility. Large numbers of challenges are still evident to spread organic farming. Many of them act as constrains to further strengthen and popularize the adoption of organic farming. Few such challenges and suggestions to tackle such challenges are

- The exports of certified products are meager compared to the organic production compelling the organic producers to market their produce in domestic market with no premium price. Proper domestic market channels big enough to handle organic products for domestic consumption are yet to be established. The channels presently available are not able to handle even 1% the present level of organic production.
- Government of India should promote market infrastructure and certification process for domestic consumption of organic foods. This will benefit the consumers as well as encourage mass scale adoption of organic agriculture
- The spread of organic farming is not uniform throughout the country. All the state Governments should draw up the plans to promote organic farming
- No much work is done in the field of processing and packing organic goods. Unless these are developed concurrent with the production, the domestic as well as export market will not respond favorably.
- It is necessary to have Indian standards for organic goods rather than following standards adopted in other countries.
- Programmes to create awareness to consume organic foods must be conceived and implemented to further strengthen the organic market
- The inputs used for organic production are not standardized, allowing the input manufacturers to claim high quality for every product that is available in the market
- Quality evaluation tests need to be standardized and strictly imposed at all the levels of production, processing, packing and marketing.
- Perishable organic products require special attention in respect of handling and marketing
- All the state agricultural universities should be allotted with adequate money to carry out the research on all the aspects organic agriculture.

REFERENCES

Helga Willer and Julia Lernoud, 2013. Organic Agriculture World Wide-Statistics and Emerging Trends pp: 2-48

FAO country profiles of different countries

FAO STAT, 2005-06 http://faostat.fao.org/site/424/default.aspx#ancor

FiBL* and IFOAM** 2013. Data on organic agriculture world-wide. In: FiBL& IFOAM (2013): The World of Organic Agriculture. Statistics and Emerging Trends. Frick and Bonn

ITC, 2006. Country Trade Profiles pp 118-140

NCOF, 2009. Organic Status in India Ghaziabad pp:1-5

Sahota, Amarjit 2013. The Global Market for Organic Food and Drink. In: FiBL & IFOAM (2013): The World of Organic Agriculture. Statistics and EmergingTrends. Frick and Bonn.

INDEX